Radiative Corrections
for e⁺e⁻ Collisions

J. H. Kühn (Ed.)

Radiative Corrections for e⁺e⁻ Collisions

Proceedings of the International Workshop
Held at Schloß Ringberg
Tegernsee, FRG, April 3-7, 1989

With 134 Figures

Springer-Verlag Berlin Heidelberg GmbH

Dr. Johann H. Kühn
Max-Planck-Institut für Physik und Astrophysik
Werner-Heisenberg-Institut für Physik
D–8000 München 40, Fed. Rep. of Germany

ISBN 978-3-642-74927-8   ISBN 978-3-642-74925-4 (eBook)
DOI 10.1007/978-3-642-74925-4

2156/3150 – 543210 – Printed on acid-free paper

Preface

In recent years the Standard Model of electroweak interactions has successfully passed a number of crucial tests, most notably in neutral current reactions and through the observation of W- and Z-bosons in proton–antiproton collisions. However, experiments are only beginning to verify one of the most basic consequences of its theoretical formulation as a local quantum field theory: quantum corrections as calculated in perturbation theory. Measurements that will be carried out at electron-positron colliders at Stanford and CERN in the very near future will improve the accuracy by more than an order of magnitude. Thus either these crucial elements of the present theoretical framework will be confirmed or the road to physics beyond the Standard Model will be opened.

A huge amount of theoretical work has been invested during the past few years to match the envisaged experimental precision. *QED corrections*, in particular from initial state radiation, will play a dominant role in the interpretation of measurements and have to be understood at a hitherto unrivalled level of accuracy. Analytical calculations – either to a fixed order in α or by summing large logarithms to arbitrary order – are complementary to recent developments of Monte Carlo techniques in the simulation of events with multiple photon emission. Measurements with hadronic final states evidently require the understanding of hadronic corrections to high accuracy. Even purely leptonic reactions are influenced by hadronic interactions through vacuum polarization.

Purely weak corrections have been calculated independently by various groups for e^+e^- annihilation into both fermion–antifermion and boson pairs. To allow for the interpretation of possible deviations from the firm predictions of the Standard Model, *alternative theories* have been studied – ranging from gauge theories based on larger symmetry groups to composite models predicting unconventional Z decays. *Experimental strategies* have been developed aiming at a fast and optimal analysis of the wealth of data expected for the second half of 1989, and preparing at the same time for the long-term future at SLC and LEP.

To allow for a critical assessment of the present status of this field shortly before the start up of experiments, and to stimulate the discussion between theory and experiment, a workshop on *Radiative Corrections for e^+e^- Collisions* was organized from April 3 to 7, 1989 at Ringberg Castle in Upper Bavaria. The intention was to provide an opportunity for the presentation of the most recent results, to allow for ample discussion time and to reach a consensus about which questions are solved and which problems are still open.

Since experimental results are expected during the second half of 1989 it was thought that a comprehensive presentation of the results of this workshop should be available to the public as quickly as possible in the form of proceedings. Part I deals with QED and QCD corrections relevant for precision measurements and with Monte

Carlo programs. Part II is concerned with electroweak corrections close to the Z-peak and above the W-pair threshold. Physics beyond the Standard Model is presented in Part III. Part IV deals with experimental strategies for precision measurements under conditions ranging from low luminosities to extremely high counting rates, and even with polarized beams.

On behalf of all the participants I would like to acknowledge the optimal working conditions provided by the locality and by Mr. Hörmann and his staff at Ringberg Castle. I am grateful to Mrs. Huber and Mrs. Paxon for their help with the preparation of the workshop. We are indebted to the Max Planck Society for the opportunity to organize this workshop and the Ernst-Rudolf-Schloeßmann Foundation for generous financial support. I wish to express my gratitude to all authors and to the Springer-Verlag, in particular to Dr. H.-U. Daniel, for having facilitated prompt publication by adhering to the stringent deadlines.

Munich, June 1989   *Johann H. Kühn*

Contents

Part IV: Experimental Strategies

Part I

**QED and Hadronic Corrections,
Monte Carlo Programs**

Radiative Corrections for e⁺e⁻ Collisions Editor: J.H. Kühn
© Springer-Verlag Berlin, Heidelberg 1989

Applications of Renormalization Group Methods to Radiative Corrections*

W. Beenakker, F.A.Berends and W.L. van Neerven

Instituut-Lorentz, University of Leiden, P.O.B. 9506, 2300 RA Leiden, The Netherlands

*Talk presented by W.L. Van Neerven.

1. INTRODUCTION

In recent years many radiative corrections were calculated to various quantities [1] in the standard model of electroweak interactions. Examples are total and differential cross sections of processes like $e^+e^- \to \mu^+\mu^-$ [1] and $e^+e^- \to W^+W^-$; $e^+e^- \to ZZ$ [2] which will be observed at LEP1 and LEP2 energies respectively. From these distributions one can derive interesting quantities like the forward backward asymmetry and the polarization asymmetry. Another interesting process is deep inelastic electron proton scattering (DIS) [3] $e+p \to \ell+'X'$ which proceeds via the exchange of the vector bosons γ, Z and W of the standard model.

A study of the analytic results obtained for the above processes reveals the presence of large logarithmic corrections of the type

$$L_m = \ln Q^2/m^2 \quad \text{with} \quad Q^2 \gg m^2 . \tag{1.1}$$

Here Q^2 stands either for a large kinematical variable like s (CM energy squared), t (momentum transfer) or a big mass like M_W, M_Z. The quantity m represents the mass of a light particle e.g. the leptons e and μ or the quarks u and d. The origin of these large logarithmic radiative corrections can be traced back to the presence of collinear divergences or mass singularities (MS) in perturbative quantum field theory [4-6]. In QED they arise when photons or fermions are radiated off in the direction of the incoming or outgoing light particle provided the momentum of the latter is kept fixed (exclusive). These divergences show up in virtual as well as in radiative graphs. Collinear divergences also arise if one performs a so-called mass singular renormalization [5]. Suppose the external momentum of a loop graph is represented by q and the unrenormalized Green's function by $F(q)$ then the renormalization is mass singular (MS) if $F^R(q^2) = F(q^2)-F(0)$. Here $F^R(q^2)$ denotes the renormalized Green's function where the subtraction has been performed at an exceptional momentum (i.e. $q^2=0$). The mass singularity present in $F^R(q^2)$ enters via $F(0)$ and shows up in the limit $-q^2 \gg m^2$. In this case $F^R(q^2)$ behaves like

$$F^R(q^2) \to \ln^i(-q^2/m^2) \tag{1.2}$$

As has been very often demonstrated in the literature collinear divergences show up when the propagator P denoted by

$$P = \frac{1}{(p_2-k_1)^2 - m_2^2} \tag{1.3}$$

gets on shell. This happens if (see fig. 1) $m_1=m_2=0$ (which implies $k_1^2=p_2^2=0$ and $\vec{k}_1 \parallel \vec{p}_2$) so that the angle θ goes to zero. In this case $(k_1-p_2)^2 - m_2^2 \rightarrow -2|\vec{k}_1||\vec{p}_2|(1-\cos\theta) \rightarrow 0$. If we now integrate over the momenta k_1 or p_2 the denominator in P (1.3) goes to zero and the expression for the Feynman graph in which this propagator appears becomes singular. Strictly speaking collinear divergences only occur if massless quanta are coupled to each other. This situation arises in QCD where one deals with massless gluons. However in the standard model of the electroweak interactions there are in principle no collinear divergences but at extremely high energies physical quantities become almost mass singular due to low mass fermions (like e,μ,u,d present in the theory. One can see this as follows. In the high energy limit the propagator in (1.3) behaves like

$$sP = \frac{s}{(p_2-k_1)^2 - m_2^2} \xrightarrow[\theta \rightarrow 0]{} \frac{-1}{\theta^2/4 + m_2^2/s} \quad , \tag{1.4}$$

where s is the CM energy of the process under study. If m_2 is of the order of the heavy vector boson masses W and Z i.e. $m_2=M_W,M_Z$ and $s \sim M_W,M_Z$ then $P \sim O(1)$ for $\theta=0$. However if m_2 stand for the electron mass m and $s \gg m^2$ then $P \gg 1$ when $\theta=0$. In this case the physical quantities depend logarithmically on m (like $\ln s/m^2$) and the latter can be considered as a kind of regulator mass for collinear singularities.

The conclusion is that the γ-e-e vertex is the most important one which appears in the standard model of elektroweak interactions. We can therefore conclude that the pure photonic (QED) corrections constitute the bulk of the radiative corrections at high energies which show up at LEP and HERA. We have to emphasize that the vanishing of the denominator in (1.3) in some regions of phase space is a necessary but not sufficient condition to have collinear singularities. In all renormalizable theories compensating terms can appear in the numerator of the Feynman integrals. For the presence of collinear and also infrared (IR) divergences in Feynman integrals we refer to the work of N. Nakanishi and T. Kinoshita [4,5]. However in gauge theories there exists a rule of thumb. Collinear divergences show up in the various graphs constituting an amplitude squared (probability) if at least two equal propagators are present provided the Feynman diagrams are computed in a physical (e.g. axial) gauge. However this is not the whole story. Collinear divergences can cancel while calculating a specific quantity. This cancellation is ruled by the KLN theorem [5,6]. It tells us that collinear divergences originating from a set of states which are degenerate in invariant mass disappear if one sums (integrates) over these states. This means that they become inclusive.

All that has been sketched above is well known in the field of perturbative QCD [7] and everything that will be presented below is borrowed from that theory. In the following we will determine the large logarithmic corrections arising in differential as well as total cross sections. Subsequently one can then derive other interesting quantities like the forward backward asymmetry, polarization asymmetry etc.

2. FORMALISM

The work on the determination of the large logarithmic corrections in QED was started by two groups [8,9]. They applied the rules of perturbative QCD to the process $e^+e^- \to \mu^+\mu^-$ (LEP1) where the total cross section was determined in the leading log (LL) approximation. The method was extended by various other groups [10,11] to obtain the next to leading logarithms (NLL) and the non log constant terms. Here we will present the most general formalism which is borrowed from QCD [12] to compute all logarithmic corrections in single as well as double differential cross sections. From the latter one can then derive the various experimentally interesting quantities. Examples are the line shape of the Z resonance, forward backward asymmetry, total cross sections for muon pair, Z-pair and W-pair production.

Since we are mainly concerned with LEP processes the incoming particles are electrons and positrons. Therefore we study processes like

$$e^+(p_1) + e^-(p_2) \to M(q_1) + {}'X' \ , \tag{2.1}$$

where X denotes the inclusive state which is allowed by quantum number conservation and M stands for the detected particle, like $e^{\pm}$, $\mu^{\pm}$, $W^{\pm}$, Z. Further we shall limit ourselves to electromagnetic radiation only. However notice that the formalism which will be developed below can also be extended to other processes where the incoming e^+e^- will be replaced by other particles. One example of this will be given in the next section. The process given in (2.1) including all photonic radiation can be described in a pictorial way see fig. 2. Let us denote $d^2\sigma/dt_1du_1$ as the double differential cross section which can be determined order by order in perturbation theory. It depends on the large logarithm $L_m \equiv \ln Q^2/m^2 (Q^2 \gg m^2)$ where $m = m_e$, M ($M = m_e, m_\mu$). Neglecting the electron mass the kinematical variables s, t_1 and u_1 are denoted by

$$s = (p_1+p_2)^2 \sim 2p_1 p_2 \qquad t_1 = (p_1-q_1)^2 - M^2 \sim -2p_1 q_1 \qquad u_1 = (p_2-q_1)^2 - M^2 \sim -2p_2 q_1 \, . \tag{2.2}$$

The shaded blob in fig. 1 represents the 'hard' process which is described by the reduced cross section $d^2\hat{\sigma}/d\hat{t}_1 d\hat{u}_1$ which <u>does not</u> contain the large logarithms L_m. Further it depends on the factorization scale Q^2 (see below). The kinematical variables corresponding to the reduced cross sections are denoted by

$$\hat{s} = (\hat{p}_1+\hat{p}_2)^2 \sim 2\hat{p}_1 \hat{p}_2 \qquad \hat{t}_1 = (\hat{p}_1-\hat{q}_1)^2 - M^2 \sim -2\hat{p}_1 \hat{q}_1 \qquad \hat{u}_1 = (\hat{p}_2-\hat{q}_1)^2 - M^2 \sim -2\hat{p}_2 \hat{q}_1 \, . \tag{2.3}$$

Here $\hat{p}_1, \hat{p}_2$ stand for the momenta of the incoming particles after the photons have been emitted by the incoming $e^+ e^-$ beams.

The outgoing momentum $\hat{q}_1$ corresponds to the outgoing particle M before the photons have been radiated off. The relation between the momenta with and without the hat is given by

$$\hat{p}_1 = x_1 p_1 \qquad \hat{p}_2 = x_2 p_2 \qquad \hat{q}_1 = \frac{1}{x_3} q_1 \, . \tag{2.4}$$

In this way the invariants in (2.2) and (2.3) are related by

$$\hat{s} = x_1 x_2 s \qquad \hat{t}_1 = \frac{x_1}{x_3} t_1 \qquad \hat{u}_1 = \frac{x_2}{x_3} u_1 \, . \tag{2.5}$$

The mass finite and mass singular cross sections $d\hat{\sigma}$ and $d\sigma$ are related via the mass factorization theorem [12]

$$s^2 \frac{d^2\sigma_{e^+e^- \to MX}}{dt_1 du_1}(s,t_1,u_1,L_m) = \int_0^1 \frac{dx_1}{x_1} \int_0^1 \frac{dx_2}{x_2} \int_0^1 \frac{dx_3}{x_3^2} \, \Gamma_{ie^+}(x_1,L_{e^+},Q^2)$$

$$\Gamma_{je^-}(x_2,L_{e^-},Q^2) \, \hat{s}^2 \frac{d^2\hat{\sigma}_{ij \to kX}}{d\hat{t}_1 d\hat{u}_1}(\hat{s},\hat{t}_1,\hat{u}_1,Q^2) D_{kM}(x_3,L_M,Q^2) \tag{2.6}$$

Γ_{ij} and D_{ij} denote the initial and final state splitting functions which depend on the mass singularities corresponding to the lines from where the photons are radiated off. Like the reduced cross section $d\hat{\sigma}$ they also depend on the mass factorization scale Q^2. Since the latter can be chosen in an arbitrary way Γ, D and $d\hat{\sigma}$ satisfy renormalization group equations [7]. These equations can be solved perturbatively. Knowledge of Γ, D and $d\hat{\sigma}$ leads to the construction of $d\sigma$. Now Γ and D do not depend on the specific process under consideration. This is the consequence of the universality of mass singularities. This theorem implies that if the splitting functions can be inferred from some process then they can be used for all other reactions. Historically Γ and D were not even calculated for some specific process but were obtained from the operator matrix elements [13]. The latter arise in the operator product expansions of the forward Compton scattering process

$$\gamma^*(q) + e(p) \rightarrow \gamma^*(q) + e(p) \ , \tag{2.7}$$

which is described by the structure tensor [3]

$$W^{\mu\nu}(q,p) = \frac{1}{4\pi} \int d^4x \ e^{iqx} <p|J^\mu(x)J^\nu(0)|p> \tag{2.8}$$

The quantities $d\sigma$, $d\hat{\sigma}$, Γ and D can be expanded as a series in the electro-magnetic coupling constant α as follows

$$d\sigma = \sum_{n=0}^{\infty} (\frac{\alpha}{\pi})^n \ d\sigma^{(n)} \ \text{with} \ d\sigma^{(n)} = \sum_{i=0}^{n} a_{ni} L_m^i \tag{2.9a}$$

$$d\hat{\sigma} = \sum_{n=0}^{\infty} (\frac{\alpha}{\pi})^n \ d\hat{\sigma}^{(n)} \tag{2.9b}$$

$$\Gamma_{ij} = \delta_{ij} + \sum_{n=1}^{\infty} (\frac{\alpha}{\pi})^n \ \Gamma_{ij}^{(n)} \ \text{with} \ \Gamma_{ij}^{(n)} = \sum_{i=0}^{n} b_{ni} L_m^i \tag{2.9c}$$

$$D_{ij} = \delta_{ij} + \sum_{n=1}^{\infty} (\frac{\alpha}{\pi})^n \ D_{ij}^{(n)} \ \text{with} \ D_{ij}^{(n)} = \sum_{i=0}^{n} c_{ni} L_m^i \tag{2.9d}$$

Starting from the Born cross section which is denoted by $d\sigma^{(0)} = d\hat{\sigma}^{(0)}$ one can determine the logarithmic terms in $d\sigma$ in an iterative way. Up to a given order in α one can calculate all logarithms in $d\sigma$ provided one knows the reduced cross section $d\hat{\sigma}$ in the previous order exactly. Notice that $d\hat{\sigma}$ can be only obtained via an explicit calculation.

The most important splitting functions are the diagonal ones since they corres-pond to multiple photon radiation and therefore contain the very important soft photon terms. The general form for Γ_{ij} (and also D_{ij}) reads as

$$\Gamma_{ij}(x,L_m,Q^2) = \delta_{ij}\delta(1-x) + (\frac{\alpha}{2\pi})P_{ij}^{(0)}(x)L_m + \frac{1}{2}(\frac{\alpha}{2\pi})^2[(P_{ik}^{(0)} \otimes P_{kj}^{(0)})(x)L_m^2 + P_{ij}^{(1)}(x)L_m]$$

$$+ \text{higher orders,} \tag{2.10}$$

where $P_{ij}^{(n)}$ are the Altarelli-Parisi splitting functions [13,14] which have been calculated in the literature up to order α^2 [13]. The convolution symbol $\otimes$ is defined by

$$(f \otimes g)(x) = \int_0^1 dx_1 \int_0^1 dx_2 \ \delta(x-x_1x_2)f(x_1)g(x_2) \ . \tag{2.11}$$

The diagonal splitting function can be split into a soft + virtual and a hard photon part as follows

$$P_{ii}(x) = \delta(1-x)P_{ii}^{S+V}(\delta) + \theta(1-\delta-x)P_{ii}^{H}(x) \qquad (2.12)$$

The parameter δ has been introduced so that one can distinguish between soft $x > 1-\delta$ and hard $x < 1-\delta$ photon contributions. Remember that $\sum_{i=1}^{n} k_i = (1-x)p_\ell$ ($\ell=1,2$) where k_i denotes the photon momentum.

If the incoming particles are electrons and positrons we have $(P_{e^+e^+} = P_{e^-e^-})$

$$P_{ee}^{(0)}(x) = \delta(1-x)(3/2 + 2\ln\delta) + \theta(1-\delta-x)\frac{1+x^2}{1-x} \qquad (2.13a)$$

$$\tfrac{1}{2}(P_{ee}^{(0)} \otimes P_{ee}^{(0)})(x) = \delta(1-x)(2\ln^2\delta + 3\ln\delta + 9/8 - 2\zeta(2)) + \theta(1-\delta-x)$$

$$\left\{ \frac{1+x^2}{1-x} (2\ln(1-x) - \ln x + 3/2) + \tfrac{1}{2}(1+x)\ln x - (1-x) \right\} . \qquad (2.13b)$$

The $P_{ee}^{(1)}(x)$ is also known [13] but it only contributes in the next to leading log approximation (NLLA) to $d\sigma$. In the following we shall limit ourselves to the leading log approximation (LLA). If we combine this with the exact non logarithmic term of the order α correction to $d\sigma$ one gets already an excellent approximation for the total radiative correction. This procedure can be even improved by re-summing all soft photon terms to all order in perturbation theory. In the LLA one only needs the lowest order term in the expansion for $d\hat{\sigma}$ i.e. $d\hat{\sigma}^{(0)}$ which is just the Born cross section. In this case we can write

$$s^2 \frac{d^2\hat{\sigma}}{d\hat{t}_1 d\hat{u}_1} (\hat{s}, \hat{t}_1, \hat{u}_1) = \delta(\hat{s} + \hat{t}_1 + \hat{u}_1)\sigma^B(\hat{s}, \hat{t}_1, \hat{u}_1) . \qquad (2.14)$$

Notice that the Born contribution is explicit scale (Q^2) independent. Our next step is that one has to choose a suitable scale Q^2 in $L_m = \ln Q^2/m^2$ (1.1), so that the next to leading log terms are as small as possible. For the differential cross section $s^2 d^2\sigma/dt_1 du_1$ it turns out that Q^2 has to be proportional to the momentum transfer to the detected particle.

$$Q^2 = \frac{t_1 u_1}{s} = \frac{\hat{t}_1 \hat{u}_1}{\hat{s}} = p_t^2 + m^2 . \qquad (2.15)$$

Notice that in order to obtain the next to leading terms one has to be very careful with the choice of the scale. Consider for example the cross section $d\sigma/ds'$ which describes the process $e^+e^- \to \mu^+\mu^-$ where s' denotes the dimuon pair mass. Here one has to choose the factorization scale s' instead of s (see ref. 11). Another choice will lead to alien terms which are not reproduced by an exact calculation. From the double differential cross section in (2.6) one can derive all other cross sections of interest. They are

1) The double differential cross section $d^2\sigma/dEdc$ $(c \equiv \cos\theta)$, where E and θ

are the energy and angle of the outgoing particle M respectively

$$\frac{d^2\sigma}{dEdc} = 2s\sqrt{E^2-M^2}\ \frac{d^2\sigma}{dt_1 du_1}\ . \tag{2.16}$$

2) From $d^2\sigma/dEdc$ one can either derive $d\sigma/dc$ or $d\sigma/dE$ including cuts on
energy or angles. Here we will give an expression for $d\sigma/dc$ only because we
need it for the applications presented in section 3. In the LLA it is very simple
to calculate $d\sigma/dc$ since we can use the delta function appearing in the Born
cross section (2.14). Using this delta function we obtain after integration over
the energy E the expression

$$\frac{d\sigma}{dc} = \frac{2}{s}\int\limits_{y(1)}^{1} dx_1 \int\limits_{y(x_1)}^{1} dx_2\ \Gamma_{ee}(x_1,L_e,s)\Gamma_{ee}(x_2,L_e,s)F(x_1,x_2,c)$$

$$+\ \frac{2}{s}\int\limits_{\frac{2M}{\sqrt{s}}}^{1} dx_1 \int\limits_{z(x_1)}^{y(x_1)} dx_2\ \Gamma_{ee}(x_1,L_e,s)\Gamma_{ee}(x_2,L_e,s)\Big[\theta(c)\{F(x_1,x_2,c) + F(x_1,x_2,-c)\}$$

$$+\ \theta(-c)\{F(x_2,x_1,c) + F(x_2,x_1,-c)\}\Big]\ . \tag{2.17}$$

The function $F(x_1,x_2,c)$ is defined by

$$F(x_1,x_2,c) = \frac{1}{x_1 x_2}\ \frac{1}{\sqrt{D}}\ \frac{[x_1 x_2(x_1-x_2)c + (x_1+x_2)\sqrt{D}]^2}{[(x_1+x_2)^2 - c^2(x_1-x_2)^2]^2}\ \sigma_B(x_1 x_2 s, x_1 t_1^*, x_2 u_1^*) \tag{2.18}$$

with

$$t_1^* = -s\ \frac{x_1 x_2\{x_1+x_2+c^2(x_2-x_1)\} - 2x_2 c\sqrt{D}}{(x_1+x_2)^2 - c^2(x_2-x_1)^2} \tag{2.19a}$$

$$u_1^* = -s\ \frac{x_1 x_2\{x_1+x_2+c^2(x_1-x_2)\} + 2x_1 c\sqrt{D}}{(x_1+x_2)^2 - c^2(x_2-x_1)^2} \tag{2.19b}$$

$$D = x_1^2 x_2^2 - \{(x_1+x_2)^2 - c^2(x_1-x_2)^2\}\ \frac{M^2}{s} \tag{2.19c}$$

The integration boundaries $y(x)$, $z(x)$ can be derived from the integration range
for E i.e. $M<E<\frac{1}{2}\sqrt{s}$ and the equation $D=0$ respectively. The result is

$$y(x) = \frac{x M/\sqrt{s}}{x - M/\sqrt{s}} \tag{2.20a}$$

$$z(x) = \frac{x M/\sqrt{s}}{x^2 - (1-c^2)M^2/s}\ [M/\sqrt{s}(1+c^2) + \{(1-c^2)x^2 + 4M^2 c^2/s\}^{\frac{1}{2}}]. \tag{2.20b}$$

In (2.17) we only took the diagonal splitting functions into account since they constitute the bulk of the radiative correction. Notice that in $d\sigma/dc$ the fragmentation function D_{MM} has disappeared. Since D_{MM} is nothing but the probability distribution in the final state energy E it equals one after integration over E. This follows from the KLN theorem [5,6]. If one integrates over E the final state becomes inclusive and the corresponding collinear divergences will disappear as $M \to 0$.

3) Finally by integrating over c in (2.17) one can derive the total cross section which has already been presented in the literature [8-11]

$$\sigma_{tot}(s) = \int_0^1 dx_1 \int_0^1 dx_2 \ \theta(x_1 x_2 s - 4M^2) \Gamma_{ee}(x_1, L_e, s) \Gamma_{ee}(x_2, L_e, s) \hat{\sigma}_{tot}(x_1 x_2 s, M^2),$$
(2.21)

where $4M^2$ denotes the threshold of the process $e^+ e^- \to MM$ including all radiative corrections. The total reduced cross section is obtained by

$$\hat{\sigma}_{tot}(\hat{s}, M^2) = \int_{(\hat{s}-\bar{s})/2}^{(\hat{s}+\bar{s})/2} d(-\hat{t}_1) \int_{-\frac{\hat{s}m^2}{\hat{t}_1}}^{\hat{s}+\hat{t}_1} d(-\hat{u}_1) \ \frac{d^2\hat{\sigma}}{d\hat{t}_1 d\hat{u}_1}$$
(2.22)

with $\quad \bar{s} = \hat{s}\sqrt{1 - 4M^2/\hat{s}}$.

Expression (2.21) can be rewritten as follows

$$\sigma_{tot}(s) = \int_{4M^2/s}^{1} dz \ \Phi_{tot}(z) \hat{\sigma}_{tot}(zs, M^2) \ ,$$
(2.23)

where $\Phi_{tot}(z)$ is the incoming particle flux which is defined by

$$\Phi_{tot}(z) = \int_0^1 dx_1 \int_0^1 dx_2 \ \delta(x_1 x_2 - z) \Gamma_{ee}(x_1, L_e, s) \Gamma_{ee}(x_2, L_e, s) \equiv (\Gamma_{ee} \otimes \Gamma_{ee})(z).$$
(2.24)

For the diagonal splitting functions it takes the form

$$\Phi_{tot}(z) = \delta(1-z) + \frac{\alpha}{\pi} P_{ee}^{(0)}(z) L_e + \frac{1}{2}\left(\frac{\alpha}{\pi}\right)^2 (P_{ee}^{(0)} \otimes P_{ee}^{(0)})(z) L_e^2 \ .$$
(2.25)

4) From (2.17) we can also obtain the forward backward asymmetry for the process $e^+ e^- \to \mu^+ \mu^-$. Putting $M = m_\mu = 0$ $d\sigma/dc$ becomes very simple. In this case we can write for the Born cross section [9]

$$\sigma_B(s, t_1, u_1) = \pi \alpha^2 \left[2W_1(s) \frac{t_1^2 + u_1^2}{s^2} + W_2(s) \frac{t_1 - u_1}{s} \right],$$
(2.26)

where $W_i(s)$ is given by [9]

$$W_1(s) = 1 + \frac{2(s-M_Z^2)s}{|Z(s)|^2} c_V^2 + \frac{s^2}{|Z(s)|^2} (c_V^2+c_A^2)^2 \qquad (2.27a)$$

$$W_2(s) = \frac{4(s-M_Z^2)s}{|Z(s)|^2} c_A^2 + \frac{8s^2}{|Z(s)|^2} c_V^2 c_A^2 . \qquad (2.27b)$$

The coupling constants c_A, c_V are denoted by

$$c_A = \frac{-1}{2\sin 2\theta_W} \qquad c_V = c_A (1 - 4\sin^2\theta_W) \qquad (2.28)$$

and the denominator of the Z-boson propagator in the narrow width approximation reads

$$Z(s) = s - M_Z^2 + iM_Z\Gamma_Z . \qquad (2.29)$$

The forward backward asymmetry A_{FB} is defined by

$$A_{FB} = [\int_0^1 dc \frac{d\sigma}{dc} - \int_{-1}^0 dc \frac{d\sigma}{dc}] / \sigma_{tot} . \qquad (2.30)$$

Inserting $\frac{d\sigma}{dc}$ (2.17) in the above expression for M=0 we obtain

$$A_{FB} = \frac{2\pi\alpha^2}{s\sigma_{tot}} \int_0^1 dx_1 \int_0^1 dx_2 \frac{1}{(x_1+x_2)^2} \Gamma_{ee}(x_1,L_e,s)\Gamma_{ee}(x_2,L_e,s)W_2(x_1 x_2 s)$$

$$= \frac{2\pi\alpha^2}{s\sigma_{tot}} \int_0^1 dz \frac{1}{(1+z)^2} \Phi_{FB}(z)W_2(zs) \qquad (2.31)$$

with $\quad \Phi_{FB}(z) = (1+z)^2 \int_0^1 dx_1 \int_0^1 dx_2 \delta(x_1 x_2 - z) \frac{1}{(x_1+x_2)^2} \Gamma_{ee}(x_1,L_e,s)\Gamma_{ee}(x_2,L_e,s).$

$$\qquad (2.32)$$

We will define the quantity Φ_{FB} as the "forward backward flux". We would like to emphasize that Φ_{FB} is not equal to Φ_{tot} as is sometimes stated in the literature [15,16]. Actually they are related as follows

$$\Phi_{FB}(z) = \Phi_{tot}(z) + \tilde{\Phi}_{FB}(z) , \qquad (2.33)$$

where Φ_{tot} (see 2.25) appears in the total cross section given by

$$\sigma_{tot}(s) = \frac{4\pi\alpha^2}{3s} \int_{\frac{4m_\mu^2}{s}}^{1} \frac{dz}{z} \Phi_{tot}(z)W_1(zs) \ . \tag{2.34}$$

An explicit calculation reveals that Φ_{FB} and Φ_{tot} start to deviate from each other at order α^2. Using (2.10) and (2.13) one can derive

$$\tilde{\Phi}_{FB}(z) = (\frac{\alpha}{2\pi})^2 L_e^2 [\frac{(1-z)^3}{2z} + \frac{(1-z)^2}{\sqrt{z}} (\text{arctg} \frac{1}{\sqrt{z}} - \text{arctg} \sqrt{z}) - (1+z)\ln z + $$
$$2(1-z)] + O(\alpha^3) \ . \tag{2.35}$$

3. APPLICATIONS

As a first example we would like to discuss the photonic corrections [17-19] to the lepton line in the following process (see fig. 3)

$$e^- + p \rightarrow e^- + 'X' \ , \tag{3.1}$$

which will be studied at HERA [3]. For convenience we shall limit ourselves to the one photon exchange contribution to (3.1). The large logarithmic terms which appear in the radiatively corrected cross section can be obtained from the master formula in (2.6). Using a shorthand notation this can be written as follows

$$d\sigma_{eq\rightarrow eX} = \Gamma_{ie}(L_e,q^2)\Gamma_{jq}(L_q,q^2)d\hat{\sigma}_{ij\rightarrow kX} \ D_{ke}(L_e,q^2) \ . \tag{3.2}$$

Notice that in practice the radiative corrections are performed for the process (fig. 3)

$$e + q \rightarrow e + 'X' \ . \tag{3.3}$$

The corrections to process (3.1) are simply obtained by folding the differential cross section (3.3) with the input parton (quark) distribution functions. If we now omit the electromagnetic corrections to the quark line one can set $\Gamma_{qq}=1$ up to all order in α. However we are not allowed to put $\Gamma_{\gamma q}=0$. This can be easily seen as follows. If one integrates over the momenta of the outgoing photons and the inclusive state X the exchanged photon in fig. 3 can become parallel to the incoming quark. In this case we get a collinear divergence which can be regulated by giving the initial state quark a mass m_q. Up to order α the radiative corrections to process (3.3) receive contributions from the diagrams given in fig. 4. The mass singularities showing up in $d\sigma_{eq\rightarrow eX}$ originate from two different production mechanisms contributing to $d\hat{\sigma}_{ij\rightarrow kX}$. They are given by the Coulomb scattering process (fig. 5a) and the Compton scattering process (fig. 5b) which are described by the cross sections $d\hat{\sigma}_{eq\to eX}^{(0)}$ and $d\hat{\sigma}_{e\gamma\to eX}^{(0)}$ respectively. Up to order α (3.2) reads

$$d\sigma_{eq\to eX} = d\hat{\sigma}_{eq\to eX}^{(0)} + \Gamma_{ee}^{(1)}(L_e,q^2)d\hat{\sigma}_{eq\to eX}^{(0)}$$

$$\text{fig. 5a} \qquad\qquad \text{fig. 6a}$$

$$+ d\hat{\sigma}_{eq\to eX}^{(0)}D_{ee}^{(1)}(L_e,q^2) + \Gamma_{\gamma q}^{(1)}(L_q,q^2)d\hat{\sigma}_{e\gamma\to eX}^{(0)} \; . \qquad (3.4)$$

$$\text{fig. 6b} \qquad\qquad \text{figs. 5, 6c1,c2} \; .$$

The last process in (3.4) contains the quark mass singularity. Therefore one cannot do the calculation without introducing a quark mass even if one computes the electromagnetic corrections to the electron line only. From the above discussion it is now very easy to obtain the coefficients of the logarithmic enhanced terms. The radiative corrections are applied to the following cross section [17] (fig.4)

$$\frac{d^2\sigma}{dxdy} = \frac{2\pi\alpha^2}{q^4} \; s[\; \{1 + (1-y)^2\} \; F_2(x,q^2) - xy^2 F_L(x,q^2)] \; , \qquad (3.5)$$

where F_2, F_L denote the deep inelastic structure functions in the case of one virtual photon exchange. Assuming the LLA in QCD we have $F_L(x,q^2) = 0$ and $F_2(x,q^2) = \sum_i e_i^2 x f_i(x,q^2)$ where $f_i(x,q^2)$ stands for the parton distribution functions. The kinematical variables are defined by (see fig. 7)

$$x = -\frac{q^2}{2p_1 q} \qquad y = \frac{p_1 q}{p_1 k_1} \qquad s = 2k_1 p_1 \qquad t_1 = q^2 = -sxy \qquad u_1 = -2k_2 p_1 = -s(1-y) \qquad (3.6)$$

The lowest order reduced cross sections are denoted by (see the notation in (2.14))

$$\hat{\sigma}_{eq_i \to eX}^B = 2\pi\alpha^2 e_i^2 \; \frac{\hat{s}^2 + \hat{u}_1^2}{\hat{t}_1^2} \qquad \text{(fig. 5a)} \qquad (3.7)$$

$$\hat{\sigma}_{e\gamma \to eX}^B = -2\pi\alpha^2 \; \frac{\hat{s}^2 + \hat{u}_1^2}{\hat{s}\hat{u}_1} \qquad \text{(fig. 5b)} \qquad (3.8)$$

The splitting functions Γ_{ee}, D_{ee} and $\Gamma_{\gamma q}$ are given by

$$D_{ee} = \Gamma_{ee} = \delta(1-x) + \frac{\alpha}{2\pi} \ln \frac{-t_1}{m_e^2} \{\delta(1-x)(2\ln\delta + \frac{3}{2}) + \theta(1-\delta-x) \frac{1+x^2}{1-x}\} \; . \quad (3.9)$$

$$\Gamma_{\gamma q} = \frac{\alpha}{2\pi} \ln \frac{-t_1}{m_q^2} \; \frac{1 + (1-x)^2}{x} \; . \qquad (3.10)$$

Here the mass factorization scale is chosen to be equal to $-q^2=t_1$. Up to order α the three contributions to the radiative corrections are

$$\Gamma_{ee}^{(1)}(L_e,q^2)d\hat{\sigma}_{eq\to eX}^{(0)} + \frac{d^2\sigma^I}{dxdy} = s\,\frac{\alpha^3}{t_1^2}\,\ln\frac{-t_1}{m_e^2}\left[\int_{x+\frac{x(1-y)\delta}{y}}^{1}\frac{dz}{z^3}\,F_2(z,q^2)\right.$$

$$\left.\frac{\{(z-xy)^2 + z^2(1-y)^2\}\{z^2 + (z-xy)^2\}}{(1-y)(z-x)(z-xy)} + (\tfrac{3}{2}+2\ln\delta)\{1 + (1-y)^2\}F_2(x,q^2)\right].$$

$$(3.11)$$

$$d\hat{\sigma}_{eq\to eX}^{(0)}D_{ee}^{(1)}(L_e,q^2) + \frac{d^2\sigma^{II}}{dxdy} = s\,\frac{\alpha^3}{t_1^2}\,\ln\frac{-t_1}{m_e^2}\left[\int_{x+\frac{x\delta}{y}}^{1}\frac{dz}{z^3}\,F_2(z,q^2)\right.$$

$$\frac{\{z^2 + (xy + z(1-y))^2\}\{(xy+ z(1-y))^2 + z^2(1-y)^2\}}{(z-x)(xy + z(1-y))} +$$

$$\left.(\tfrac{3}{2}+2\ln\delta)\{1 + (1-y)^2\}F_2(x,q^2)\right].$$

$$(3.12)$$

$$\Gamma_{\gamma q}^{(1)}(L_q,q^2)d\hat{\sigma}_{e\gamma\to eX}^{(0)} + \frac{d^2\sigma^{III}}{dxdy} = s\,\frac{\alpha^3}{t_1^2}\sum_i e_i^2\,\ln\frac{-t_1}{m_i^2}\int_x^1\frac{dz}{z^2}\,f_i(z,q^2)$$

$$\frac{y^2\{1 + (1-y)^2\}\{z^2 + (x-z)^2\}}{(1-y)},$$

$$(3.13)$$

where e_i and m_i denote the charge and mass of the quark i respectively. The above expressions agree with the logarithmic terms presented in eq. 26 of ref. 18. These authors obtained them via an explicit Feynman diagram calculation. This example gives a nice demonstration how to get mass singular logarithms without an explicit diagrammatic calculation.

Further we would like to remark that the authors in ref. 18 have replaced m_i^2 in the logarithm of (3.13) by the proton mass. To our opinion this term should be removed via mass factorization and has to be absorbed in the input quark distribution function. In fig. 8 we show the ratio

$$\delta(x,y) = \frac{d^2\sigma^{(1)}}{dxdy}\bigg/\frac{d^2\sigma^{(0)}}{dxdy}$$

$$(3.14)$$

for the exact calculation obtained from eq. 26 in ref. 18 and the LLA given above. Notice that we have omitted in the exact as well as in the approximate expression for δ the contribution of (3.13). It turns out that the latter only affects the size of δ near the region y=1. Fig. 8 reveals that the LLA agrees very well with the exact calculation of ref. 16. This in particular holds in the small y and large x region where the results of refs. 16 and 19 disagree.

The next example we want to show is the application to Z- and W-pair production at LEP 2 [20]. The processes under study are $e^+e^- \to ZZ$ and $e^+e^- \to W^+W^-$ including photonic corrections in LLA up to order α^2.

In figs. 9, 10 we show the total cross section for Z and W production respectively. The chosen parameters are $M_Z = 93$ GeV, $M_W = 83$ GeV and $\cos^2\theta_W = M_W^2/M_Z^2$. Using the same parameters we have plotted the differential cross section $d\sigma/dc$ (c=cosθ) in figs. 11,12. From these figures we can conclude the following.

a) Near threshold the radiative corrections are very large. This effect can be wholly attributed to soft photon radiation. Further the order $\bar{\alpha}$ corrections turn out to be negative whereas the order α^2 contributions are positive.

b) The order α^2 contribution is only noticeable if soft photon radiation dominates the radiative correction. Far away from threshold soft photon contributions become less important and the corrections beyond order α become unobservable. This can be seen in fig. 12 where the difference between order α and order α^2 is unnoticeable.

c) Variation of M_W between 81 and 83 GeV in fig. 11 shows that a knowledge about radiative corrections is indispensable for an accurate determination of the vector boson masses at LEP 2.

Notice that the figures and the above conclusions are based on the zero width approximation. Inclusion of the width will modify the threshold behaviour. However the qualitative effect of the photonic corrections on the finite width cross sections will be the same as in the case of the zero width. If soft photons are dominating the radiative corrections we can resum them (see e.g. refs. 5, 8, 10, 11) as follows:

$$\sigma_{tot}^s(s,M^2) = \frac{e^{\frac{3}{2}\frac{\alpha}{\pi}L_e} e^{-\beta\gamma}}{\Gamma(1+\beta)} \int_{\frac{4M^2}{s}}^{1} dz\ \beta(1-z)^{\beta-1}\ \sigma_{tot}(sz, M^2) + \text{hard photon}$$

$$\text{corrections.} \qquad (3.15)$$

$$\beta = \frac{2\alpha}{\pi}(L_e-1)\ . \qquad (3.16)$$

where γ is the Euler constant and Γ denotes the gamma function. In the above formula we have also included the NLL soft photon terms. The hard photon corrections have to be included if the soft photon contribution does not dominate the correction. In practice it turns out that the order α^2 corrected cross section is very close to the result in (3.15).

Summarizing our findings we conclude the following. Using the renormalization group and mass factorization techniques we have developed a general formalism to determine the large logarithmic corrections present in radiative corrections. The logarithms originate from both the virtual plus soft and hard photonic contributions. This technique can be applied to single as well as double differential cross sections from which one can derive quantities as σ_{tot}, A_{FB}, A_{LR} etc. Notice that in this talk we have limited ourselves to single particle exclusive processes. However the above method can be generalized to include more complicated processes where two or more particles are detected. The exact leading and next to leading log results for all these processes can be used as a consistency check on Monte Carlo programs. Notice that these programs allow for cuts on energy and angles of the radiated photons which in general cannot be imposed on the splitting functions Γ and D. Using these techniques one can also improve perturbation theory in order to construct soft photon resummation techniques as is shown in (3.15). It is very important to choose the appropriate mass factorization scale Q^2 so that the next to leading terms become as small as possible. Finally as is demonstrated in our first example one can use this method as a check on the exact calculations.

REFERENCES

1. "Physics at LEP", eds. J. Ellis and R. Peccei, CERN Report 86-02, Vols. 1 and 2.

2. "ECFA Workshop on LEP 200", eds. A. Böhm and W. Hoogland, CERN Report 87-08, Vols. 1 and 2.

3. "Proceedings of the HERA Workshop", ed. R.D. Peccei, Hamburg, October 12-14, 1987, Vols. 1 and 2.

4. N. Nakanishi, Progr. Theor. Phys. $\underline{19}$ (1958) 159.

5. T. Kinoshita, J. Math. Phys. $\underline{3}$ (1962) 650.

6. T.D. Lee and M. Nauenberg, Phys. Rev. $\underline{133}$ (1964) B 1549.

7. G. Altarelli, Phys. Rep. $\underline{81}$ (1982) 1.

8. E.A. Kuraev and V.S. Fadin, Sov. J. Nucl. Phys. $\underline{41}$ (1985) 466.

9. See ref. 1) G. Altarelli and G. Martinelli, Vol. 1, 47.

10. O. Nicrosini and L. Trentadue, Phys. Lett. $\underline{196B}$ (1987) 551.

11. F.A. Berends, W.L. van Neerven and G.J.H. Burgers, Nucl. Phys. $\underline{B297}$ (1988) 429.

12. J.C. Collins and G. Sterman, Nucl. Phys. $\underline{B185}$ (1982) 172.
J.C. Collins, D.E. Soper and G. Sterman, Phys. Lett. $\underline{134B}$ (1984) 263; Nucl. Phys. $\underline{B261}$ (1985) 104.
G.T. Bodwin, Phys. Rev. $\underline{D31}$ (1985) 2616 and references therein.

13. E.G. Floratos, D.A. Ross and C.T. Sachrajda, Nucl. Phys. $\underline{B129}$ (1977) 66, Erratum Nucl. Phys. $\underline{B139}$ (1978) 545.
A. González-Arroyo, C. López and F.J. Yuduraïn, Nucl. Phys. $\underline{B153}$ (1979) 161.
A. González-Arroyo and C. López, Nucl. Phys. $\underline{B166}$ (1980) 429.
E.G. Floratos, C. Kounnas and R. Lacaze, Phys. Lett. $\underline{98B}$ (1981) 89; 285.

14. V.N. Gribov and L.N. Lipatov, Sov. J. Nucl. Phys. $\underline{15}$ (1972) 438, 675.
G. Altarelli and G. Parisi, Nucl. Phys. $\underline{B126}$ (1977) 298.

15. J.E. Campagne and R. Zitoun, LPN HEP-88-06.

16. L. Trentadue and O. Nicrosini, "Structure functions techniques in e^+e^- Collisions", talk presented at the Workshop on "Electroweak Radiative Corrections", Ringberg Castle (FRG), 2-8 April, 1989.

17. See ref. 3) D. Yu. Bardin et al., Vol. 2, 577.

18. D. Yu. Bardin, C. Burdik, P.Ch. Christova, T. Riemann, JINR Dubna preprint E2-87-595.

19. See ref. 3) H. Spiesberger, Vol. 2, 605.

20. See ref. 1) G. Barbiellini et al., Vol. 2, 1.
     See ref. 2) J. Bijnens et al., Vol. 1, 49.

FIGURE CAPTIONS

Fig. 1: Momentum diagram showing the collinear region.

Fig. 2: Diagram depicting the mass factorization as given in (2.6). The shaded blob represents the hard process.

Fig. 3: Diagrams depicting the mass factorization for the final and initial state bremsstrahlung from the electron in the process $e+p \rightarrow e+'X'$ and $e+q \rightarrow e+X$ respectively.

Fig. 4: Photonic correction to the electron line as calculated in refs.17-19.

Fig. 5: Lowest order contributions to the reduced cross section of the reaction $e+q \rightarrow e+'X'$.

Fig. 6: Momentum diagrams showing the collinear regions in the process $e+q \rightarrow e+\gamma+'X'$.

Fig. 7: Kinematics of the process $e+p \rightarrow e+'X'$.

Fig. 8: Order α correction $\delta(x,y)$ (%) to the process $e+p \rightarrow e+'X'$.

Fig. 9: Total cross section in pb of the process $e^+e^- \rightarrow ZZ$; dashed line: Born; dashed dotted line: $O(\alpha)$ corrected; solid lines: $O(\alpha^2)$ corrected.

Fig. 10: Total cross section in pb of the process $e^+e^- \rightarrow W^+W^-$; dashed line: Born; dashed dotted line: $O(\alpha)$ corrected; solid line: $O(\alpha^2)$ corrected.

Fig. 11: $d\sigma/d \cos\theta$ [pb] for the reaction e^+e^- ZZ; dashed line: Born; dashed dotted line: $O(\alpha)$ corrected; solid line: $O(\alpha^2)$ corrected.

Fig. 12: $d\sigma/d \cos\theta$ [pb] for the reaction $e^+e^- \rightarrow W^+W^-$; dashed line: Born; dashed dotted line: $O(\alpha)$ corrected.

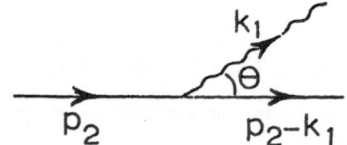

Fig. 1

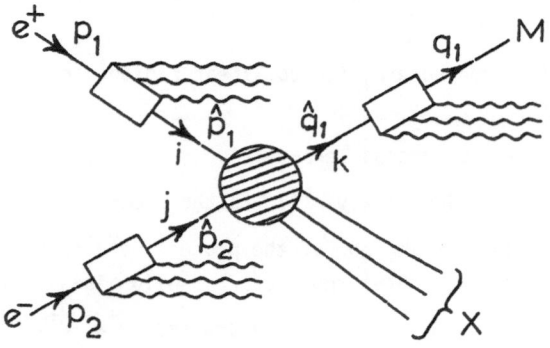

Fig. 2

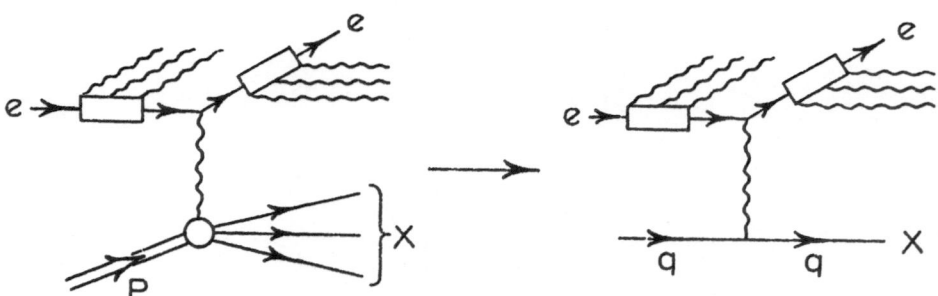

Fig. 3

21

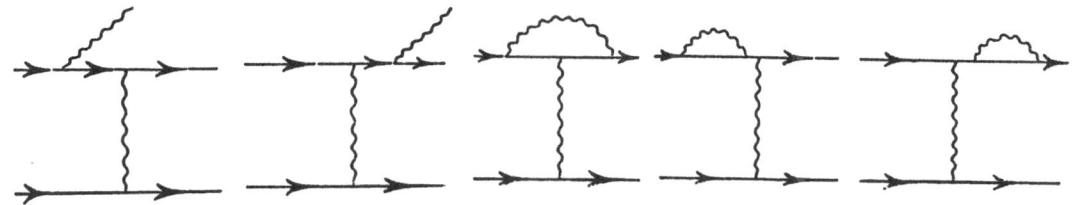

Fig. 4

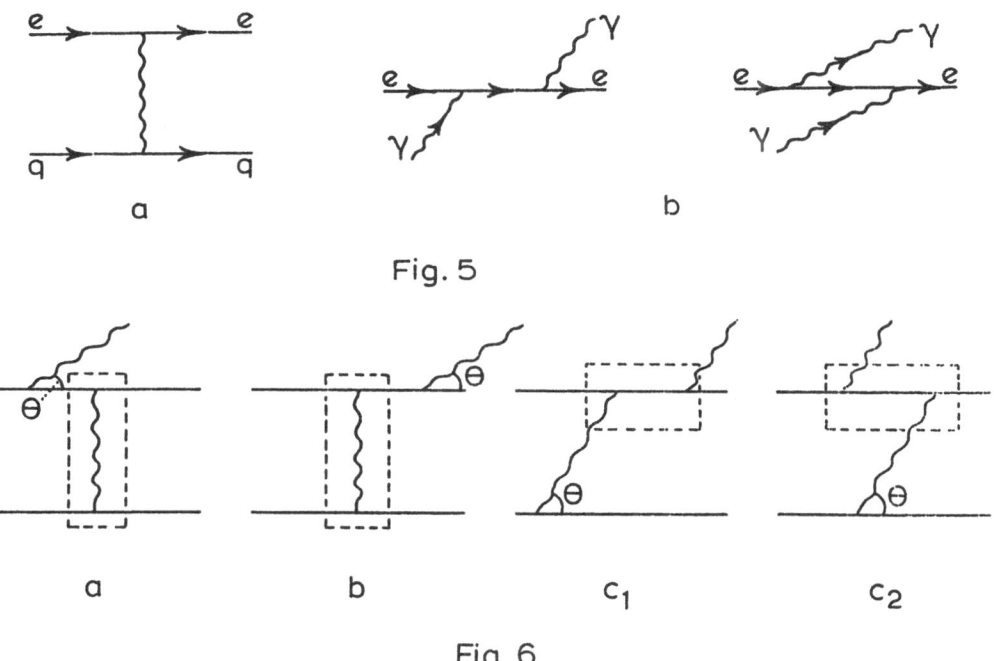

a

b

Fig. 5

a   b c_1 c_2

Fig. 6

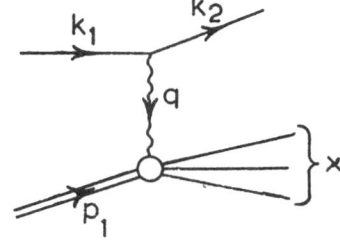

Fig. 7

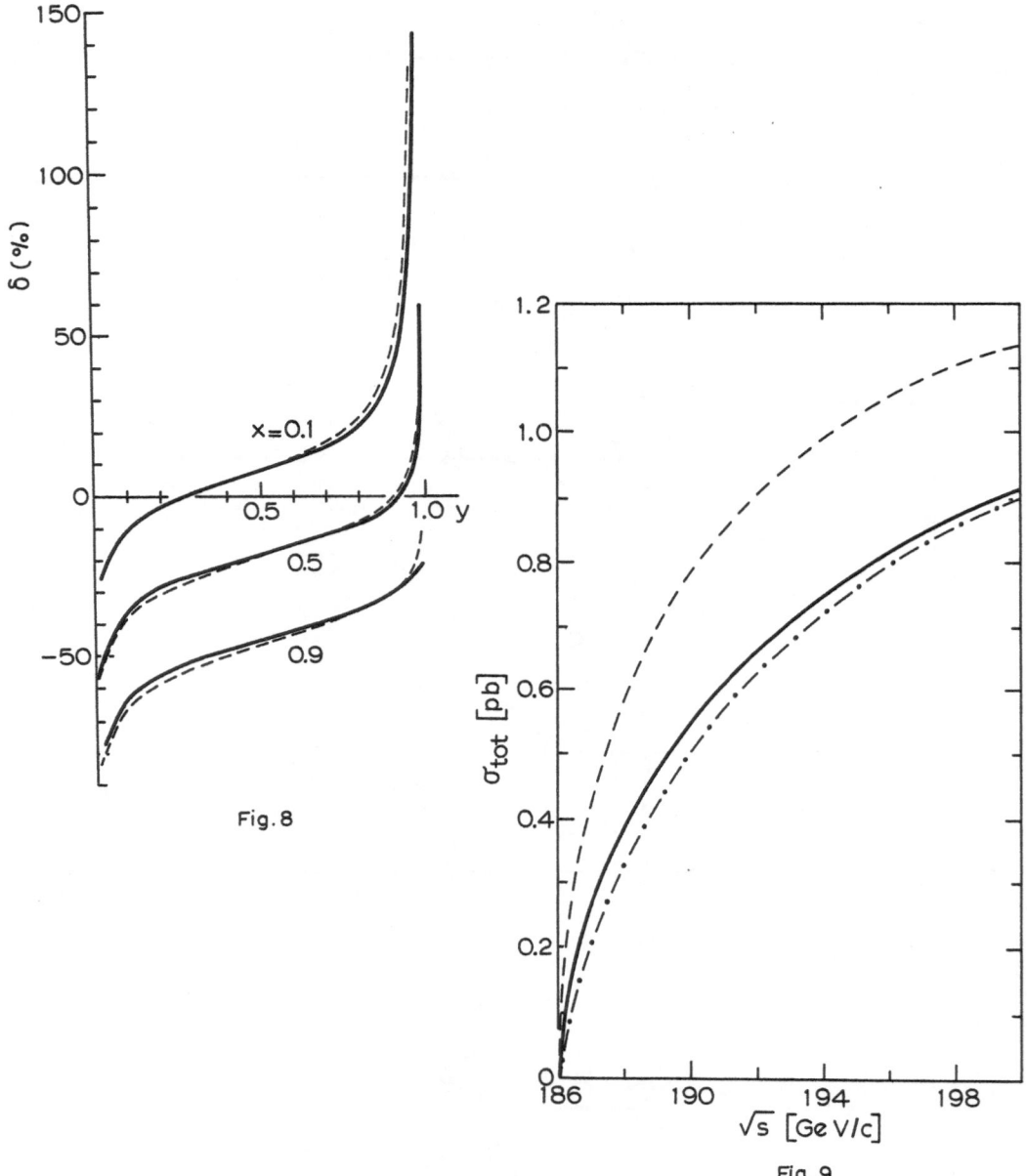

Fig. 8

Fig. 9

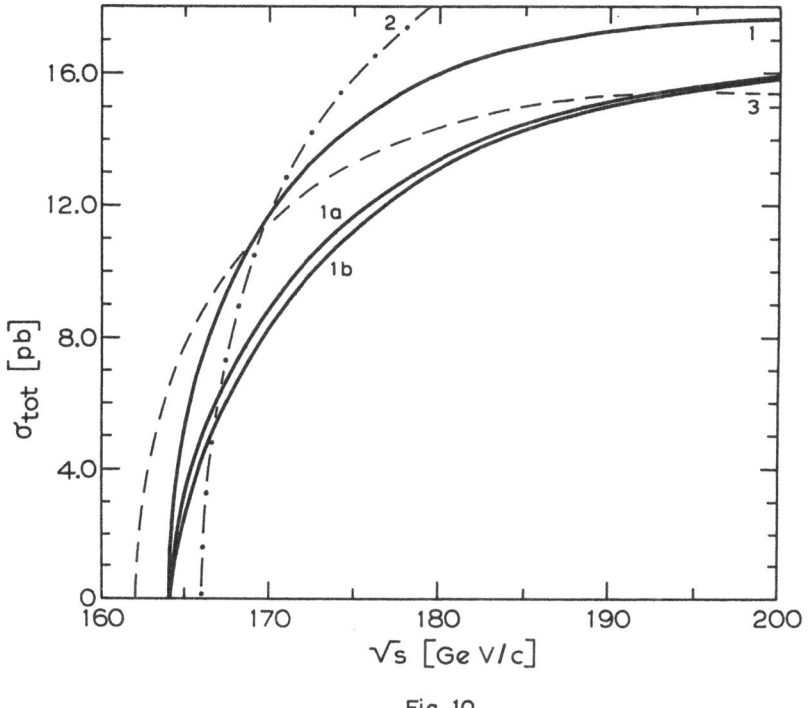

Fig. 10

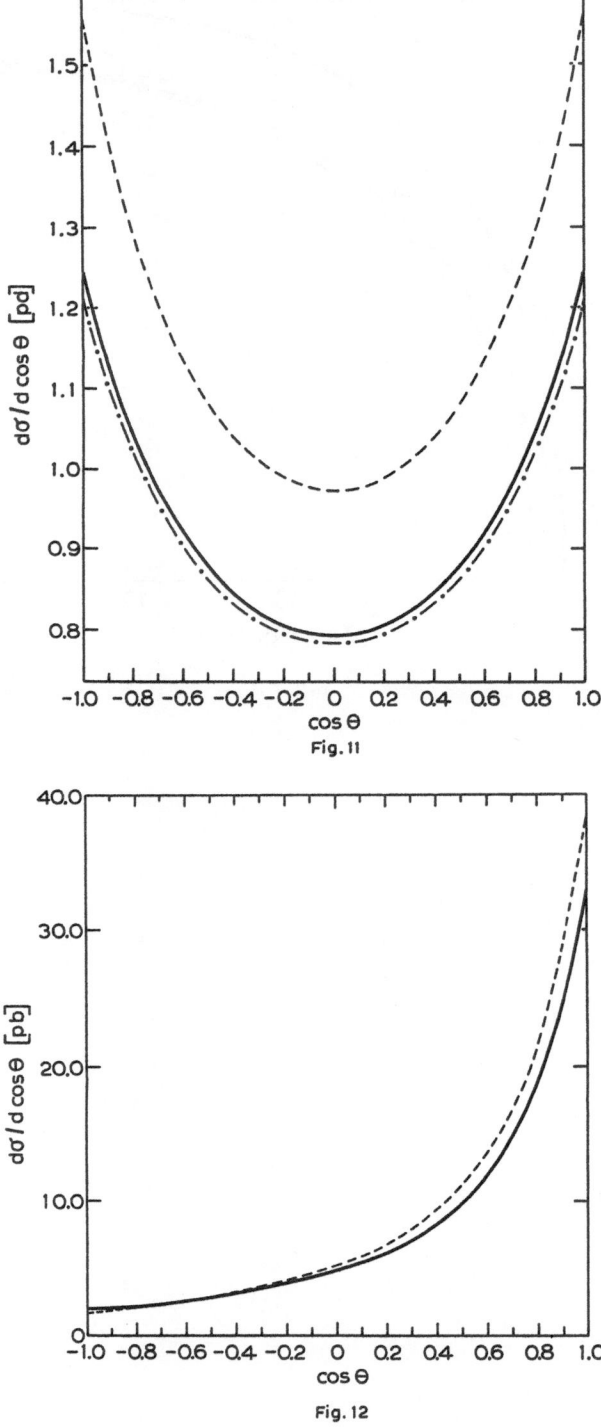

Fig. 11

Fig. 12

Radiative Corrections for e⁺e⁻ Collisions   Editor: J.H. Kühn
© Springer-Verlag Berlin, Heidelberg 1989

STRUCTURE FUNCTIONS TECHNIQUES IN e^+e^- COLLISIONS

O. Nicrosini

Istituto Nazionale di Fisica Nucleare, Sezione di Pavia, and
Dipartimento di Fisica Nucleare e Teorica dell'Università, Pavia, Italy

and

L. Trentadue

Dipartimento di Fisica dell'Università, Parma,
Istituto Nazionale di Fisica Nucleare,
Gruppo Collegato di Parma, Sezione di Milano, Italy, and
CERN
Geneva, Switzerland

Abstract

The use of the structure functions formalism to describe the electromagnetic radiative corrections in e^+e^- colliders is described. Examples are given of specific applications to physical quantities. The total cross section with both initial and final state radiative corrections is derived. p_t dependent structure functions are introduced and applied to the radiative neutrino counting problem. Expressions are derived for the forward-backward asymmetry A_{FB}. The problem of the next-to-leading factorization and of the use of structure functions to construct a Monte Carlo type algorithm is also briefly sketched

1. Introduction

A careful study of the properties of Standard Model and an accurate testing of its parameters are among the goals of the new electron-positron colliders LEP and SLc..

Precision tests of the electroweak theory around the Z^0 are in this context a relevant issue [1]. A significant role is played by the electromagnetic radiative corrections to the initial electron-positron state.

This has been an argument familiar, since long time, to those working at the electron-positron accelerators also in the last few years actively investigated [2]. The formalism of the structure functions, widely applied to describe the interactions of partons within the Quantum Chromodynamics QCD theory of the strong forces, has been recently used [3-5] for a new series of applications in the framework of the Quantum Electrodynamics QED theory. The main purpose of this new development has been the evaluation of the electromagnetic radiative corrections to leptonic processes.

Historically structure function evolution equations have been first applied to describe the interactions of fermions within vector theories [6]. The more complex case of the interaction within non-Abelian gauge theories has been also considered [7]. This formalism has been proposed as a new tool [3-5] to deal with the problems of describing the dynamics also of electron and photon states by using the corresponding evolution equations. One of the main motivations for this approach is the relevance that at high energies, particularly around peaked resonances, have the infrared and collinear singularity structure of the radiative corrections [2,3-5]. Structure Functions are with this respect able to deal with the mass singularity sector in a simple and effective way. The problem of evaluating two-loop corrections within this approach has been also studied [3,5] thus allowing, within the formalism, an evaluation of these perturbative contributions accurate to $O(\alpha^2)$.

2. General Formalism

In order to introduce the formalism of the structure functions for the Quantum Electrodynamics case it is useful to follow the analogy with the corresponding QCD case. A parton 'electron' can be defined as consisting of a cloud of real and virtual particles quarks and gluons 'electrons and photons'. The probability of finding within this give state a an electron or a photon b at a given scale or virtuality $k^2 = s$ with a fraction of longitudinal momentum $x = \frac{k_L}{E}$ with $E = \sqrt{s}/2$ can be given by defining a certain distribution $D_{ab}(x, s)$.

In general 'matrix' coupled equations should be considered connecting the different branching probabilities among the various channels. In fact if we call $D_{ab}(x, s)$ the probability distribution we may have that $a, b = e^+, e^-, \gamma$ and $D_{ab} = D_{ee}, D_{e\gamma}, D_{\gamma e}, D_{\gamma\gamma}$, see Fig.1 ,as well as $P(z) = P_{ee}, P_{e\gamma}, P_{\gamma e}, P_{\gamma\gamma}$. The correspondig coupled 'matrix' equations are:

$$s\frac{\partial D_{ee}(x, s)}{\partial s} = \frac{\alpha}{2\pi} \int_x^1 \frac{dz}{z} [D_{ee}(\frac{x}{z}, s) P_{ee}(z) + D_{e\gamma}(\frac{x}{z}, s) P_{e\gamma}(z)]$$

$$s\frac{\partial D_{e\gamma}(x, s)}{\partial s} = \frac{\alpha}{2\pi} \int_x^1 \frac{dz}{z} [D_{e\gamma}(\frac{x}{z}, s) P_{\gamma\gamma}(z) + (D_{ee+}(\frac{x}{z}, s) + D_{ee-}(\frac{x}{z}, s)) P_{e\gamma}(z)] \qquad (1)$$

with

$$P_{ee}(z) = P_{e^+e^-}(z) = P_{\gamma e}(1 - z) = P_{\gamma e+}(1 - z)$$

and

$$P_{e^+\gamma}(z) = P_{e\gamma}(z) = P_{e\gamma}(1 - z) = \frac{1}{2}(z^2 + (1 - z)^2)$$

We will be mainly interested in the radiative corrections to the cross section for photon emission. From a QCD analogy these contributions correspond to the Non-Singlet (NS) channel in the evolution equations i.e. the electron or positron line is continuously connected to the annihilation vertex * and all the coupled contributions of the Singlet type are neglected. The corresponding evolution equations can be therefore written with the electron and positron distributions that satisfy the 'master' evolution equation:

$$D_{ee}(x,s) = D(x,s) = \delta(1-x) + \int_{m_e^2}^s \frac{dk^2}{k^2} \frac{\alpha(k^2)}{2\pi} \int_x^1 \frac{dz}{z} P^{NS}(z) D_{ee}(\frac{x}{z},k^2) \qquad (2)$$

with

$$\alpha(k^2) = \frac{\alpha}{1 - \frac{\alpha}{3\pi}\ln(k^2/m_e^2)}$$

where

$$P^{NS}(z) = P(z) = \frac{1+z^2}{1-z} - \delta(1-z)\int_0^1 dx \frac{1+x^2}{1-x}$$

is the regularized *electron* $\rightarrow$ *electron + photon* vertex where the first term represents the 'real' photon radiation and the second the 'virtual' corrections, see Fig.2. Here the $\delta(1-x)$ represents a Born source term within the evolution equation and the regularization is obtained with the inclusion of the 'virtual' self-energy type contributions. The 'master' equation eq.(2) is graphically represented in Fig.3.

3. Initial State Radiative Corrections in e^+e^- Around the Z^0

The physics involved in the problem of the radiative corrections to the initial state in e^+e^- annihilation is related to the emission of quanta from the annihilating electron and positron states and to their effect on the size and shape of the resonance peak. It can be explained as follows : In the production of a resonance of mass M with colliding beams of total center of mass energy $\sqrt{s} = 2E$ emission of soft and hard photons occours. For $\sqrt{s} > M$, in the energy region above the resonance mass, this has the effect of decreasing the total center of mass energy to M. These states, therefore, effectively contribute to the resonance production. As a result a radiative tail arises on the right hand side of the resonance peak and, being the area under the peak not affected by radiative corrections, it follows a lowering of the maximum of the peak, see Fig.4 .

The determination of the Z^0 resonance cross section requires the evaluation of the radiative corrections to the electron-positron vertex as described in Fig.5. These should include both soft and hard photon radiation. At LEP/SLC it is expected that the accuracy on the determination of the mass and width of the Z^0 will be of the order of few dozen of MeV [1]. The size of the sistematic errors therefore can be reduced by obtaining a comparable accuracy on the estimates of the radiative electromagnetic corrections to the initial state.

For an electron or positron state the radiative corrections are characterized in the perturbative series by terms of the type $(\alpha/\pi)^n \ln^p(s/m_e^2) \ln^q(E/\lambda)$ with m_e the electron mass and λ a scale on which there is a sizeable variation of the cross section. In our case $\lambda = \delta E$ is the difference $\sqrt{s} - M$ i.e. the energy radiated away by the annihilating states. At the mass of the Z^0 the fact that $L = \ln(s/m_e^2) \simeq 24$ and the possibility that also the soft logarithms $l = \ln(E/\delta E)$ are also large implies that the effective expansion parameter $(\alpha/\pi)Ll$ is large too and therefore these terms in the perturbative series must be taken into account and summed.

Whereas the sum of collinear logarithms, due to the factorization property, is straightforward this is not the case for the soft logaritmic contributions. Techniques to perform these sums at the level of leading and non-leading terms have been developed to deal with the analogous problem in Quantum Chromodynamics. There, the combination $(\alpha_s/\pi)Ll$ assumes, due to the strong

---

* Since in this note only photon emission cross section will be considered we will consistently neglect all the contributions related to the appearence of virtual as well as real fermionic pairs.

running coupling constant, in most cases, a even larger value. We will apply these techniques to the present problem.

The process we are considering is the e^+e^- annihilation into the Z^0. According to a QCD analogy this scattering might be seen as a Drell-Yan process, see Fig.6, and, taking into account the well known theorems on the factorization of the mass and infrared singularities [9], its cross section can be written in the following form :

$$\sigma(s) = \int dx_1 \int dx_2 \; D_{e^-}(x_1, s) \; D_{e^+}(x_2, s) \; \sigma_R(\dot{x}_1 x_2 s) \tag{3}$$

$D_{e^-(+)}(x, s)$ represents the electron (positron) structure function giving the probability of finding an electron (positron) within an electron (positron) with a longitudinal momentum fraction $x = p_{e^-(+)}/E$ and virtualness s in a electron (positron). $\sigma_R(x_1 x_2 s)$ represents the resonance cross section at the reduced energy $x_1 x_2 s = s'$.

In order to evaluate the cross section one has to solve the evolution equations for the electon and positron distributions. Various procedures are possible at different approximation levels :

a) One possible method is to simply iterate Eq.(2) and write the distribution in terms of a series which for the Non-Singlet case, i.e. by taking into account only the graphs containing photon emission in eq.(2) see Fig.7, we are considering is:

$$D_{e^-(+)}(x, s) = \delta(1 - x) + \int \frac{dk^2}{k^2} \frac{\alpha(k^2)}{2\pi} \left(P(x) + \int \frac{dl^2}{l^2} \frac{\alpha(l^2)}{2\pi} \int \frac{dz}{z} P(z) P(\frac{x}{z}) \right) + ...$$

The series can be truncated to the required level of approximation. Here we will neglect all the terms of the order $(\alpha/\pi)^3$. The perturbative expansion contains terms of the form :

$(\alpha/\pi)^n L^n l^n \; (dominant)$; $(\alpha/\pi)^n L^n l^{n-j} \; (L - dominant)$;

$(\alpha/\pi)^n L^{n-j} l^n \; (l - dominant)$; $(\alpha/\pi)^n L^{n-j} l^{n-i} \; (non - dominant)$;

with $n \geq 1$ and $n \geq i, j \geq 1$. A first classification of the various terms according to their decreasing importance can be made by considering that the largest contribution will be given by those containing at least a logarithm $L = \ln(s/m_e^2)$ for each power of α/π, i.e. $(\alpha L)^n l^m$, $n \geq m \geq 0$.

b) A second solution can be obtained in the soft, large-x, limit. Here the large logarithm can be taken into account and summed by using the method developed by Gribov and Lipatov [10]. With the use of the running coupling constant the solution to the Eq.(2) reads :

$$D(x, s) = \frac{e^{\frac{\eta}{4}(\frac{3}{2} - 2\gamma_E)}}{\Gamma(\frac{\eta}{2})} (1 - x)^{\frac{\eta}{2} - 1} \tag{4}$$

with $\eta = -6 \ln(1 - \frac{\alpha}{3\pi} \ln(\frac{s}{m_e^2}))$ where γ_E is the Euler's constant.

By taking the product $D(x_1, s) D(x_2, s)$ in the expression for the cross section in Eq.(3) and substituting the solution of Eq.(4) and expanding in series in α/π one has, with the substitution $1 - x = E/\delta E$, that Eq.(3) becomes in the large-x soft limit:

$$\sigma(s) = \sigma_R(s)(1 + \sum_{n=1}^{2} \sum_{m=0}^{n} g_{nm}(l)(\frac{\alpha}{\pi})^n L^m) \tag{5}$$

with

$$g_{11}(l) = -2l + \frac{3}{2}; \; g_{22}(l) = 2l^2 - \frac{10}{3}l + \frac{11}{8} - \frac{\pi^2}{3}$$

It is now interesting to compare this expression with a finite second order result. We use the results of the work of Barbieri, Mignaco and Remiddi [8] on the electron form factor corresponding to the set of diagrams in Fig.8.

The cross section, with emission of only photons by the electron-positron state, is :

$$\sigma^{(\gamma)}(s) = \sigma_R(s)(1 + \sum_{n=1}^{2} \sum_{m=0}^{n} a_{nm}(l)(\frac{\alpha}{\pi})^n L^m)$$ (6)

with

$$a_{10}(l) = 2l + \frac{\pi^2}{3} - 2; \; a_{11}(l) = -2l + \frac{3}{2};$$

$$a_{20}(l) = 2l^2 + (\frac{2}{3}\pi^2 - 4)l - \frac{6}{5}(\zeta(2))^2 - \frac{9}{2}\zeta(3) - 6\zeta(2)\ln 2 + \frac{3}{8}\zeta(2) + \frac{57}{12};$$

$$a_{21}(l) = -4l^2 - \frac{45}{16} + (7 - \frac{2}{3}\pi^2)l + \frac{11}{2}\zeta(2) + 3\zeta(3);$$

$$a_{22}(l) = 2l^2 - 3l + \frac{9}{8} - 2\zeta(2);$$

This expression contains all the logarithmic contributions and in $a_{20}(l)$ the constant terms. By comparing Eq.(6) with Eq.(5) it appears that the coefficients of the large $(\alpha L)^n l^n$ leading logarithmic terms are exactly reproduced by the Gribov-Lipatov solution.

In Eq.(4), however, other terms are not reproduced which do appear in the second order solution Eq.(6). These contributions, which are of the type $(\alpha/\pi)^n L^j$ with $n > j$, can be included by applying the iterative method to solve Eq.(2). Here we use a fixed coupling constant and factorize the summed Gribov-Lipatov solution. From the second order result [3-5,13] one can also extract in fact the remaining virtual terms. This procedure gives:

$$D_{e^-(+)}^{NS}(x,s) = \frac{\beta}{2}(1-x)^{\frac{\beta}{2}-1}\Delta^{\frac{1}{2}} - (1+x)\frac{\alpha L}{2\pi} + (1+x)\frac{\alpha}{2\pi}$$

$$+\frac{1}{2}(\frac{\alpha L}{2\pi})^2[(1+x)(-4\ln(1-x) + 3\ln x) - \frac{4}{1-x}\ln x - 5 - x]$$

$$-\frac{1}{2}(\frac{\alpha}{2\pi})^2(2L-1)[(1+x)(-4\ln(1-x) + 3\ln x) - \frac{4}{1-x}\ln x - 5 - x]$$ (7)

where $\beta = \frac{2\alpha}{\pi}(L-1)$ and Δ is given by the expression [3,5] :

$$\Delta = 1 + \frac{\alpha}{\pi}(\frac{3}{2}L + \frac{\pi^2}{3} - 2) + (\frac{\alpha}{\pi})^2[(\frac{9}{8} - 2\zeta(2))L^2 + (-\frac{45}{16}$$ (8)

$$+\frac{11}{2}\zeta(2) + 3\zeta(3))L - \frac{6}{5}(\zeta(2))^2 - \frac{9}{2}\zeta(3) - 6\zeta(2)\ln 2 + \frac{3}{8}\zeta(2) + \frac{57}{12}].$$

By substituting the result for $D_{e^-(e^+)}(x,s)$ into Eq.(1) we have by defining $(1-\chi)s = s'$ that the cross section becomes:

$$\sigma(s) = \int d\chi \, \sigma_R((1-\chi)s) \, H(\chi,s)$$ (9)

with

$$H(\chi,s) = \Delta \left(\beta\chi^{\beta-1} - \frac{\beta^2\chi^\beta}{4}\right) - \frac{\beta\chi^{\frac{\beta}{2}}\Delta_{(1)}^{\frac{1}{2}}}{4}\left[(2-x)(1 + (1-\chi)^{-\frac{\beta}{2}})\right.$$

$$\left. - \frac{\beta}{(2+\beta)}\chi(1-(1-\chi)^{-\frac{\beta}{2}})\right] + \frac{\beta^2}{16}\left[(\chi - 2)(4\ln\chi - 3\ln(1-\chi))\right.$$

$$-\frac{4}{\chi}\ln(1-\chi)-6+\chi+2\chi-(2-\chi)\ln(1-\chi)\Big],$$

where $\Delta_{(1)} = 1 + \frac{\alpha}{\pi}(\frac{3}{2}L + \frac{\pi^2}{3} - 2)$.

In the expression above not only the dominant $(\alpha L)^n$ terms are summed but also the less dominant $\alpha^n L^m$ with $n \geq m \geq 0$ are taken into account up to $n = 2$. In Eq.(7) the first term $\frac{\beta}{2}(1-x)^{\frac{\beta}{2}-1}$ corresponds to the soft photon approximation. The terms proportional to $(1+x)$ and the last term contain contributions which modify the result accounting for emission of hard photons. The Δ factor contains terms of the type $(\frac{\alpha}{\pi})^n L^m, n \geq m \geq 0$.

Let us compare our expression Eq.(9) with previous results :

This expression [5] differs from the one obtained in ref.[3] for the inclusion of the terms $(\frac{\alpha}{\pi})^2 L^2, (\frac{\alpha}{\pi})^2 L, (\frac{\alpha}{\pi})^2 constant$ in the factor Δ. It agrees with the result obtained in ref.[11] by an exact $O(\alpha^2)$ calculation apart from terms that are relevant only within the 'hard' $x \to 0$ limit.

These facts all show that, being $O(\alpha^3)$ corrections really negligible, the accuracy reached on the cross section in Eq.(9) is below the one per cent level.

In order to represent the effect of the corrections we have numerically integrated Eq.(9). By using the cross section for $e^+e^- \to \mu^+\mu^-$ with $\Gamma = 2.6\,GeV$ and $M = 93.2\,GeV$.

$$\sigma_R(s) = \frac{4\pi\alpha^2}{3s}(1 + \frac{2(s-M^2)sc_v^2}{|s-M^2+iM\Gamma|^2} + \frac{s^2(c_v^2+c_A^2)^2}{|s-M^2+iM\Gamma|^2})$$

with $c_A = \frac{-1}{2sin2\theta_W}$, $c_v = c_A(1-4sin^2\theta_W)$ and $sin^2\theta_W = 0.223$. Fig.9 shows the effect of the radiative corrections for this cross section.

In Fig.10 is represented the fully corrected cross section in Eq.(9) as compared with the finite second and first order results. These are obtained by substituting to the exponentiated factor $\beta\chi^{\beta-1}$, representing the sum of the logarithmic terms in Eq.(9), its finite order forms, obtained by expanding it to the corresponding finite order.

4. Final State Radiative Corrections

As we have seen before radiation from the initial states strongly affects the the behaviour of the production cross section around the Z^0 peak by modifying both its shape and size. Also final state corrections [12] must be taken into account, since, together with initial final interference, are needed to reproduce physical quantities such as acollinearities and asymmetries. Any separation between radiation arising from initial as compared to the one from final states is, in processes involving charged particles, of no physical content. Radiative corrections should be considered therefore as a whole set of corrections for any given process.

Since finite $O(\alpha)$ results are highly inadequate to describe the many-particle nature of the electromagnetic radiation [2], expressions to all orders of the perturbative expansion are necessary. In ref.[13] it has been considered the general case of including the electromagnetic radiative corrections to the entire process. To this purpose the structure function formalism developed in [3-5] has been used and extended to both initial and final states. This will be done by carring an $O(\alpha^2)$ calculation which takes into account the resummation to all orders of the dominant and next to dominant [3] logarithmic contributions and uses finite order expressions [12] for the box and the interference contributions. In order to evaluate the final state radiation it is useful to start with a classification of the set of the diagrams involved. We will study the annihilation process $e^+e^- \to \mu^+\mu^-$ in an inclusive sense, i.e. with all the radiated photons summed over

The "bare" process is represented in Fig.11 a). The bulk of the radiative corrections is represented by the blob in Fig.11 b).

As shown before we can factorize initial state radiation and write the cross section as a convolution of a "bare" cross section with a "radiator" representing initial state radiation eq.(9). By applying the Kinoshita-Lee-Nauenberg factorization theorem [9], the radiatively corrected cross section can be written as:

$$\sigma(s) = \int dx \, \sigma_0((1-x)s) \, H_e(x,s)$$

According to the theorem on the factorization of infrared and mass singularities, the evaluation of the radiator gives a quantity independent from the process itself. Analogously also for the final state, a radiator can be defined that does factorize as the initial one does.

By taking the graph in Fig.11 c) as a representation of the initial state radiator $H_e(x,s)$, the final state radiator can be described by the symmetrical one in Fig.11 d). This as far as initial or final state radiation alone are concerned. Let us now consider corrections not belonging to the above classes, i.e. those obtained by joining initial with final states by means of photon lines. These, non-factorizable, corrections correspond to both interference terms and box diagrams. If one calls σ_k the set of the factorizable (Born) plus non factorizable (box diagrams and initial-final state interference) elementary cross sections the totally radiatively corrected process in Fig.11 b) can be effectively decomposed as described in Fig.11 e). The calculation of the total cross section can be performed by computing the graphs corresponding to the structure described in Fig.11 e). The accuracy that might be obtained depends on the various dominant and non-dominant contributions that are properly included within the three factors $H_e(x,s)$, $H_\mu(x,s)$ and σ_k. Initial, final states and the kernel σ_k can be evaluated with any desired accuracy, provided this is done consistently order by order in the perturbation expansion.

The radiator $H_e(x,s)$ is defined as :

$$H_e(x,s) = \int_{1-x}^{1} \frac{dz}{z} \, D_e(z,s) \, D_e(\frac{1-x}{z},s). \tag{10}$$

H_e contains contributions at all orders in perturbation theory at dominant and non-dominant level.

The generalization of eq.(9) to take into account also final state radiation is, for the factorized part of the cross section,

$$\sigma_f(s) = \int dx_1 dx_2 dy_1 dy_2 \, \sigma_0(s') \, D_e(x_1,s) D_e(x_2,s) D_\mu(y_1,s'') D_\mu(y_2,s''). \tag{11}$$

where $s' = x_1 x_2 s$ and $s'' = y_1 y_2 s'$, with $x_{1,2}$ and $y_{1,2}$ fractions of longitudinal momentum of the electrons and muons respectively. $D_\mu(x,s)$ is the structure function for the muon, obtained from $D_e(x,s)$ with the substitution $m_e \to m_\mu$. Note that the scale s'' at which D_μ are evaluated is the invariant mass squared of the final real muon pair. By making the substitution $1-x = x_1 x_2$ and $1-y = y_1 y_2$ and recalling eq.(2) for the initial state radiator one has:

$$\sigma_f(s) = \int_0^{r_{max}} dx \, \sigma_0(s') \, H_e(x,s) \, F_\mu(r_{max} - x, s') \tag{12}$$

where the upper limit r_{max} is the maximum fraction of radiation emitted and can be properly choosen and $F_\mu(z,s) = \int_0^z dx \, H_\mu(x,(1-x)s)$. The final state radiation kernel $H_\mu(x,s)$ is defined as:

$$H_\mu(x,s) = \int_{1-x}^{1} \frac{dy}{y} \, D_\mu(y,s) \, D_\mu(\frac{1-x}{y},s). $$

H_μ contains the same set of contributions as $H_e(x,s)$. Eq.(12) is in a factorized form. σ_0 corresponds to the Born cross section. To take into account also the non factorizable corrections having photon lines connecting initial with final state, box diagrams and initial-final state interference contributions should be also included. Let us define the effective kernel σ_k of the central blob in Fig.11 e). It contains all these non factorizable diagrams. A special care must be devoted to the cancellation of infrared singularities among initial, final and non factorizable contributions. By closely following the recipe sketched above, the cross section becomes:

$$\tilde{\sigma}(s) = \sigma_f(s) + \sigma_{box}(s) + \sigma_{int}(s) =$$

$$= \int_0^{r_{max}} dx\, \sigma_0(s')\, H_e(x,s)\, F_\mu(r_{max} - x, s') + \sigma_{box}(s) + \sigma_{int}(s) \qquad (13)$$

where x is the energy fraction radiated away.

$H_e(x,s)$, the initial state radiator as calculated by means of eq.(11), is [3]

$$H_e(x,s) = \Delta_e(s)\, \beta_e\, x^{\beta_e - 1} - \frac{1}{2}\, \beta_e\, (2-x)$$

$$+ \frac{1}{8}\, \beta_e^2 \left[(2-x)\, [3\ln(1-x) - 4\ln x] - \frac{4\ln(1-x)}{x} - 6 + x \right] \qquad (14)$$

$F_\mu(t,s)$, the final state radiator, can be written as:

$$F_\mu(t,s) = \int_0^t dy\, H_\mu(y, (1-y)s) \approx$$

$$\approx \int_0^t dy\, H_\mu(y,s) = \Delta_\mu(s)\, t^{\beta_\mu(s)} - \beta_\mu(s)(t - \frac{1}{4}t^2)$$

$$- \frac{1}{2}\beta_\mu^2(s) \left[Li_2(1-t) + \ln(1-t)\ln t + \frac{3}{2}\ln(1-t)\left[(1-t) + \frac{1}{4}(t^2 - 1)\right] \right.$$

$$\left. + 2\ln t(t - \frac{1}{4}t^2) - \frac{1}{16}t^2 + \frac{5}{8}t - \zeta(2) \right] \qquad (15)$$

The second approximate equality corresponds to neglecting the y dependence of $L_\mu((1-y)s) = \ln\frac{(1-y)s}{m_\mu^2}$ in Δ_μ and β_μ. With this approximation we neglect hard $y \approx 1$ radiation from the final state and this corresponds to set the scale of the final state to be s' instead of s''.

$\sigma_{box}(s)$ and $\sigma_{int}(s)$ respectively represent the contributions of non factorizable virtual and real interference terms. Following the notation of the last paper of ref.[12] they can be written as:

$$\sigma_{box}(s) = \int d\Omega \left[\frac{d\sigma^Q}{d\Omega} \delta_{\gamma\gamma}^Q + \frac{d\sigma^I}{d\Omega} (\delta_{\gamma\gamma}^I + \delta_{\gamma Z}^I) + \frac{d\sigma^Z}{d\Omega} \delta_{\gamma Z}^Z \right] \qquad (16)$$

$$\sigma_{int}(s) = \int d\Omega \frac{d\sigma_0}{d\Omega} \frac{2\alpha}{\pi} \left[4\ln\tan\frac{\theta}{2} \ln\frac{2E}{\lambda} + 2\ln^2(\sin\frac{\theta}{2}) \right.$$

$$\left. - 2\ln^2(\cos\frac{\theta}{2}) - Li_2(\sin^2\frac{\theta}{2}) + Li_2(\cos^2\frac{\theta}{2}) \right]$$

$$+ \frac{\alpha^3}{s} \int_0^{r_{max}} \frac{dz}{z} \left[\left[C(s,(1-z)s) - D(s,(1-z)s) \right](2-z) + 2\,W_2(s) \right] - \frac{2\alpha^3}{s} W_2(s) \ln r_{max} \qquad (17)$$

where λ is the parameter choosen to regularize the infrared divergence. Finally, it must be observed that both $\sigma_{box}(s)$ and $\sigma_{int}(s)$ are O(α) contributions with respect to $\sigma_0(s)$. So, consistently to the O(α) in the corrections, one can rewrite eq.(13) as

$$\sigma(s) = \int_0^{r_{max}} dx\, \sigma_k(s')\, H_e(x,s)\, F_\mu(r_{max} - x, s') \tag{18}$$

where $\sigma_k(s)$ is the effective integration kernel as defined above and given by

$$\sigma_k(s) = \sigma_0(s) + \sigma_{box}(s) + \sigma_{int}(s).$$

By comparing $eq.(18)$ with $eq.(13)$ we see that further corrections are taken into account

$$\sigma(s) = \sigma_f(s) + \int_0^{r_{max}} dx\, \Big[\sigma_{box}(s') + \sigma_{int}(s')\Big] H_e(x,s)\, F_\mu(r_{max} - x, s')$$

$\sigma(s)$ reduces to $\tilde\sigma(s)$ if only the $O(\alpha^0)$ in H_e and F_μ in the second term are taken into account. To better understand the content of $eq.(18)$, let's focus our attention on the first term in the r.h.s. of $eq.(13)$. By substituting in $eq.(13)$ the definition of $F_\mu(t,s)$ of $eq.(15)$ and by properly splitting the integration domain in the $x - y$ plane by using a cut-off ϵ, three contributions can be defined.

a) The *soft* cross section

$$\sigma_{soft}(s) = \int_0^\epsilon dx \int_0^{\epsilon-x} dy\, \sigma_0((1-x)s)\, H_e(x,s)\, H_\mu(y,(1-y)(1-x)s)$$

reduces to the following expression:

$$\sigma_{soft}(s) \approx \sigma_0(s) \int_0^\epsilon dx\, H_e(x,s) \int_0^\epsilon dy\, H_\mu(y,s) =$$

$$\sigma_0(s)\left[1 + 2\frac{\alpha}{\pi}\Big[[-1 + \frac{\pi^2}{6} + \frac{3}{4}L_e + (L_e - 1)\ln\epsilon] + [-1 + \frac{\pi^2}{6} + \frac{3}{4}L_\mu + (L_\mu - 1)\ln\epsilon]\Big]\right]$$

which represents the factorized part of the contribution of soft radiation from both initial and final states.

b) The *hard − initial* cross section

$$\sigma_e(s) = \int_\epsilon^{r_{max}} dx \int_0^{r_{max}-x} dy\, \sigma_0((1-x)s)\, H_e(x,s)\, H_\mu(y,(1-y)(1-x)s)$$

by freezing the final state radiation ($H_\mu(y,s) = \delta(y)$) and expanding to $O(\alpha)$ reduces to

$$\sigma_e(s) \approx \int_\epsilon^{r_{max}} dx\, \frac{d\sigma_e}{dx}$$

with

$$\frac{d\sigma_e}{dx} = \sigma_0((1-x)s)\frac{\alpha}{\pi}(L_e(s) - 1)\frac{1 + (1-x)^2}{x}$$

which is precisely the initial state bremsstrahlung spectrum.

c) The *hard − final* cross section

$$\sigma_\mu(s) = \int_0^\epsilon dx \int_{\epsilon-x}^{r_{max}-x} dy\, \sigma_0((1-x)s)\, H_e(x,s)\, H_\mu(y,(1-y)(1-x)s)$$

by freezing initial state radiation ($H_e(x,s) = \delta(x)$) and expanding to $O(\alpha)$ as for the previous case becomes

$$\sigma_\mu(s) \approx \int_\epsilon^{r_{max}} dy\, \frac{d\sigma_\mu}{dy}$$

where

$$\frac{d\sigma_\mu}{dy} = \sigma_0(s) \frac{\alpha}{\pi} \left[L_\mu\big((1-y)s\big) - 1 \right] \frac{1 + (1-y)^2}{y}$$

is nothing but the final state hard bremsstrahlung spectrum. The result in eq.(18) is nothing but a factorized correction to initial and final states of an effective one loop kernel. It reproduces the bulk of second order contributions to the Born cross section, resumming in addition the leading and next to leading singularities.

Recalling the definition of $F_\mu(z, s)$, one can see that the integrand in eq.(12) is the double differential cross section

$$\frac{d^2\sigma}{dx\,dy} = \sigma_k((1-x)s)\, H_e(x,s)\, H_\mu(y, (1-x)(1-y)s) \tag{19}$$

with x and y fractions of energy radiated away from initial and final state respectively. From eq.(19) one can derive the initial, final and total radiation spectra, resummed to all orders:

$$\frac{d\sigma_e}{dx} = \sigma_0((1-x)s)H_e(x,s) \tag{20}$$

$$\frac{d\sigma_\mu}{dy} = \sigma_0(s)H_\mu(y, (1-y)s) \tag{21}$$

$$\frac{d\sigma}{dz} = \int_0^z dx\, \sigma_k((1-x)s)\, H_e(x,s)\, H_\mu(z-x, (1-z+x)(1-x)s) \tag{22}$$

Eq.(20) and (21) are obtained by freezing final and initial radiation respectively. Eq.(22) is obtained by integrating eq.(19) over x and y with the constraint $x + y = z$. With the following choice of the standard model parameters: mass of the Z^0 $M_{Z^0} = 93.2\,GeV$, width $\Gamma = 2.6\,GeV$ and $\sin^2\theta_W = .222$.

The effect of the full set of the radiative corrections on the total cross section in eq.(11) is to lower the peak, to shift its position and to raise a radiative tail at high energy with respect to the Born approximation $\sigma_0(s)$. This effect can be seen in Fig.12.

Moreover a correct all orders treatment of initial state radiative corrections alone is almost exaustive, as far as total cross sections are concerned, as can be seen in the blow up Fig.13.

In order to use eq.(18) to analyze real experiments let's remark that detectors hardly distinguish between charged particles and the accompaining radiation. Calorimetric cross sections [14] have to be used by integrating over defined angles and ranges of energies of the emitted radiation surrounding the final particles.

5. Transverse Momentum Structure Functions

In order to deal with the problem of taking into account also the angles and the transverse momentum of the emitted photons, some extended distributions must be introduced. As discussed in the introduction, for the neutrino counting problem a somewhat less-inclusive quantity than the total cross section has to be evaluated, i.e. the rate for the process $e^+e^- \to \gamma, Z^0 \to \gamma\nu\bar\nu$.

For the QCD case, in ref.[15] (see also refs. therein) eqs. (1),(2) have been generalized to take into account also the transverse degrees of freedom. The evolution equation for $D(x, p_t; s)$ in the case of space-like kinematics, which is appropriate for the annihilation process, has the form:

$$D(x, p_t; s) = \delta(1-x)\delta^{(2)}(p_t) + \frac{\alpha}{2\pi} \int_{m^2}^s \frac{dk^2}{k^2 + m^2} \int_x^1 \frac{dz}{z} P(z)$$

$$\int \frac{d^2 q_t}{\pi} \delta((1-z)k^2 + z(1-z)m^2 - q_t^2) D(\frac{x}{z}, p_t - \frac{x}{z} q_t; k^2). \tag{23}$$

$D(x, p_t; s)$ represents the probability of finding inside a parent electron, at the scale s, an electron with fraction of longitudinal momentum x and transverse momentum p_t with respect to the initial beam direction see Fig.14. Eq. (23) obeys transverse momentum conservation. The transverse momentum of the electron is balanced by the one of the photon in the elementary branching process.

Let us solve eq. (23) by iterating the first term on the r.h.s., analogously to what can be done for the integrated $D(x, s)$ distribution. With one iteration of the source term, the solution at order α gives the following form for the distribution function:

$$
D(x, p_t; s) = \delta(1-x)\delta^{(2)}(p_t)
$$
$$
+ \frac{\alpha}{2\pi} P(x) \frac{1}{\pi} \frac{1}{p_t^2 + (1-x)^2 m^2} \Theta\left((1-x)s - p_t^2\right) + O\left(\frac{p_t^2}{E_\gamma^2}\right), \qquad (24)
$$

where Θ is the step function. $D(x, p_t; s)$ is normalized in such a way that, neglecting the electron mass m in the denominator, one has, by integrating over the transverse momentum degrees of freedom:

$$
\int d^2 p_t \, D(x, p_t; s) = D(x, s)
$$

consistently at order α, so that the original longitudinal momentum distribution is recovered.

Following the same lines as for the x dependent distributions, we define a p_t dependent radiator $H(x, p_t; s)$ as:

$$
H(x, p_t; s) = \int_{1-x}^{1} \frac{dz}{z} \int d^2 k_t \, D_{e^-}(z, k_t; s) D_{e^+}(\frac{1-x}{z}, p_t - k_t; s),
$$

where the indices e^- and e^+ label radiation from electron and positron respectively. The expression for $H(x, p_t; s)$ is, at order α,

$$
H^{(\alpha)}(x, p_t; s) = \delta(x)\delta^{(2)}(p_t) + \frac{\alpha}{2\pi} P(1-x)
$$
$$
\frac{1}{\pi} \left[\frac{1}{p_{te^+}^2 + x^2 m^2} + \frac{1}{p_{te^-}^2 + x^2 m^2} \right] \Theta\left(xs - p_t^2\right) + O\left(\frac{p_t^2}{E_\gamma^2}\right). \qquad (25)
$$

This defines the radiator for the neutrino counting problem. By using the p_t dependent radiator an angle dependent radiator can be defined $H^{(\alpha)}(x, \cos\theta; s)$. The transverse momentum and the angle of the emitted photon, measured with respect to the incoming electron, are linked by the relations

$$
p_{te^-}^2 = 2EE_\gamma (1 - \cos\theta) \qquad p_{te^+}^2 = 2EE_\gamma (1 + \cos\theta).
$$

One obtains, for the angle dependent radiator, the expression:

$$
H^{(\alpha)}(x, \cos\theta; s) = \frac{\alpha}{\pi} \frac{1 + (1-x)^2}{x} \frac{1}{1 + \frac{4m^2}{s} - \cos^2\theta} + O\left(\frac{m^2}{s}\right).
$$

The matrix element in the last equation does contain the leading soft and collinear singularities respectively represented by the $\frac{1}{x}$ and $\frac{1}{1-\cos^2\theta}$ poles. We can include also other less dominant (less singular) contributions by using the exact matrix element in the radiator, i.e. $H(x, \cos\theta; s)$ can be rewritten as [16]:

$$
H^{(\alpha)}(x, \cos\theta; s) = \frac{\alpha}{2\pi} \frac{1}{x} \left[2 \frac{1 + (1-x)^2}{1 + \frac{4m^2}{s} - \cos^2\theta} - x^2 \right] + O\left(\frac{m^2}{s}\right).
$$

This expression is more appropriate in the case when very hard photons ($x \approx 1$) are emitted at large angles ($\theta \approx \pi/2$). Moreover, for photons of energy $E_\gamma \approx 10$ GeV at $\sqrt{s} = 100$ GeV, the difference between the exact matrix element and the one given in eq.(25) is less than the 1% level, so that, to this accuracy, one can use the following form for the radiator:

$$H^{(\alpha)}(x, \cos\theta; s) = \frac{\alpha}{\pi} \frac{1 + (1-x)^2}{x} \frac{1}{1-\cos^2\theta} \Theta \left(1 - \frac{4m^2}{s} - |\cos\theta|\right).$$ (26)

6. Neutrino Counting

It has been proposed [17] that the number of neutrino families is a quantity that can be determined by counting photons produced in electron-positron annihilation around the Z^0 resonance peak. Th corresponding cross section can be computed by using the method of the structure functions for the initial fermions [20].

As for the total cross section case this formalism has the main advantage that it is able to take into account and resum, in a compact and straightforward way, initial as well as final state configurations that contain soft and collinear radiation.

The neutrino counting problem is related to different final states, i.e. those containing one or more isolated photons [17]. These, radiated only by the initial electrons, are detected. Those configurations, see Fig.15, define a quantity which is less inclusive than the total cross section and, for the neutrino counting problem, states with one or more radiated photons should also be accounted for.

With respect to the reaction $e^+e^- \to \gamma, Z^0 \to \gamma\nu\bar{\nu}$, there are various sources of background contamination: one is due to the background originated from Bhabha processes, but also the reactions $e^+e^- \to e^+e^- + \mu^+\mu^-\gamma, e^+e^- \to e^+e^- + R_{hadronic\ resonance} \to n\gamma$ and $e^+e^- \to \gamma\gamma\gamma$ give rise to a sizeable background.

Let us now consider the direct channel process and its physics. The problem is to determine the energy and the angle of emission of photons radiated by initial electrons and positrons accompanying the production of Z^0's. When a Z^0 decays into a $\nu\bar{\nu}$ pair, the photon is the only observed state, see Fig.15. In the average event, however, together with this process also hard photons radiated in the very forward direction and soft ones radiated all over the solid angle are likely to be produced. These change the effective process kinematics see Figs.16,17. The former photons are lost in the beam pipe or rejected according to the geometrical set-up of the apparatus; the second ones will not be observed for energies below the detector threshold.

Let us consider the cross section for producing a single photon by an electron-positron pair, $e^+e^- \to Z^0, W \to \gamma\nu\bar{\nu}$. The bare spectrum has been computed by many authors [17]. The corresponding cross section, obtained by evaluating the graphs in Fig. 18 in the limit $M_W = \infty$, can be written as follows:

$$\frac{d^2\sigma_0}{dx\,dy} = \frac{G_F^2 \alpha s(1-x)\left[(1-\frac{x}{2})^2 + \frac{x^2y^2}{4}\right]}{6\pi^2 x(1-y^2)} \cdot$$

$$\cdot \left(2 + \frac{N_\nu\left(g_v^2 + g_a^2\right) + 2\left(g_v + g_a\right)\left[1 - \frac{s(1-x)}{M_Z^2}\right]}{\left[1 - \frac{s(1-x)}{M_Z^2}\right]^2 + \Gamma_Z^2/M_Z^2}\right),$$ (27)

where G_F is the Fermi coupling constant, $x = 2E_\gamma/\sqrt{s}$ is the fraction of energy carried away by the emitted photon, $y = \cos\theta$, with θ the angle of the photon see Fig.19, N_ν is the number of neutrinos and $g_v = -\frac{1}{2} + 2\sin^2\theta_W$ and $g_a = -\frac{1}{2}$. In eq. (27) the first term in the last factor comes from the square of the W exchange amplitude the term containing N_ν from the square of the Z amplitude and the last one from the $Z - W$ interference Fig.18.

Radiative corrections to $e^+e^- \to \gamma\nu\bar{\nu}$ have been evaluated with standard Feynman-diagram tecniques [18,19] to the order α and the corresponding diagrams are in Fig.20. We will use the structure function approach in both these cases, firstly by defining structure functions that give probabilities of radiating photons at a particular angle with a given energy, and subsequently convoluting fermion lines with inclusive radiation of soft and collinear nature.

By using the radiator obtained in eq. (26), the cross section for the process under consideration is

$$\frac{d^2\sigma_0}{dx\,dy} = H^{(\alpha)}(x,y;s)\,\sigma_0\left((1-x)s\right), \qquad (28)$$

where σ_0 is the "reduced" cross section for the process $e^+e^- \to Z, W \to \nu\bar{\nu}$:

$$\sigma_0(s) = \frac{G_F^2 s}{12\pi}\left(2 + \frac{N_\nu\left(g_v^2 + g_a^2\right) + 2\left(g_v + g_a\right)\left[1 - \frac{s}{M_Z^2}\right]}{\left[1 - \frac{s}{M_Z^2}\right]^2 + \Gamma_Z^2/M_Z^2}\right). \qquad (29)$$

In order to analyze the various contributions to the radiative corrections let us first consider the virtual and soft ones described in Fig.20. By using the bare spectrum defined in eq. (28), we obtain the total bare cross section for producing a real photon by integrating x from $x_{min} = 2E_{\gamma min}/\sqrt{s}$ to 1, where $E_{\gamma min}$ is the minimum detectable energy, and y from minus $\cos\theta_{min}$ to plus $\cos\theta_{min}$, where θ_{min} is the minimum detectable angle (for a schematic description of the experimental set up see Fig. 19). The total bare cross section $\sigma^{(\gamma)}$ is then given by the expression:

$$\sigma^{(\gamma)}(s) = \int_{x_{min}}^{1} dx \int_{-\cos\theta_{min}}^{\cos\theta_{min}} dy\, H^{(\alpha)}(x,y;s)\,\sigma_0\left((1-x)s\right). \qquad (30)$$

The integral over y of $H^{(\alpha)}(x,y;s)$ is easily performed to give:

$$\int_{-\cos\theta_{min}}^{\cos\theta_{min}} dy\, H^{(\alpha)}(x,y;s) \approx$$

$$\frac{\alpha}{\pi}\frac{1+(1-x)^2}{x}\left[\ln\left(\frac{1+\cos\theta_{min}}{1-\cos\theta_{min}}\right)\cdot\Theta\left(1 - \frac{4m^2}{s} - \cos\theta_{min}\right) + \right.$$

$$\left. + \ln\frac{s}{m^2}\cdot\Theta\left(\cos\theta_{min} - 1 + \frac{4m^2}{s}\right)\right] = H^{(\alpha)}_{\theta_{min}}(x;s). \qquad (31)$$

In order to evaluate the $O(\alpha)$ virtual and soft radiative corrections to the bare process, it is sufficient to correct the cross section $\sigma^{(\gamma)}$ by the $O(\alpha)$ radiator $H_{(1)}(x,s)$ as given by:

$$H_{(1)}(x,s) = \frac{\beta}{x} - \frac{\beta}{2}(2-x)$$

in the following way:

$$\sigma^{(\gamma)}_{(1)}(s) = \int_{0}^{x_{min}} d\xi\, H(\xi,s)\,\sigma^\gamma\left((1-\xi)s\right)|_{(\alpha)} =$$

$$= \int_{0}^{x_{min}} d\xi\, H_{(1)}(\xi,s)\left[\sigma^{(\gamma)}\left((1-\xi)s\right) - \sigma^{(\gamma)}(s)\right]$$

$$+ \sigma^{(\gamma)}(s)\int_{0}^{x_{min}} d\xi\, H(\xi,s)|_{(\alpha)}, \qquad (32)$$

where $\sigma^{(\gamma)}_{(1)}$ is the $O(\alpha)$ corrected cross section. Since $H_{(1)}(x,s)$ is the $O(\alpha)$ expression of the the radiator $H(x,s)$ of eq. (7), the all orders soft and virtual radiative corrections [1-3] are taken into account by substituting, in the previous expression, $H(x, s)$ to $H_{(1)}(x, s)$. The the fully corrected cross section $\sigma_c^{(\gamma)}$ is given by:

$$\sigma_c^{(\gamma)}(s) = \int_0^{x_{min}} d\xi \, H(\xi, s) \, \sigma^{(\gamma)}((1 - \xi)s). \tag{33}$$

By interchanging the order of integration over ξ and x and dropping the integral over x, one obtains the differential spectrum by using both the single bremmstrahlung spectrum of eq. (26) and the radiator H_e:

$$\frac{d\sigma^{soft}}{dx} = H_{\theta_{min}}^{(\alpha)}(x, s) \int_0^{x_{min}} d\xi \, H(\xi, s) \, \sigma_0((1 - x)(1 - \xi)s). \tag{34}$$

Eq. (34) describes the x spectrum of a real photon accompanied by soft and virtual radiation. Let us now consider the case of hard corrections to the bare process. By hard corrections we mean those due to photons that are detectable but that are for various reasons lost. These photons effectively contribute to the bare process and must be taken into account. The differential spectrum is given by:

$$\frac{d\sigma}{dx_1 dx_2 dy_1 dy_2} = H^{(\alpha)}(x_1, y_1; s) H^{(\alpha)}(x_2, y_2; (1 - x_1)s)\sigma_0((1 - x_1)(1 - x_2)s).$$

By substituting to $H^{(\alpha)}(x, y; s)$ the matrix element as given in eq. (26), the double bremsstrahlung matrix element is recovered.

It is necessary to distinguish two kinds of hard corrections: a) hard photons lost in the pipe and b) hard photons parallel (within an apparatus dependent resolution angle) to the observed one.

a) These photons are lost being emitted at angles smaller than a *veto* angle (see Fig.19). The collinear singularity gives rise to logarithmic contributions of the form $\frac{\alpha}{\pi} \ln(\frac{s}{m^2})$. As for the infrared singularity, one should sum them to all orders. The production spectrum must be integrated from the forward 0 angle direction to the veto angle.

To $O(\alpha)$, the contribution coming from one photon lost in the pipe is given by:

$$\frac{d\sigma_{(1)}^{pipe}}{dx} = H_{\theta_{min}}^{(\alpha)}(x; s) \int_{\frac{x_{min}}{\sqrt{1-x}}}^{\sqrt{1-x}} dx_1 \cdot 2$$

$$\cdot \int_{\cos\theta_v}^{1 - \frac{m^2}{s}} dy_1 H^{(\alpha)}(x_1, y_1; (1 - x)s)\sigma_0((1 - x_1)(1 - x)s),$$

where $H^{(\alpha)}(x, y; s)$ is the angle dependent radiator of eq. (26) and θ_v is the veto angle. The factor 2 takes into account the backward collinear singularity.

The expression in eq. (26) can be generalized in order to handle the collinear singularity resummation problem. By defining

$$H(x, y; s) = \Delta(s) \frac{2\alpha}{\pi} \frac{1}{1 - y^2} x^{\beta(y)-1} [1 + \beta(y) \ln(x)]$$

$$- \frac{\alpha}{\pi} \frac{1}{1 - y^2} (2 - x) + \left(\frac{\alpha}{\pi}\right)^2 \frac{1}{1 - y^2} \ln\left(\frac{s(1 - y^2)}{2m^2}\right) \cdot$$

$$\cdot \left[(2 - x) \left[3 \ln(1 - x) - 4 \ln x\right] - \frac{4 \ln(1 - x)}{x} - 6 + x\right], \tag{35}$$

where $\beta(y)$ is given by

$$\beta(y) = \frac{2\alpha}{\pi} \ln\left(\frac{s(1-y^2)}{2m^2}\right).$$

The integral over y of $H(x,y;s)$ gives the resummed x dependent radiator. The contribution to the spectrum, by integrating over the angular region as discussed before, is

$$\frac{d\sigma^{pipe}}{dx} = H_{\theta_{min}}^{(\alpha)}(x;s) \int_{\frac{x_{min}}{\sqrt{1-x}}}^{\sqrt{1-x}} dx_1 \cdot 2$$

$$\cdot \int_{\cos\theta_v}^{1-\frac{m^2}{s}} dy_1 H(x_1, y_1; (1-x)s)\sigma_0\left((1-x_1)(1-x)s\right).$$

The integral over y_1 is easily performed [20] to give

$$\frac{d\sigma^{pipe}}{dx} = H_{\theta_{min}}^{(\alpha)}(x;s) \cdot$$

$$\cdot \int_{\frac{x_{min}}{\sqrt{1-x}}}^{\sqrt{1-x}} dx_1 H^{pipe}(x_1; (1-x)s)\sigma_0\left((1-x_1)(1-x)s\right), \tag{36}$$

where H^{pipe} is given by

$$H^{pipe}(x,s) \equiv 2 \cdot \int_{\cos\theta_{veto}}^{1-\frac{m^2}{s}} dy\, H(x,y;s) = \Delta(s)\,\beta_v\, x^{\beta_v-1} - \frac{1}{2}\,\beta_v\,(2-x)$$

$$+ \frac{1}{8}\,\beta_v^2\left[(2-x)\,[\,3\ln(1-x)-4\ln x\,] - \frac{4\ln(1-x)}{x} - 6 + x\right], \tag{37}$$

with $\beta_v = \frac{2\alpha}{\pi}\ln\left(\frac{s(1-c_v)}{m^2}\right)$ and $c_v = \cos\theta_v$. b) photons parallel to the observed one. The contribution to the cross section give by photons parallel to the observed one can also be computed. One has [20] :

$$\frac{d\sigma^{par}}{dx} = 2 \cdot \int_0^{c_m} dy H^{(\alpha)}(x,y;s) \int_{\frac{x_{min}}{\sqrt{1-x}}}^{\sqrt{1-x}} dx_1$$

$$\cdot \int_{yc_r-\sqrt{1-y^2}s_r}^{yc_r+\sqrt{1-y^2}s_r} dy_1 H^{(\alpha)}(x_1, y_1; (1-x)s)\,\sigma_0((1-x)(1-x_1)s), \tag{38}$$

where $c_r = \cos\theta_r$, $s_r = \sin\theta_r$, θ_r being half of the resolution angle see Fig.19.
The contribution can be written as:

$$\frac{d\sigma^{par}}{dx} = 2 \cdot \int_0^{c_m} dy H^{(\alpha)}(x,y;s)$$

$$\int_{\frac{x_{min}}{\sqrt{1-x}}}^{\sqrt{1-x}} dx_1 H_{par}^{(\alpha)}(x_1, y; (1-x)s)\sigma_0((1-x)(1-x_1)s). \tag{39}$$

The spectrum of the observed photon $\frac{d\sigma}{dx}$ is in fact given by the following sum:

$$\frac{d\sigma}{dx} = \frac{d\sigma^{soft}}{dx} + \frac{d\sigma^{pipe}}{dx} + \frac{d\sigma^{par}}{dx}, \tag{40}$$

where the three contributions of the r.h.s. are respectively given by eqs. (34) (36) and (39).
We have performed a quantitative analysis with the choice of parameters: $G_F = \frac{\pi\alpha}{\sqrt{2}\sin^2\theta_W M_W^2}$,

$M_W = 82$ GeV, $M_Z = 93.2$ GeV, $\Gamma_Z = 2.6$ GeV, $N_\nu = 3$, $E_{\gamma\,min} = 1$ GeV, $\sin^2\theta_W = 0.223$, $\theta_{min} = 20°$, $\theta_v = 3°$ and $\theta_r = .5°$.

In Figs.21,22 total cross sections are plotted. As can be seen, the effect of resummation is to slightly reduce the effect of the finite order electromagnetic radiative corrections, the resummed peak being a little higher than the one at order α. In Fig.23 ($\sqrt{s} = 98$ GeV) a direct comparison between the Born-approximation energy spectrum and the order α and fully corrected ones is shown: the effect of resummation is to sensibly modify the shape of the distribution.

7. Forward-Backward Asymmetries

Let us confine ourselves to the process $e^+e^- \rightarrow \gamma, Z^0 \rightarrow \mu^+\mu^-$. The zeroth order differential cross section is given by

$$\frac{d\sigma_0}{d\Omega_\mu} = \frac{\alpha^2}{4s}\left[W_1(s)(1 + c^2) + W_2(s)c\right], \tag{41}$$

where $c = \cos\vartheta$, ϑ being the angle between the momenta of e^- and μ^-, and the functions $W_1(s)$ and $W_2(s)$ are given by

$$W_1(s) = 1 + \frac{2(s - M^2)sc_v^2}{|Z(s)|^2} + \frac{s^2(c_v^2 + c_a^2)^2}{|Z(s)|^2},$$

$$W_2(s) = \frac{4(s - M^2)sc_a^2}{|Z(s)|^2} + \frac{8s^2c_v^2c_a^2}{|Z(s)|^2}, \tag{42}$$

$$Z(s) = s - M^2 + iM\Gamma.$$

In eq. (42) M and Γ are the mass and width of the Z^0 respectively. c_a and c_v are given by

$$c_a = -\frac{1}{2\sin(2\vartheta_W)}, \qquad c_v = -c_a(4\sin^2\vartheta_W - 1).$$

The forward-backward asymmetry is defined as

$$A_0(s) = \frac{\sigma_0^{(+)}(s) - \sigma_0^{(-)}(s)}{\sigma_0(s)}, \tag{43}$$

where $\sigma_0^{(+)}(s)$ and $\sigma_0^{(-)}(s)$ are the forward and backward hemisphere cross sections respectively, and $\sigma_0(s)$ is the total cross section. An explicit calculation gives

$$\sigma_0^{(+)}(s) = \frac{\pi\alpha^2}{4s}\left[\frac{8}{3}W_1(s) + W_2(s)\right],$$

$$\sigma_0^{(-)}(s) = \frac{\pi\alpha^2}{4s}\left[\frac{8}{3}W_1(s) - W_2(s)\right], \tag{44}$$

$$\sigma_0(s) = \frac{4\pi\alpha^2}{3s}W_1(s).$$

Inserting eq. (44) into (43) one obtains the expression for the born approximation asymmetry:

$$A_0(s) = \frac{3}{8}\frac{W_2(s)}{W_1(s)}. \tag{45}$$

The calculation of the QED radiative corrections is reduced to the evaluation of the structure function for the initial states [3-5]. In particular the corrected cross section can be written as a convolution of a "bare" cross section and a "radiator" H. The structure function approach can be generalized to include final state radiative corrections too [13]. The corrected cross section can be written in the form [21]

$$\sigma(s) = \int_0^\varepsilon dx \, H_e(x,s) F_\mu\left(\varepsilon - x, (1-x)s\right) \sigma_0\left((1-x)s\right) + \int d\Omega_\mu \frac{d\sigma_B}{d\Omega_\mu}, \qquad (46)$$

where $\frac{d\sigma_B}{d\Omega_\mu}$ is is the box and interference contribution (see for example ref.[12]). H_e and F_μ represent initial and final state radiation respectively [13]. Radiative corrections to the born approximation asymmetry are implemented by simply applying eq. (46) to the forward and backward hemisphere born approximation cross sections $\sigma_0^{(+)}$ and $\sigma_0^{(-)}$ of eq. (44)

$$\sigma^{(+)}(s) = \int_0^\varepsilon dx \, H_e(x,s) F_\mu\left(\varepsilon - x, (1-x)s\right) \sigma_0^{(+)}\left((1-x)s\right) + \int_0^1 d\Omega_\mu \frac{d\sigma_B}{d\Omega_\mu},$$

$$\sigma^{(-)}(s) = \int_0^\varepsilon dx \, H_e(x,s) F_\mu\left(\varepsilon - x, (1-x)s\right) \sigma_0^{(-)}\left((1-x)s\right) + \int_{-1}^0 d\Omega_\mu \frac{d\sigma_B}{d\Omega_\mu}, \qquad (47)$$

and then combining the results according to the definition

$$A(s) = \frac{\sigma^{(+)}(s) - \sigma^{(-)}(s)}{\sigma(s)}. \qquad (48)$$

Eq. (47) takes into account the following contributions to QED radiative corrections:
- exact $O(\alpha^2)$ real and virtual photonic radiation from initial and final state
- $O(\alpha)$ box and interference contributions
- resummation to all orders of perturbation theory of leading and part of next to leading logarithms both form initial and final state radiation Finite order results can be obtained by simply expanding the initial and final state radiators to the desired order. Pure initial state results are obtained by putting $F_\mu = 1$ everywhere. At the peak soft photon contributions represent the largest effect of the radiation. As for the line shape case the factorized form in eq.(46) is justified [22].

In order to quantify the effects of QED radiative corrections, we have numerically evaluated eqs. (47) and (48) with the following choice for the parameters: $M = 93.2$ GeV, $\Gamma = 2.6$ GeV, $\sin^2 \vartheta_W = 0.223$.

In Fig.24 we have plotted the forward-backward asymmetry as a function of the centre of mass energy $\sqrt{s}$. The continuous and dash-dotted lines represent the resummed and $O(\alpha)$ results respectively, obtained by using eqs. (47) and (48) with the resummed radiators or their $O(\alpha)$ expression. As is well known, the effect of radiative corrections is to shift the position of the zeroes and to sensibly change the shape of the asymmetry under and above the resonance energy.

In the same situation as before (no photon-energy cut and no box-interference contribution), in Fig.25 we have plotted the forward-backward asymmetry as a function of the centre of mass energy in a very strict region around the Z^0 mass. The position of the zeroes can be evaluated. In particular, the dashed line corresponds to Born approximation; the contribution of the initial-final-state radiation to $O(\alpha)$ asymmetry is represented by the dashed-dotted curve; according to eq. (47), resummation is included in the solid line which lies between the results for born and $O(\alpha)$ asymmetries. The correction due to resummation of the initial-state radiation only gives a curve which is, in the scale considered, coincident with the total-resummation solid line and for this reason doesn't appear in the figure. In Fig.26 the box and interference contribution are taken into account when the energy radiated by the photon is less than 1 GeV ($\varepsilon = 2/\sqrt{s}$). It has already been pointed out, in fact, that initial-final interference diagrams are important if photon-energy cuts are considered [23]. Our results support this conclusion. The sparsely dotted line is the $O(\alpha)$ asymmetry without box-interference terms while the closely dotted line corresponds to $O(\alpha)$ asymmetry with the inclusion of box-interference corrections. The effect on the shape is remarkable: more precisely, the inclusion of box and interference effects gives larger values for the asymmetry at the same center-of-mass energy and the position of the zero is changed towards lower energy of about 100 MeV. It is worth noting that this correction is of the same order of magnitude and sign as the one due to resummation only (dash-dotted line). Expanding to $O(\alpha)$ and using the proper parameters the results given in ref.[12] are reproduced. We also agree with the results given in ref.[23], where the same formalism, but with slightly different structure functions and paremeters, has been adopted. The effect of resummation is important in the calculation of the asymmetry, causing both a sizable shift in the location of the zero and a modification of the shape, expecially above the Z^0 resonance. Initial state resummation is the bulk of the effect. In imposing cuts, the effect of resummation becomes of the same order of magnitude of the correction due to $O(\alpha)$ box and interference contributions so that, to be consistent, one has to take both into account.

8. Monte Carlo

Structure Functions Formalism has been applied to develop a Monte Carlo code [24]. the main features of this code are:
- Use of p_t dependent structure functions as defined in Sect.4 which allow to describe both the longitudinal and the transverse degrees of freedom in the electron evolution.
- The radiatively corrected cross sections have the simple, factorized, structure given by the convolution of a Radiator with the corresponding Born cross section.
- Multiple photon configurations with the explicit four momenta are generated according to a Poissonian distribution.
- Multiphoton exponentiation and next to leading factorization are implemented.
- By integrating exclusive final configurations the Z^0 line shape as well as photon spectrum in the neutrino counting process are reproduced according to the analytical results within the statistical errors of the Monte Carlo [24].

References

[1] G. Altarelli, in "Physics at LEP", CERN - Yellow Report, 86-02, J. Ellis and R. Peccei Editors, Geneva, February 1986; F. Gilman, SLAC-PUB-4002, June 1986, Talk presented at the Seventh Vanderbilt Conference on High Energy Physics, Nashville, Tennessee, May 15-17, 1986,.

[2] E. Etim, G. Pancheri, B. Touschek, Nuovo Cimento 51B (1967) 276; G. Bonneau, F. Martin, Nucl. Phys. B27 (1971) 381; M. Greco, G. Pancheri-Srivastava, Y. Srivastava, Nucl. Phys. B101 (1975) 11, B171 (1980), 118; J. D. Jackson, D. L. Scharre, Nucl. Instruments and Methods 128 (1975) 13; F. A. Berends, R. Kleiss, Nucl. Phys. B178 (1981) 141; V. Baier, V. S. Fadin, V. Khoze, E. A. Kuraev, Phys. Rep. 78, 294 (1981); Y. S. Tsai, SLAC-PUB-3129(1983), Presented at the Asia Pacific Conference, Singapore, June 12-18 1983; V. S. Fadin, V. A. Khoze, Yad. Fiz., 47 (1988) 1693.

[3] E. A. Kuraev, V. S. Fadin, Yad. Fiz. 41, 753(1985) [Sov. J. Nucl. Phys. 41 (3), 1985, 466].

[4] G. Altarelli, G. Martinelli, in "Physics at LEP", CERN-Yellow Report, 86-02, J. Ellis and R. Peccei Editors, Geneva, February 1986.

[5] O. Nicrosini, L. Trentadue - Phys. Lett. 196B (1987) 551.

[6] L. Lipatov, Yad. Fiz. 20, 181 (1974)[Sov. J. of Nucl. Phys. 20, 94 (1975)]; V. Baier, V. S. Fadin, V. A. Khoze, Nucl. Phys. B65 (1973) 381; J. Kogut, L. Susskind, Phys. Rev. D9 (1974) 693, 3391.

[7] G. Altarelli, G. Parisi, Nucl. Phys. B126 (1977) 298.

[8] R. Barbieri, J. A. Mignaco, E. Remiddi, Nuovo Cimento 11A (1972) 824.

[9] T. Kinoshita, J. Math. Phys., 3, 650 (1962); T. D. Lee, M. Nauenberg, Phys. Rev. 133 (1964) 1549.

[10] V. Gribov, L. Lipatov, Yad. Fiz. 15, 781, 1218 (1972) [Sov. J. Nucl. Phys, 15, 938, 675 (1972)].

[11] F. A. Berends, G. J. H. Burgers, and W. L. van Neerven, Phys. Lett. 185B (1987) 395; G. J. Burgers, Phys. Lett. 164B (1985) 167; F. A. Berends, G. Burgers, W. Hollik and W. L. van Neerven, Phys. Lett. B203 (1988) 177; J. P. Alexander, G. Bonvicini, P. S. Drell and R. Frey, Phys. Rev. D37 (1988).

[12] M. Greco, G. Pancheri, Y. Srivastava , Nucl. Phys. B101 (1975) 11, B171 (1980) 118, B197 (1982) 543; M. Greco - "Physics at LEP", CERN - Yellow Report, 86-02, J. Ellis and R. Peccei Editors, Geneva, February 1986; F. A. Berends, R. Kleiss, S. Jadach, Nucl. Phys. B202 (1982) 63.

[13] O. Nicrosini, L. Trentadue, Zeitsch. Phys. C39 (1988) 479.

[14] G. Sterman, S. Weinberg, Phys. Rev. Lett. 39 (1977) 1436; G. Curci, M. Greco, Phys. Lett. 79B (1978) 406; M.Greco, "Physics at LEP", CERN - Yellow Report, 86-02, J. Ellis and R. Peccei Editors, Geneva, February 1986;

[15] A. Bassetto, M. Ciafaloni, G. Marchesini, Phys. Rep. 100 (4) 1983.

[16] G. Bonneau and F. Martin, Nucl. Phys. B27 (1971) 381; see also F. A. Berends and R. Kleiss, Nucl. Phys. B260 (1985) 32.

[17] A. D. Dolgov, L. B. Okun and V. I. Zacharov, Nucl. Phys. B41 (1972) 197; V. S. Fadin, V. Khoze, ZhETF, Pis Red 17, 8, 438 (1973); E. Ma and J. Okada, Phys. Rev. Lett. 41 (1978) 287; K. J. F. Gaemers, R. Gastmans and F. M. Renard, Phys. Rev. D19 (1979) 1605; G. Barbiellini, B. Richter and J. L. Siegrist, Phys. Lett. 106B (1981) 414.

[18] F. A. Berends, G. J. H. Burgers, C. Mana, M. Martinez and W. L. van Neerven, Nucl. Phys. B301 (1988) 583.

[19] M. Igarashi and N. Nakazawa, Nucl. Phys. B288 (1987) 301.

[20] O. Nicrosini, L. Trentadue, Nucl. Phys. B318 (1989) 1.

[21] G. Montagna, O. Nicrosini, L. Trentadue, Universita' di Pavia, preprint in preparation.

[22] See also F.Berends and W.van Neerven, at this Workshop.

[23] J. E. Campagne, R. Zitoun, LPNHE 88-06, 88-08.

[24] G. Bonvicini, L. Trentadue, UMHEP preprint 1989 (Nucl.Phys.B in press).

44

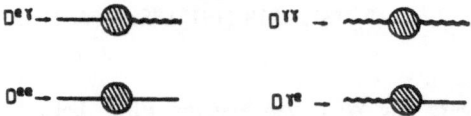

Fig.1 Coupled photon and electron distribution function.

Fig.2 Real and virtual radiative emission.

Fig.3 Graphical representation of the 'master' equation.

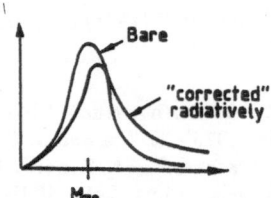

Fig.4 The effect of lowering of the peak hight and the emerging radiative tail due to radiative corrections.

Fig.5 The diagrams correcting the Born electron-positron annihilation process.

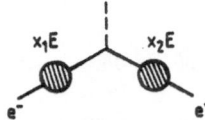

Fig.6 The Drell-Yan type dressed electron and positron states.

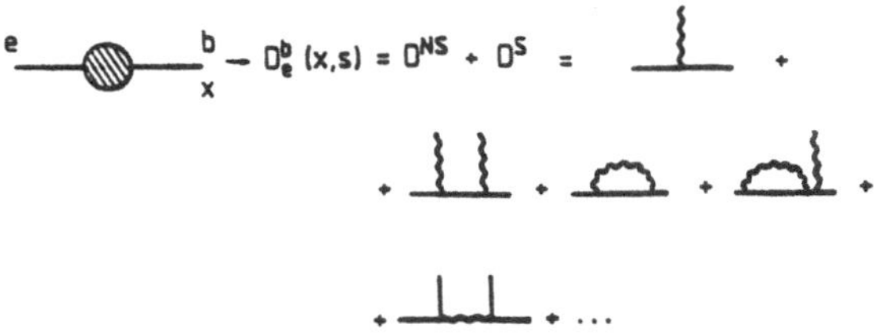

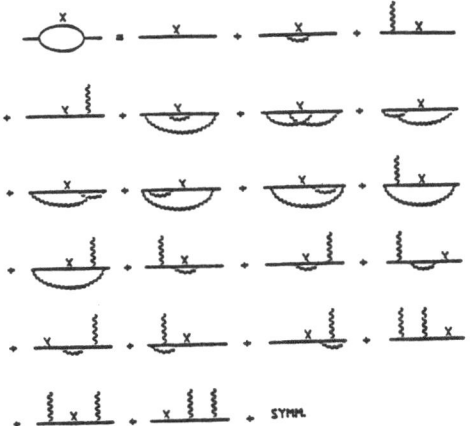

Fig.7 The full electron distribution function: Non-Singlet and Singlet.

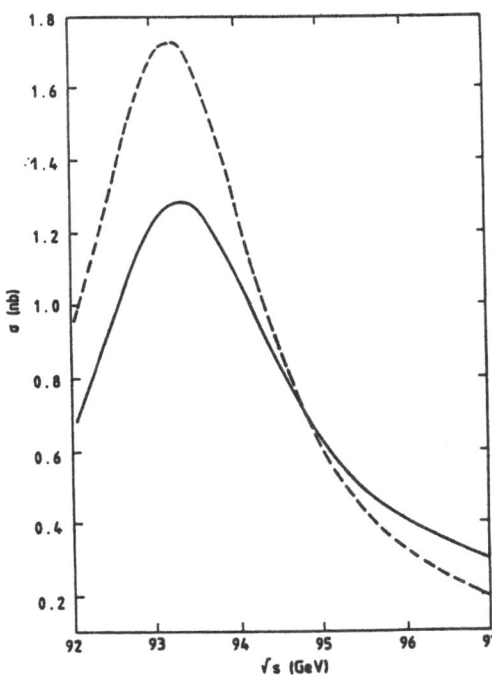

Fig.8 Diagrams corresponding to the electron form factor as evaluated in ref.[8] and contained in the radiator $H_e(x, s)$.

Fig.9 Fully corrected (solid-line) and uncorrected (dashed-line) cross section for the cross section $e^+e^- \rightarrow \mu^+\mu^-$ with $\Gamma = 2.6\,GeV$ and $M = 93.2\,GeV$.

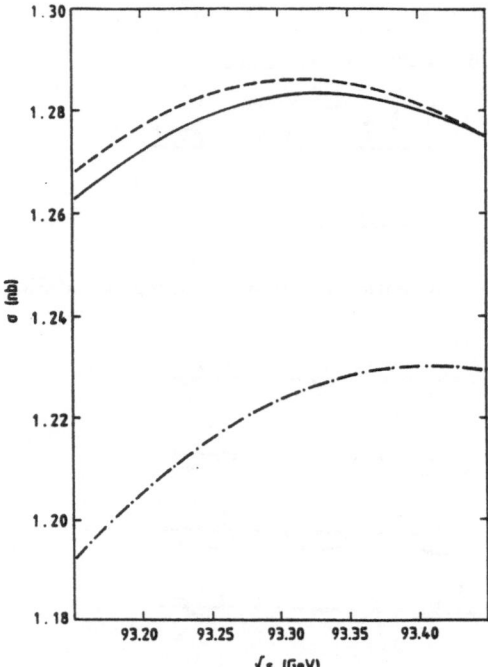

Fig.10 First (dashed-dotted line) and second (dashed-line) order finite cross sections and fully corrected (solid-line) all-orders cross section for $e^+e^- \rightarrow \mu^+\mu^-$.

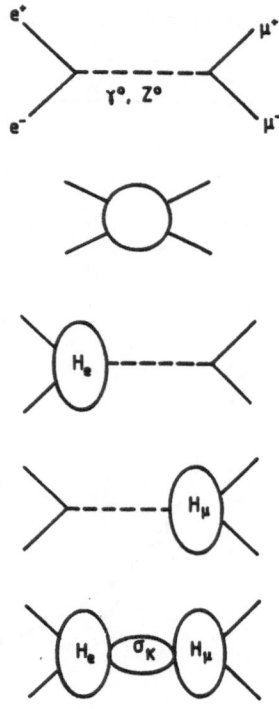

Fig.11 a)Born approximation b)Fully corrected c)Initial state corrected d)Final state corrected e)Effective decomposition of corrections.

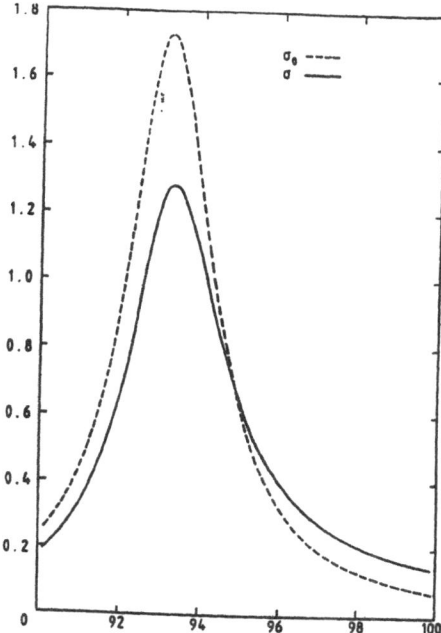

Fig.12 Born approximation (dashed line) and fully corrected $eq.(18)$ (solid line) cross sections for the process $e^+e^- \rightarrow \gamma^*, Z^0 \rightarrow \mu^+\mu^-$.

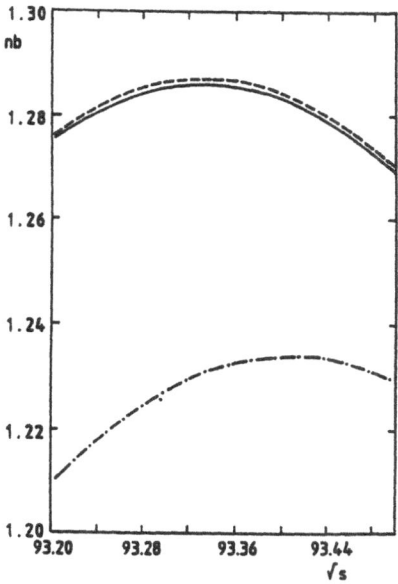

Fig.13 First order (dash-dotted line), initial state (dashed line) and fully corrected (solid line) cross sections.

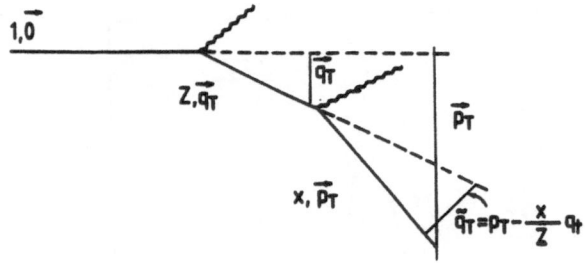

Fig.14 Transverse momentum decomposition in the electron evolution equation.

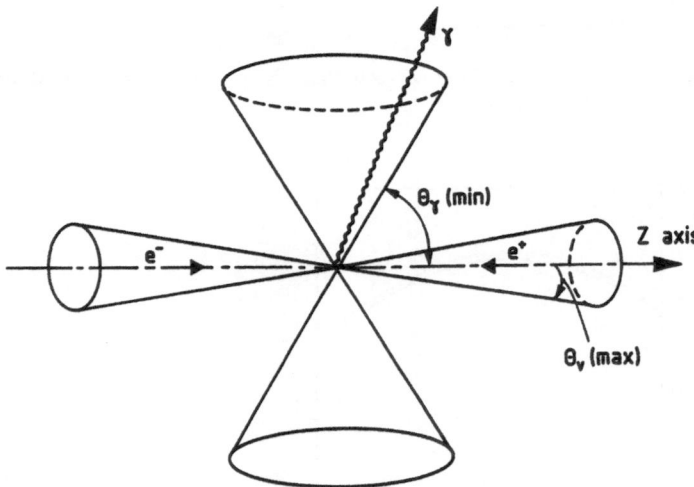

Fig.15 Experimental set-up for the radiative neutrino counting reaction.

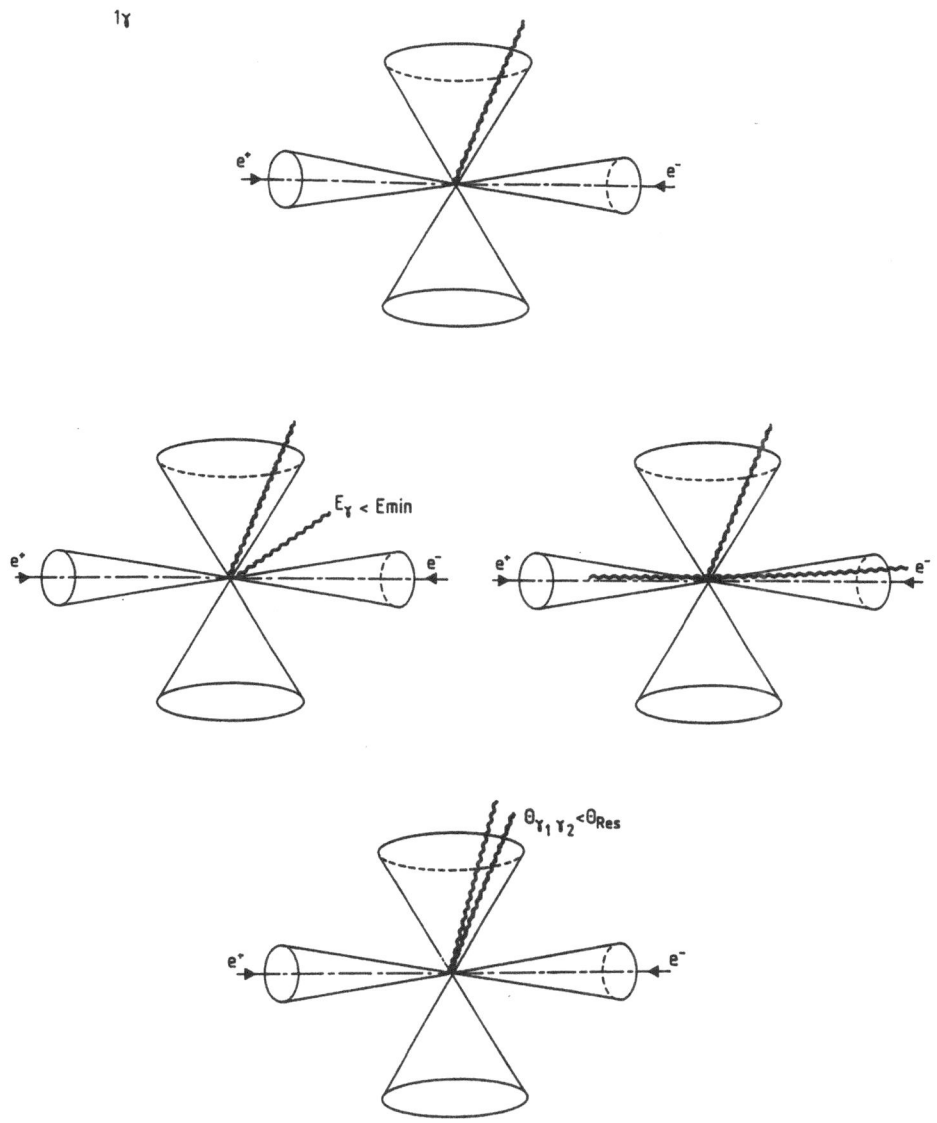

Fig.16 Different combinations of soft, collinear and parallel photons accompaining the observed one.

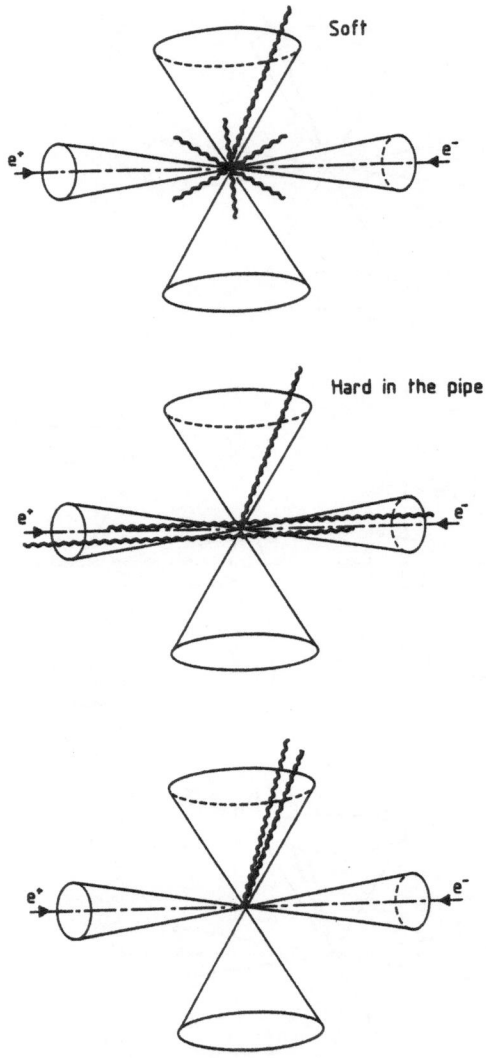

Fig.17 As before with many-photon states.

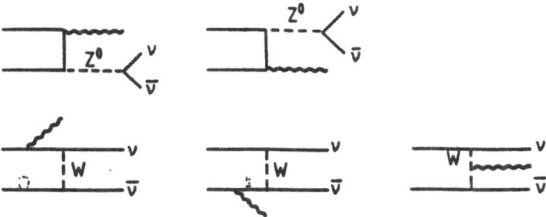

Fig.18 Feynman diagrams corresponding to the bare process. Z^0 and W contributions.

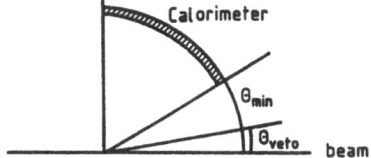

Fig. 19 Experimental set-up: θ_{veto} is the veto angle and θ_{min} is the minimum detectable angle.

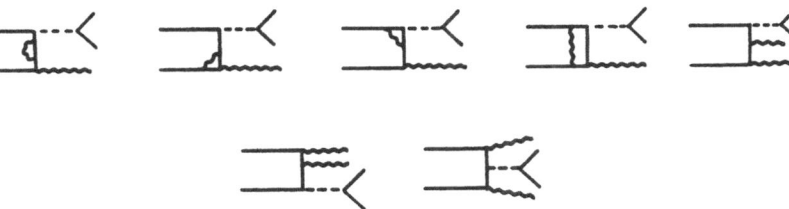

Fig.20 Feynman graphs corresponding to virtual and soft $O(\alpha)$ radiative corrections. Their contribution is reproduced by the application of the initial state radiator $H(\xi, s)$ in eq.(34) together with resummation to all orders.

52

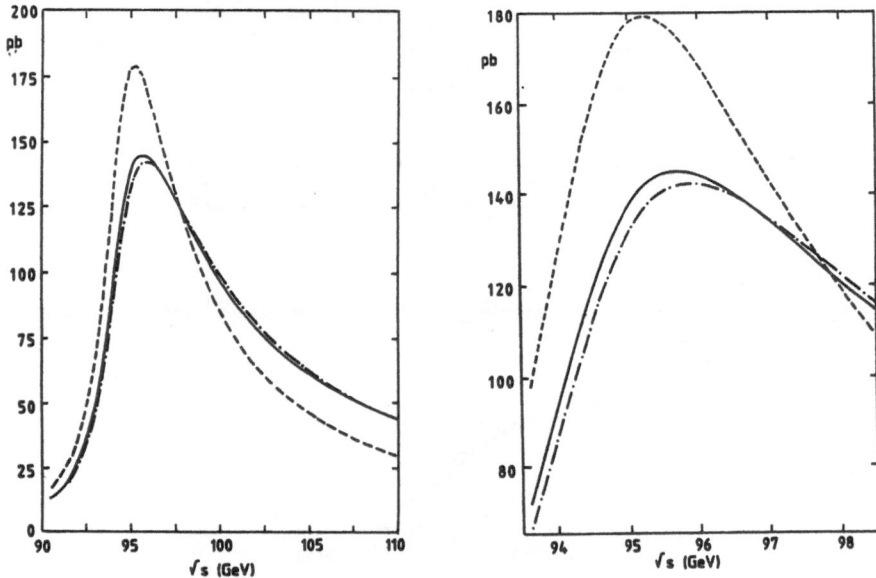

Fig.21-22 Total cross sections 21 and blow-up 22 in the Born approximation (dashed-line) obtained by integrating eq. (28) over x and y, fully corrected (solid-line) and order α corrected (dash-dotted line) obtained by integrating eq. (40) over x without and with expanding the radiators to $O(\alpha)$ respectively.

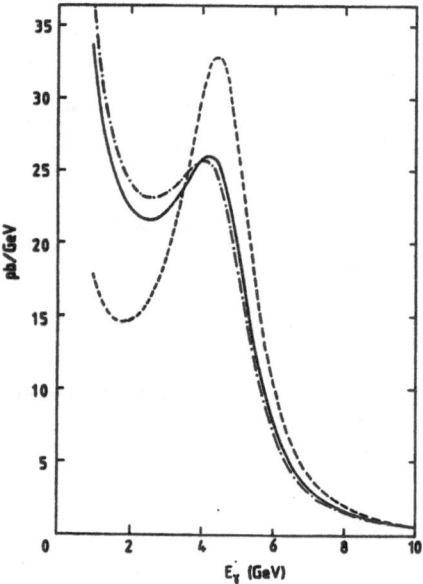

Fig.23 Energy spectrum of the observed photon. Born approximation (dashed line), fully corrected (solid-line) eq.(40) and order α corrected (dash-dotted line) obtained expanding eq.(26) to $O(\alpha)$ at $(\sqrt{s} = 98 \text{ GeV})$

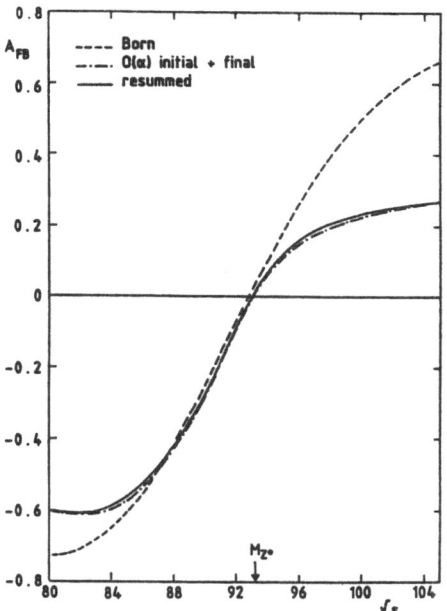

Fig.24 Forward-backward asymmetry as a function of the centre of mass energy. Born approximation of eq. (45) (dashed line), all order result of eqs. (47) and (48) (solid line) and O(α) result obtained by expanding the radiators to this order (dash-dotted line).

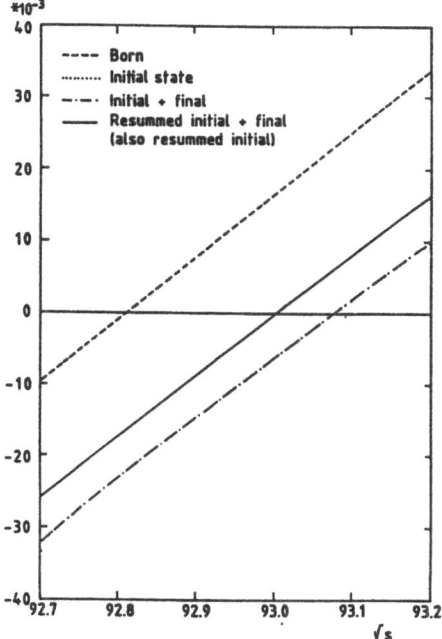

Fig. 25 The same as Fig.1 (blow up)

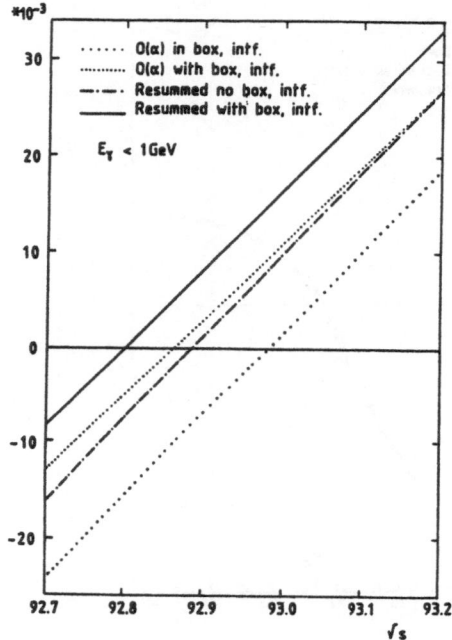

Fig.26 O(α) corrected asymmetry without (sparsely-dotted line) and with (closely-dotted line) box and interference contributions; all order corrected asymmetry without (dash-dotted line) and with (solid line) box and interference contributions. Cut on the photon energy: $E_\gamma <$ 1 GeV.

Radiative Corrections for e⁺e⁻ Collisions   Editor: J.H. Kühn
© Springer-Verlag Berlin, Heidelberg 1989

The Z Line Shape

F.A. Berends

Instituut-Lorentz, University of Leiden,

P.O. Box 9506, 2300 EA Leiden, The Netherlands

1. INTRODUCTION

During the past few years much progress has been made in the evaluation of various effects which contribute to the line shape of the Z. Some time ago the first order in α corrections [1,2] to the Z line shape were studied, even in such a way that experimental cuts could be incorporated [2]. It was clear from these studies that a theoretical assessment of higher order corrections would become necessary for an accurate comparison between theory and experiment. It is the purpose of this report to review the present situation and to give the most recent actual numbers for the line shape following from the standard model. This review can be considered as a continuation of work presented elsewhere [3,4]. We restrict ourselves to the ideal case of no experimental cuts. The motivation is that we first want to be sure what the expectations are for this case, before we introduce the additional problem of cuts. The formulae and numbers of this report can serve as a guideline for event generators, which should at least reproduce these results.

The precision aimed for is for the shape of the Z an accuracy of less than 10 MeV, i.e. the peak and half width positions should be calculated to this level. The overall size of the line shape should be evaluated within 0.3% precision. This objective requires the inclusion of some effects but allows for the omission of others, as will become clear below.

The line shape is discussed for three channels

$$e^+e^- \rightarrow \bar{\nu}\nu , \tag{1.1}$$

$$e^+e^- \rightarrow \mu^+\mu^- , \tag{1.2}$$

$$e^+e^- \rightarrow \text{hadrons} , \tag{1.3}$$

the latter made up from the five different quark pair production reactions. In this discussion we assume that the top quark mass is at least above 60 GeV. Although (1.1) will be only measured indirectly in neutrino counting a proper inclusion of the Z-shape matters for this measurement [5]. A fourth channel, that of Bhabha scattering is left out of consideration, since its structure with an additional t-channel diagram is essentially different.

The ingredients which go into the line shape are non-photonic (non-QED, or electroweak) and photonic (QED) corrections. This terminology is not used by everyone in exactly the same sense. Here we mean by the former the loop corrections which do not involve photons and by the latter those loop contributions which contain photons. Also bremsstrahlung of photons and fermion pairs belong to the second category.

For the calculations in the standard model we use the on-shell (OS) renormalization scheme [6]. In this scheme there still is freedom in the treatment of field renormalization. The results from the literature which will be reviewed here in connection with the line shape use somewhat different versions of the OS scheme [7,8]. Also the gauge choice is different in these papers: 't Hooft-Feynman versus the unitary gauge [7]. In the OS scheme the input parameters are the fine structure constant α and the physical masses of the particles. For the quark masses values are used which fit the numerical results for the polarization [9]. The actual numbers used are

$$m_\mu = m_d = 0.041 \text{ GeV,}$$
$$m_c = 1.5 \text{ GeV,} \quad m_s = 0.15 \text{ GeV ,} \tag{1.4}$$
$$m_b = 4.5 \text{ GeV ,}$$

whereas m_t is a free parameter. The other mass parameters are the vector boson masses M_Z, M_W and the Higgs boson mass M_H. In order to reduce the number of parameters the muon decay constant G_μ is used to derive M_W. In the calculation of hadronic widths one more parameter is required, the QCD coupling constant α_s. We use a value of 0.12 ± 0.02 determined from e^+e^- collisions [10], which introduces at the Z-mass a correction of $(4 \pm 0.7)\%$ to massless quark pair production and $(4.5 \pm 0.7)\%$ to $b\bar{b}$ production [11]. Unless stated differently the Z-mass is taken to be 92 GeV. All results will then depend on m_t and M_H, which we restrict to the ranges 60-230 GeV and 10-1000 GeV.

The outline of this review is as follows. In section 2 the non-photonic corrections will be discussed, whereas in section 3 the photonic ones are considered. Besides the exact evaluation of both types of corrections, we discuss approximate analytic formulae. This will lead to an explicit formula for the line shape which is accurate to at least 0.4% in the Z-region. For quick results and fitting purposes such a formula is very useful.

2. TOTAL CROSS SECTION WITH NON-PHOTONIC CORRECTIONS

2.1. The outline of the section

Firstly the lowest order predictions for the total cross sections and widths will be discussed. The expressions can be rewritten in terms of total and partial widths, which will be useful for the development of approximate formulae. Secondly, the corrections to the widths are described and numerical comparisons between different calculations are given. Then the non-photonic corrections to the lowest order cross section are taken into account. The resulting total cross section can be approximated by a simple formula, which is a generalization of a particular Born term expression.

2.2. Lowest order widths and cross sections

As discussed in section 1 the unknown input parameters in the actual calculations are M_Z, m_t and M_H. From the muon decay constant G_μ or in other words from

$$A = \frac{\pi\alpha}{\sqrt{2}\, G_\mu} = (37.281 \text{ GeV})^2 \tag{2.1}$$

one determines M_W. In lowest order the required relation reads

$$M_W^2 \sin^2\theta_W = A \ , \tag{2.2}$$

but the inclusion of electroweak corrections in muon decay modifies this relation to (cf. last paper of ref. [6])

$$M_W^2 \sin^2\theta_W = \frac{A}{1-\Delta r} \ . \tag{2.3}$$

The actual value of Δr and its dependence on the parameters of the theory (amongst others M_Z, M_W, M_H and m_t) is discussed for instance by Hollik in these proceedings and elsewhere [12]. In the OS scheme one uses as definition of the weak mixing angle

$$s_W^2 \equiv \sin^2\theta_W = 1 - \frac{M_W^2}{M_Z^2} \ . \tag{2.4}$$

This relation combined with eq. (2.2) gives one prediction for M_W whereas the combination with eq. (2.3) gives another. The latter is the realistic one, since it includes the electroweak corrections.

In lowest order the Z decays only into fermion pairs. The Born term expression for the partial width is

$$\Gamma^{(0)}_{Z \to \bar{f}f} = \frac{\alpha}{6} N_c M_Z \sqrt{1 - \frac{4m_f^2}{M_Z^2}} \left((g_f^-)^2 + (g_f^+)^2 + \frac{m_f^2}{M_Z^2} (6g_f^- g_f^+ - (g_f^-)^2 - (g_f^+)^2) \right)$$

(2.5)

with

$$g_f^- = (I_f^3 - Q_f s_W^2) / (s_W c_W)$$

(2.6)

$$g_f^+ = -Q_f s_W^2 / (s_W c_W) ,$$

(2.7)

where $c_W = \cos\theta_W$, N_c represents the number of colours whereas Q_f and I_f^3 refer to the charge and weak isospin of the fermion.

Making use of the tree level relation (2.2) one may rewrite eq. (2.5) in terms of G_μ

$$\bar{\Gamma}^{(0)}_{Z \to \bar{f}f} = N_c \frac{G_\mu M_Z^3}{24\pi\sqrt{2}} \sqrt{1 - \frac{4m_f^2}{M_Z^2}} \left(1 - \frac{4m_f^2}{M_Z^2} + \left(2I_3^f - 4Q_f s_W^2\right)^2 \left(1 + \frac{2m_f^2}{M_Z^2}\right) \right) .$$

(2.8)

For a fixed M_Z, M_H and m_t one can determine from eqs. (2.3) and (2.4) M_W and s_W^2. Inserting the latter into eqs. (2.7) and (2.8) gives different values for the widths. They would be the same when the lowest order relation (2.2) is used to determine s_W^2.

The total lowest order width is denoted by

$$\Gamma_Z^{(0)} = \sum_f \Gamma^{(0)}_{Z \to \bar{f}f} .$$

(2.9)

For massless fermions (in practice all but the b quark) the partial width simplifies to

$$\Gamma^{(0)}_{Z \to \bar{f}f} = \Gamma_-(f) + \Gamma_+(f) = \frac{\alpha}{6} N_c M_Z \left(|g_f^-|^2 + |g_f^+|^2 \right) ,$$

(2.10)

where the - and + signs refer to the helicity of f, the helicity of $\bar{f}$ being opposite.

The total cross section in lowest order for

$$e^+ e^- \to \bar{f}f$$

(2.11)

with massless fermions reads

$$\sigma_o(s) = \frac{4\pi\alpha^2}{3s} \frac{N_c s^2}{4} \left\{ \left| \frac{g_e^- g_f^-}{s - M_Z^2 + iM_Z \Gamma_Z} + \frac{Q_e Q_f}{s} \right|^2 \right.$$

$$+ \left| \frac{g_e^- g_f^+}{s - M_Z^2 + iM_Z \Gamma_Z} + \frac{Q_e Q_f}{s} \right|^2 + \left| \frac{g_e^+ g_f^-}{s - M_Z^2 + iM_Z \Gamma_Z} + \frac{Q_e Q_f}{s} \right|^2$$

$$\left. + \left| \frac{g_e^+ g_f^+}{s - M_Z^2 + iM_Z \Gamma_Z} + \frac{Q_e Q_f}{s} \right|^2 \right\} . \tag{2.12}$$

The four terms in the cross section correspond to the following helicity combinations for $e^+ e^-$, $\bar{f}$ and f: (+- +-), (+- -+), (-+ +-) and (-+ -+). It is clear that in the neutrino case only the first and third terms contribute. For Γ_Z the lowest order total width (2.9) is used. For massive fermions the expression is somewhat more involved and can be found in the literature (see e.g. [13]).

Another representation for the total lowest order cross section will turn out to be useful as basis for an approximate expression to the corrected total cross section. It uses the partial (helicity) widths and follows from eqs. (2.10) and (2.12). Firstly we rewrite eq. (2.10) in the form

$$\sigma_o(s) = \frac{N_c s}{48\pi} \sum_{\lambda_e \lambda_f} \left| \frac{R_{\lambda_e \lambda_f}}{s - M_Z^2 + iM_Z \Gamma_Z} + \frac{Q_e Q_f e^2}{s} \right|^2 , \tag{2.13}$$

with

$$R_{\lambda_e \lambda_f} = \pm \lambda_e \lambda_f \frac{24\pi \, \Gamma_{\lambda_e}^{\frac{1}{2}} \Gamma_{\lambda_f}^{\frac{1}{2}}}{N_c^{\frac{1}{2}} M_Z} , \tag{2.14}$$

where the $\pm$ sign corresponds to $I_3 = \mp \frac{1}{2}$ of the final state and λ_e, λ_f refer to the helicities of the e^- and f. Carrying out the helicity summations one finds

$$\sigma_o(s) = \frac{s N_c}{(s - M_Z^2)^2 + M_Z^2 \Gamma_Z^2} \left[\frac{12\pi \, \Gamma_e \Gamma_f}{M_Z^2 N_c} + I \, \frac{s - M_Z^2}{s} \right] + \frac{4\pi Q_f^2 \alpha^2 N_c}{3s} , \tag{2.15}$$

where

$$I = \pm \frac{4\pi Q_e Q_f \alpha}{N_c^{\frac{1}{2}} M_Z} \, (\Gamma_+^{\frac{1}{2}}(e) - \Gamma_-^{\frac{1}{2}}(e))(\Gamma_+^{\frac{1}{2}}(f) - \Gamma_-^{\frac{1}{2}}(f)) . \tag{2.16}$$

The first term in eq. (2.15) is the Breit-Wigner form for a spin 1 resonance, the last term is the pure QED cross section. The latter term still is of relevance for mupair production. The interference term is positive for realistic s_W^2 values and is less relevant for mupair production than for quark pair production.

Although we have not indicated this explicitly in the expressions (2.10)-(2.16) all partial and total widths are here lowest order expressions.

At this point some qualitative features of the lowest order total cross section (2.15) may be noticed. For neutrino pair production only the pure Breit-Wigner formula contributes

$$\sigma_o(s) = \frac{12\pi \ \Gamma_e \Gamma_f}{M_Z^2} \ \frac{s}{(s-M_Z^2)^2 + M_Z^2 \Gamma_Z^2} \ . \tag{2.17}$$

The peak value of

$$\sigma_{max} = \frac{12\pi \ \Gamma_e \Gamma_f}{M_Z^2 \Gamma_Z^2} \ (1+\gamma^2)^{\frac{1}{2}} \ , \tag{2.18}$$

where

$$\gamma = \Gamma \ /M_Z \tag{2.19}$$

is obtained for

$$(\sqrt{s})_{max} = M_Z(1+\gamma^2)^{\frac{1}{2}} \ . \tag{2.20}$$

The right and left half maxima positions are at

$$(\sqrt{s})_{\pm} = (1+\gamma^2)^{\frac{1}{8}} \ [\ M(1+\gamma^2)^{\frac{1}{4}} \pm \frac{\Gamma}{2} \ ] \ . \tag{2.21}$$

Thus the peak position is about 17 MeV above M_Z, whereas the average value of the half maxima positions is greater than (2.20). The distance between the half maxima positions is the same as the width Γ i.e. approximately within 0.3 MeV.

When one photon exchange is present the shift in the peak position remains but the distance between the half maxima becomes larger. This is most noticeable for mupair production, since the pure QED term is most important for this channel. It contributes about 0.5% to the peak value and forms a slowly varying background to the Breit-Wigner resonance.

These qualitative effects will be somewhat modified after the inclusion of electroweak corrections. This will be illustrated by actual numbers and formulae in subsections 2.4 and 2.5.

2.3. The corrected partial and total widths

As a first step to the corrections to the total cross section we discuss corrections to the partial width $\Gamma^{(0)}_{Z \to f\bar{f}}$. The corrections can be divided into four classes:

1. Non-photonic loop corrections to the decay into fermion pairs.
2. Photonic loop corrections and radiative decay.
3. QCD corrections.
4. Decay into three or more particles.

In this list the second class may look out of place since photonic corrections are dealt with in section 3. The emphasis in that section will however be on initial state radiation for the reactions (1.1)-(1.3) and not so much on final state radiation, which is also responsible for the QED correction to the partial width. The second class of corrections has to be included here since it contributes to the total width, which through the Z-propagator affects the total cross section. Moreover it is similar to the QCD corrections.

The photonic loop correction and radiative decay gives in first order the correction

$$\delta_{QED} = \frac{3\alpha}{4\pi} Q_f^2 = 0.0017 \, Q_f^2 \, , \tag{2.22}$$

which is a tiny effect. The QCD correction is larger due to the size of α_s. As mentioned in the introduction one obtains for massless quarks

$$\delta_{QCD} = 0.040 \pm 0.007 \tag{2.23}$$

and for b-quarks, using the expressions of ref. [11]

$$\delta_{QCD} = 0.045 \pm 0.007. \tag{2.24}$$

The errors will eventually lead to an error of about 10 MeV on the total width.

The decay into three or more particles has been considered in the literature [14, 15, 17]. The most important decay is

$$Z \to H f\bar{f} \, , \tag{2.25}$$

but it is only relevant for $M_H \lesssim 10$ GeV. Using the formulae of ref. [16] and adding all fermion combinations the partial width for the decay (2.25) is of the order of 5 MeV. The sum of all other decays e.g.

$$Z \to f\bar{f} \, f\bar{f} \tag{2.26}$$

where at least two fermions are leptons, is expected to have a width of at most a few MeV [15]. It may be that smaller phase space decays like

$$Z \to \pi^\circ \bar{\ell}\ell \qquad (2.27)$$

which belong to the class (2.26) could give slight deviations from the expectations. On the other hand the decay $Z\to\pi^\circ\gamma$ is already included in the above mentioned QED corrections, just like the decays into four quarks is a QCD correction.

In view of the required accuracy we neglect the above rare decays, keeping in mind however that for small M_H the decay (2.25) should be taken into account.

We now turn to the non-photonic loop corrections, which have been discussed extensively in the literature [18, 19, 20]. These corrections come from vertex corrections and wave function renormalizations. In ref. [18] the correction to eq. (2.5) is sizeable, mainly due to the wave function renormalization of the Z. In the scheme of ref. [19] the correction to eq. (2.8) is considered which is small. The corrected partial and total widths are however the same and are given in table 1. These values include the effects of the first three classes of corrections.

Table 1

Partial and total Z widths for M_Z=92 GeV and α_s=0.12. The first line lists the results of ref. 18, the second of ref. 19. In the second line only the digits differing from line 1 are printed. In the cases where M_H=10 the decay (2.25) has a width of 5.3 and 5.8 (m_t=230) respectively, which is not included in the listed value for Γ_Z.

m_t (GeV)	M_H (GeV)	$\sin^2\theta_W$	Γ_Z (MeV)	$\Gamma_{Z\to\nu\bar\nu}$ (MeV)	$\Gamma_{Z\to e^+e^-}$ (MeV)	$\Gamma_{Z\to u\bar u}$ (MeV)	$\Gamma_{Z\to d\bar d}$ (MeV)	$\Gamma_{Z\to b\bar b}$ (MeV)
60	100	0.2296	2562	170.5	85.8	306.3	394.4	392.1
			1			.1	.2	.0
90	10	0.2242	2564	170.4	85.8	307.1	395.0	391.9
						6.9	4.8	
90	100	0.2258	2567	170.8	86.0	307.2	395.4	392.3
		7		.7	5.9	.0	.2	.2
90	1000	0.2289	2559	170.5	85.7	305.9	394.0	390.9
			7	.3		.6	3.6	.6
150	100	0.2186	2581	171.5	86.4	309.6	398.2	391.6
			78			.4	7.9	
200	100	0.2109	2596	172.5	86.9	312.5	401.5	390.5
				.4		.3	.3	90.8
230	10	0.2035	2604	172.8	87.1	314.3	403.5	389.1
			5				.4	.8
230	100	0.2052	2607	173.2	87.2	314.5	403.9	389.6
			8				.8	90.2
230	1000	0.2086	2599	172.9	87.1	313.3	402.6	388.5
			8	.7	.0	.0	.2	.7

The agreement is within 0.1%. For all massless fermion channels also the results of [20] are in agreement with the table. The difference between the partial widths for decays into d- and b-quarks is due to three effects: phase space, the QCD corrections and the contribution of a massive top quark to the vertex correction.

2.4. Total cross section with non-photonic corrections

From the lowest order cross section for massless fermions it can be seen that outside of the resonance the cross section is of order α^2 whereas near the resonance it becomes of order α^0. When one wants to include first order corrections to the cross section one should therefore use a total width up to first order in α [21]. Since Γ_Z is related to the imaginary part of the Z self-energy it means that the one-loop corrections to the propagator are not sufficient but that two-loop corrections should be taken into account in the resonance region.

The one-loop corrections to $\bar{f}f$ production consist of three types: self-energies, vertex corrections and box diagrams. In the resonance region we can neglect the latter. This means that in eq. (2.12) the coupling constants are replaced by s-dependent form factors and the propagators are replaced by [12]

$$\frac{1}{s} \rightarrow \frac{1}{s + \Sigma_\gamma(s)} \quad , \tag{2.28}$$

$$\frac{1}{s - M_Z^2 + iM_Z\Gamma_Z} \rightarrow \frac{1}{s - M_Z^2 + \Sigma_Z(s)} \quad , \tag{2.29}$$

where

$$\Sigma_\gamma = \Sigma_{\gamma\gamma} - \frac{\Sigma_{\gamma Z}^2(s)}{s - M_Z + \Sigma_{ZZ}(s)} \quad , \tag{2.30}$$

$$\Sigma_Z = \Sigma_{ZZ} - \frac{\Sigma_{\gamma Z}^2(s)}{s + \Sigma_\gamma(s)} \quad . \tag{2.31}$$

These expressions are obtained from a Dyson series summation involving the renormalized one particle irreducible self-energies $\Sigma_{\gamma\gamma}$, Σ_{ZZ} and $\Sigma_{\gamma Z}$. Besides the above propagators one has also to include a γ-Z mixing propagator, which takes the form

$$D_{\gamma Z} \doteq \frac{-\Sigma_{\gamma Z}(s)}{s[s - M_Z^2 + \Sigma_Z(s)]} \quad . \tag{2.32}$$

In these expressions the real parts of Σ_γ and Σ_Z are taken in first order. The imaginary part of Σ_Z is considered up to second order. That is, the imaginary part of Σ_{ZZ} should be evaluated in second order. This is done by the following approximation

$$\mathrm{Im}\ \Sigma_{ZZ}^{(2)}(s) = \frac{s}{M_Z^2}\ \mathrm{Im}\ \Sigma_{ZZ}^{(2)}(M_Z^2) , \tag{2.33}$$

where the latter expression is related to the first order corrections to the width. All corrections to the width contribute to eq. (2.33) except for the wave function renormalization of the Z. This can be seen by expanding (2.29) in the resonance region

$$\frac{1}{s - M_Z^2 + \Sigma_Z(s)} = \frac{1}{1 + \Pi_Z(M_Z^2)}\ \frac{1}{s - M_Z^2 + \dfrac{\mathrm{Im}\ \Sigma_Z(M_Z^2)}{1 + \Pi_Z(M_Z^2)}} , \tag{2.34}$$

where

Table 2

The total cross section including electroweak corrections for $e^+e^-\to\bar\nu\nu$. The first line gives the results of ref. [22], the second of ref. [23]. Only the digits which deviate from the first entry are listed in the second line.

m_t (GeV)	M_H (Gev)	Γ_Z (MeV)	σ_{max} (nb)	$\sqrt{s}_{max}$ (GeV)	$\sqrt{s}_-$ (GeV)	$\sqrt{s}_+$ (GeV)
60	100	2562 1	3.861 0	91.982	90.711	93.271
90	10	2564	3.854 3	91.982	90.709	93.272
90	100	2567	3.859 8	91.982	90.708	93.274
90	1000	2559 7	3.866 70	91.982	90.712 3	93.270 69
150	100	2581 0	3.859 8	91.982	90.701 2	93.280
200	100	2596	3.863 0	91.982	90.694	93.288
230	10	2604 5	3.862 59	91.982 1	90.689	93.292
230	100	2607 8	3.866 3	91.982	90.688	93.294
230	1000	2599 8	3.871 3	91.982	90.692	93.290 89

Table 3

The total cross section including electroweak corrections for $e^+e^- \to \mu^+\mu^-$. Same conventions as in table 2.

m_t (GeV)	M_H (GeV)	Γ_Z (MeV)	σ_{max} (GeV)	$\sqrt{s}_{max}$ (GeV)	$\sqrt{s}_-$ (GeV)	$\sqrt{s}_+$ (GeV)
60	100	2562 61	1.951 50	91.983	90.704 5	93.280
90	10	2564	1.949	91.983	90.703	93.281
90	100	2567	1.951 0	91.983	90.701 2	93.283
90	1000	2559 7	1.953 5	91.983	90.706 7	93.278
150	100	2581 0	1.954 3	91.982 3	90.695	93.290
200	100	2596	1.958 7	91.982 3	90.687	93.297 8
230	10	2604 5	1.963 1	91.982	90.683 2	93.301 2
230	100	2607 8	1.963 2	91.982	90.681	93.303
230	1000	2599 8	1.963 5	91.982	90.685 6	93.299

Table 4

The total hadronic cross section including electroweak corrections. Same conventions as in table 2.

m_t (GeV)	M_H (GeV)	Γ_Z (MeV)	σ_{max} (GeV)	$\sqrt{s}_{max}$ (GeV)	$\sqrt{s}_-$ (GeV)	$\sqrt{s}_+$ (GeV)
60	100	2562 1	40.627 04	91.983 4	90.712 3	93.275 6
90	10	2564	40.634 11	91.984	90.711	93.277
90	100	2567	40.633 11	91.983 4	90.710	93.278 9
90	1000	2559 7	40.634 62	91.983 4	90.714 5	93.274
150	100	2581 0	40.678 53	91.983 4	90.703 4	93.286 5
200	100	2596	40.752 24	91.983 4	90.696	93.294
230	10	2604 5	40.833 02	91.983 4	90.692 1	93.297 8
230	100	2607 8	40.821 .789	91.983 4	90.690	93.299 .300
230	1000	2599 8	40.797 .820	91.983 4	90.694 5	93.295

$$\pi_Z(s) = \frac{\partial}{\partial s} \, \text{Re} \, \Sigma_Z(s) \tag{2.35}$$

and

$$M_Z \Gamma_Z = \frac{\text{Im} \, \Sigma_Z(M_Z^2)}{1 + \pi_Z(M_Z)} \, . \tag{2.36}$$

The denominator in eq. (2.36) represents the wave function renormalization of the Z. It gives a first order correction to Γ_Z. When one also considers the real part of the second term in eq. (2.31) one effectively takes a part of the second order correction of the real part of Σ_Z into account. The effect of this has been discussed in [12]. For high top masses ($m_t \gtrsim 150$ GeV) deviations from the results presented here occur. It should be stressed that for a conclusive discussion of this effect the full second order calculation of Re Σ_Z should be performed.

The numerical results for the reactions (1.1)-(1.3) are collected in tables 2-4. The numbers are based on the evaluations of refs. [22,23]. The former evaluation is a continuation of refs. [3] and [24], the latter of ref. [19]. The agreement between the two calculations is within 0.1%. Listed are the peak value of the cross section and the position of the maximum and half maxima. Some of the features of these exact results will be discussed in the next subsection, where approximate formulae are introduced. It should be noted that for the hadronic cross sections the QCD corrections (2.23) and (2.24) are applied to the final states. The QED correction (2.22) on the final states is however left out. The reason is that it belongs to section 3. The effect is tiny.

2.4. Approximate expressions for the electroweak corrected cross sections

As mentioned in the previous subsection the modification of the total cross section (2.12) due to electroweak corrections consists of an introduction of s-dependent form factors which replace the coupling constants $g^\pm$ and consists of changes in the propagators. In the resonance region the s-dependence of $\Sigma_Z(s)$ is crucial, the other s-dependences have a small influence. The s-dependence of Im $\Sigma_Z(s)$ near the resonance can be well approximated by the replacement

$$\frac{1}{s - M_Z^2 + iM_Z\Gamma_Z} \rightarrow \frac{1}{s - M_Z^2 + is\Gamma_Z/M_Z} \, , \tag{2.37}$$

with the total corrected width, as listed in table 1. The other s-dependent corrections can be taken at $s=M_Z^2$. Effectively we get coupling constants $g^\pm(M_Z^2)$ and

$$e(M_Z^2) = \frac{1}{1 + \pi_\gamma(M_Z^2)} \quad , \tag{2.38}$$

where

$$\pi_\gamma(s) = \frac{\partial}{\partial s} \, \text{Re} \, \Sigma_\gamma(s) \quad . \tag{2.39}$$

Note that $\text{Im} \, \Sigma_\gamma(s)$ is neglected when we introduce $e(M_Z^2)$.

As approximation to the electroweak corrected total cross section we use eq. (2.15) with the following rules: the replacement (2.37) for the Z-propagator, the partial widths $\Gamma_e, \Gamma_f \, \Gamma_\pm(e), \Gamma_\pm(f)$ are now the corrected ones and α is replaced by $\alpha(M_Z^2)$ according to eq. (2.38). It should be noted that the QED correction (2.22) should be left out from the partial widths (e.g. should be removed from table 1) since the final state QED correction is neglected in tables 2-4. As far as the QCD corrections (2.23) and (2.24) are concerned it is most practical to remove them from the partial widths as well but include them as an overall factor to eq. (2.15) in the end. Since the interference term in eq. (2.15) is valid for massless fermions, the approximation for the $\bar{b}b$ channel uses $\Gamma_\pm(b)$ for massless b quarks.

It turns out that in the region of the resonance $(M_Z - \Gamma_Z, M_Z + \Gamma_Z)$ the approximation is accurate within 0.2%. We now use the approximate formula to discuss some qualitative features, which can also be recognized from tables 2-4. For this it is convenient to rewrite the propagator [25]

$$\chi(s) = \frac{1}{s - M + is\gamma} = \frac{1}{1 + i\gamma} \, \frac{1}{s - \tilde{M}_Z + i\tilde{M}_Z\tilde{\Gamma}_Z} \quad , \tag{2.40}$$

where

$$\tilde{M}_Z = M_Z(1+\gamma^2)^{-\frac{1}{2}} \, , \quad \tilde{\Gamma}_Z = \Gamma_Z(1+\gamma^2)^{-\frac{1}{2}} \quad . \tag{2.41}$$

The results of subsection (2.2) apply when M_Z and Γ_Z are replaced by $\tilde{M}_Z$ and $\tilde{\Gamma}_Z$. For instance, the peak position is

$$\sqrt{s}_{max} = M_Z(1+\gamma^2)^{-\frac{1}{6}} \quad , \tag{2.42}$$

which is about 17 MeV below M_Z. Thus the inclusion of the energy dependence of $\text{Im} \, \Sigma_Z(s)$ is crucial to the peak position [3,25]. The distance between half maxima positions is largest for the mupair case due to the relative importance of the pure QED term. It is about 14 MeV greater than the width.

3. THE PHOTONIC OR QED CORRECTIONS

3.1. The outline of the section

In the first place the most important QED correction, that of initial state photon radiation is discussed and tables with results are given. Then an approach with approximate analytic formulae is presented.

2.2. Initial state photonic corrections

When one considers the full first order QED correction to the line shape, one finds contributions from initial state radiation, final state radiation and the interference. When no cuts on the outgoing fermions are imposed the final state radiative correction is just eq. (2.22) and is therefore small. Also the interference term is negligible [2, 26, 27]. Thus the initial state radiative corrections remain and they are sizeable due to the occurrence of large logarithms of the type

$$L = \ln \frac{s}{m_e} \ . \tag{3.1}$$

In first order in α only single photon emission has to be considered, in second order not only double photon emission occurs but also the emission of an additional fermion pair. For instance as correction to reaction (1.2) one has

$$e^+e^- \rightarrow \mu^+\mu^-e^+e^- \ . \tag{3.2}$$

In ref. [28] a discussion of the effects of photon emission and of (3.2) on $d\sigma/ds'$ is given, where s' is the square of the invariant mass of the mupair. The conclusion is that the effect of (3.2) can be neglected with respect to photon emission, except for $s'/s < 0.3$, where s is the square of the laboratory energy. In the region of low invariant mass the so-called two photon production of a mupair, which is one of the mechanisms in reaction (3.2), becomes more important than the bremsstrahlung for $d\sigma/ds'$. In the following it is assumed that the specific characteristics of the two photon mechanism allow for the removal of that type of events from the data. What is then left in $d\sigma/ds'$ comes essentially from photon emission. In case a cut on s' is used for the removal of the events one should in principle take into account a modification of the final state correction (2.22). We ignore this point and consider from now on as QED correction only multiple photon emission from the incoming e^+ and e^- and the corresponding vertex corrections.

From explicit calculations up to order α^2 and the resummation of soft photons [28], which takes a part of all higher order corrections into account the following formula for the initial state photonic corrections can be derived

$$\sigma_T(s) = \int_{z_o}^{1} dz\ \sigma(sz)G(z)\ ,\tag{3.3}$$

with $s'=sz$ and $4m_f^2/s \le z_o$. In eq. (3.3) the cross section $\sigma(s)$ is the electro-weak corrected one from section 2.4. The function $G(z)$ takes the form [28]

$$G(z) = \beta(1-z)^{\beta-1}\delta^{V+S} + \delta^H,\tag{3.4}$$

where

$$\beta = \frac{2\alpha}{\pi}\ (L-1)\ ,\tag{3.5}$$

$$\delta^{V+S} = 1 + \delta_1^{V+S} + \delta_2^{V+S}\ ,\tag{3.6}$$

$$\delta^H = \delta_1^H + \delta_2^H\ ,\tag{3.7}$$

$$\delta_1^{V+S} = \frac{\alpha}{\pi}\ (\ \frac{3}{2}\ L + 2\varsigma(2) - 2)\ ,\tag{3.8}$$

$$\delta_2^{V+S} = (\frac{\alpha}{\pi})^2\ [(\ \frac{9}{8} - 2\varsigma(2))L^2 + (-\ \frac{45}{16} + \frac{11}{2}\ \varsigma(2) + 3\varsigma(3))L$$
$$-\ \frac{6}{5}\ \varsigma(2)^2 - \frac{9}{2}\ \varsigma(3) - 6\varsigma(2)\ln 2 + \frac{3}{8}\ \varsigma(2) + \frac{19}{4}\ ]\ ,\tag{3.9}$$

$$\delta_1^H = -\ \frac{\alpha}{\pi}\ (1+z)(L-1)\ ,\tag{3.10}$$

$$\delta_2^H = (\frac{\alpha}{\pi})^2\ \left\{ X - (1+z)[2\ln(1-z)(L-1)^2 + (L-1)(\frac{3}{2}L + 2\varsigma(2) - z)]\right\},\tag{3.11}$$

$$X = \left(-\ \frac{1+z^2}{1-z}\ \ln z + (1+z)\ \frac{1}{2}\ln z + z - 1\right) L^2$$

$$+ \left[\frac{1+z^2}{1-z}\left(\mathrm{Li}_2(1-z) + \ln z\ \ln(1-z) + \frac{7}{2}\ln z - \frac{1}{2}\ln^2 z\right)\right.$$

$$\left.+ (1+z)\ \frac{1}{4}\ln^2 z - \ln z + \frac{7}{2} - 3z\right] L$$

$$+ \frac{1+z^2}{1-z}\left(-\ \frac{1}{6}\ln^3 z + \frac{1}{2}\ln z\ \mathrm{Li}_2(1-z) + \frac{1}{2}\ln^2 z\ \ln(1-z)\right.$$

$$\left.-\ \frac{3}{2}\ \mathrm{Li}_2(1-z) - \frac{3}{2}\ \ln z\ \ln(1-z) + \varsigma(2)\ln z - \frac{17}{6}\ \ln z - \ln^2 z\right)$$

$$+ (1+z)\left(\frac{3}{2}\ \mathrm{Li}_3(1-z) - 2S_{1,2}(1-z) - \ln(1-z)\mathrm{Li}_2(1-z) - \frac{1}{2}\right)$$

$$-\ \frac{1}{4}\ (1-5z)\ln^2(1-z) + \frac{1}{2}\ (1-7z)\ln z\ \ln(1-z) - \frac{25}{6}\ z\ \mathrm{Li}_2(1-z)$$

$$+ (-1 + \frac{13}{3}\ z)\varsigma(2) - (\ \frac{3}{2}+z)\ln(1-z) + \frac{1}{6}\ (11+10z)\ln z$$

$$+ \frac{2}{(1-z)^2}\ \ln^2 z - \frac{25}{11}\ z\ \ln^2 z - \frac{2}{3}\ \frac{z}{1-z}\left(1 + \frac{2}{1-z}\ \ln z + \frac{1}{(1-z)^2}\ \ln^2 z\right).\tag{3.12}$$

In these definitions the polylogarithms [29,30] $\mathrm{Li}_n(x)$ and $S_{n,p}(x)$ have been introduced and the Riemann zeta function $\varsigma(2) = \pi^2/6$ and $\varsigma(3) = 1.202$. The terms δ_i^{V+S} and δ_i^H originate respectively from virtual, soft photon, and hard photon corrections of order α^i. It should be noted that the form (3.3) was

Table 5

The total cross section including all corrections for $e^+e^- \rightarrow \bar{\nu}\nu$

m_t (GeV)	M_H (GeV)	Γ_Z (MeV)	σ_{max} (nb)	$\sqrt{s}_{max}$ (GeV)	$\sqrt{s}_-$ (GeV)	$\sqrt{s}_+$ (GeV)
60	100	2562	2.853	92.093	90.771	93.698
90	10	2564	2.849	92.094	90.770	93.699
90	100	2567	2.853	92.094	90.769	93.702
90	1000	2559	2.857	92.093	90.773	93.696
150	100	2581	2.854	92.094	90.762	93.710
200	100	2596	2.858	92.094	90.755	93.720
230	10	2604	2.859	92.095	90.751	93.725
230	100	2607	2.862	92.095	90.749	93.728
230	1000	2599	2.865	92.095	90.753	93.723

Table 6

The total cross section including all corrections for $e^+e^- \rightarrow \mu^-\mu^-$

m_t (GeV)	M_H (GeV)	Γ_Z (MeV)	σ_{max} (nb)	$\sqrt{s}_{max}$ (GeV)	$\sqrt{s}_-$ (GeV)	$\sqrt{s}_+$ (GeV)
60	100	2562	1.453	92.094	90.754	93.724
90	10	2564	1.452	92.094	90.752	93.726
90	100	2567	1.453	92.094	90.751	93.728
90	1000	2559	1.454	92.094	90.755	93.722
150	100	2581	1.456	92.095	90.745	93.737
200	100	2596	1.460	92.095	90.737	93.747
230	10	2604	1.464	92.095	90.733	93.752
230	100	2607	1.464	92.095	90.732	93.755
230	1000	2599	1.464	92.095	90.736	93.749

Table 7

The total cross section including all corrections for $e^+e^- \rightarrow$ hadrons

m_t (GeV)	M_H (GeV)	Γ_Z (MeV)	σ_{max} (nb)	$\sqrt{s}_{max}$ (GeV)	$\sqrt{s}_-$ (GeV)	$\sqrt{s}_+$ (GeV)
60	100	2562	30.043	92.095	90.772	93.705
90	10	2564	30.051	92.095	90.771	93.707
90	100	2567	30.054	92.095	90.770	93.709
90	1000	2559	30.054	92.095	90.773	93.703
150	100	2581	30.102	92.096	90.763	93.718
200	100	2596	30.174	92.097	90.756	93.729
230	10	2604	30.244	92.097	90.753	93.734
230	100	2607	30.238	92.097	90.751	93.736
230	1000	2599	30.211	92.097	90.755	93.731

first introduced in the structure function approach to this problem [31-33]. However usually only the leading log terms $(\alpha/\pi)^n L^n$ are derived and the result (3.4) goes beyond this. The leading logs found in the explicit evaluation [28] of G agree with those of the structure function approach. The subleading logs were evaluated [28] also in both methods and were found to agree.

The numerical results following from the convolution (3.3) are given in tables 5-7. The peak height decreases by a factor 0.74 and the position is shifted by about 112 MeV. Furthermore the shape is distorted, the half maxima shift by typically 60 and 430 MeV, the asymmetry being caused by the radiative tail of the resonance. The pure QED term in mupair production becomes more important after the convolution, such that the line shape becomes wider than for the other reactions.

3.3. Analytical results for the line shape

In order to obtain approximate analytical results for the line shape we successively make the following approximations. Related approaches to the Z line shape are given in refs. [35-37].

Firstly, in the integrand of eq. (3.3) we take for $\sigma(s')$ the approximation of subsection 2.4 and introduce $\tilde{M}$ and $\tilde{\Gamma}$. One finds with $s' = s(1-x)$

$$\sigma(x) = \left\{ \frac{s(1-x)(C_R+C_I) - (\tilde{M}^2+\tilde{r}^2)C_I}{s^2[(1 - x - \tilde{M}_Z^2/s)^2 + \tilde{M}_Z^2\tilde{r}_Z^2/s]} + \frac{C_Q}{s(1-x)} \right\} (1 + \delta_{QCD}) , \tag{3.13}$$

where

$$C_R = \frac{12\pi \, r^e_r r^f}{M_Z^2(1+\gamma^2)} , \quad C_I = \frac{I^N_C}{1+\gamma^2} \tag{3.14}$$

$$C_Q = 4\pi \, Q_f^2 \alpha^2(M_Z^2)N_C/3 . \tag{3.15}$$

The partial widths (also in I) contain electroweak corrections, but not the QED and QCD corrections. The latter is included as overall factor in eq. (3.13). Secondly, the function G in eq. (3.4) is taken in the first order exponentiated form, which means that δ_2^{V+S} and δ_2^H are omitted. For the convolution of the pure QED term even the exponentiation can be omitted and the first order result [38] is used.

Thirdly, the integration region for the exponential part of G is extended to $(0,\infty)$, for the other part of G $(0,1)$ is used and for the QED part the original integration region is kept. Introducing partly the same procedure as ref. [35] which is inspired by the J/ψ line shape [39] one has the integrals

$$J_\beta = \beta \int_0^\infty dx \, \frac{x^{\beta-1}}{x^2 - 2\eta x \, \cos\zeta + \eta^2} = \eta^{\beta-2} \, \phi(\cos\zeta,\beta) , \tag{3.16}$$

where

$$\phi(\cos\zeta,\beta) = \frac{\pi\beta \, \sin[(1-\beta)\zeta]}{\sin\pi\beta \, \sin\zeta} , \tag{3.17}$$

with

$$\eta = \sqrt{a^2+b^2} , \quad \cos\zeta = a/\eta , \tag{3.18}$$

$$a = \tilde{M}_Z^2/s-1 , \quad b = \tilde{M}\tilde{r}/s , \tag{3.19}$$

and from the δ_1^H part

$$J_1 = \int_0^1 dx \, \frac{2-x}{(x+a)^2 + b^2} = \frac{2+a}{b} A - \frac{B}{2} , \tag{3.20}$$

$$J_2 = \int_0^1 dx \, \frac{x(2-x)}{(x+a)^2 + b^2} = -\frac{a^2+2a-b^2}{b} A + (1+a)B - 1 , \tag{3.21}$$

with

$$A = \arctan \frac{a+1}{b} - \arctan \frac{a}{b} , \tag{3.22}$$

$$B = \ln \frac{2a+1+a^2+b^2}{a^2+b^2} . \tag{3.23}$$

The analytic approximate cross section is

$$
\begin{aligned}
\sigma(s) = \Bigg\{ &\left(\frac{C_R + C_I}{s} - \frac{\tilde{M}^2 + \tilde{\Gamma}^2}{s^2} C_I \right) \left(n^{\beta-2} \phi(\cos\zeta, \beta)(1 + \delta_1^{V+S}) - \frac{\beta}{2} J_1 \right) \\
&- \frac{C_R + C_I}{s} \left(\frac{\beta}{\beta+1} n^{\beta-1} \phi(\cos\zeta, \beta + 1)(1 + \delta_1^{V+S}) - \frac{\beta}{2} J_2 \right) \\
&+ \frac{C_Q}{s} \left[1 + \frac{\alpha}{\pi} L \left(2\ln(1 - \frac{s_m}{s}) - \ln\frac{s_m}{s} + \frac{1}{2} + \frac{s_m}{s} \right) \right. \\
&\left. + \frac{\alpha}{\pi} \left(\frac{\pi^2}{3} - 1 - 2\ln(1 - \frac{s_m}{s}) + \ln\frac{s_m}{s} - \frac{s_m}{s} \right) \right] \Bigg\} (1 + \delta_{QCD}) , \qquad (3.24)
\end{aligned}
$$

where $s_m > 4m_f^2$.

The expression (3.23) is within 0.4% a good approximation to the line shape in the region $(M_Z - 3\Gamma_Z, M_Z + \Gamma_Z)$. It applies to the lepton and quark channels. For the lepton channels we take $\sqrt{s_m}$ to be 200 MeV, for the hadrons 10 GeV.

ACKNOWLEDGEMENT

I gratefully acknowledge the fruitful collaboration with W. Beenakker, G. Burgers, W. Hollik, S. van der Marck and W.L. van Neerven. Moreover, the efforts of D. Bardin, T. Riemann and M. Sachwitz contributed greatly to obtain the numbers presented in this report.

REFERENCES

1. M. Greco, G. Pancheri-Srivastava and Y. Srivastava, Nucl. Phys. B171 (1980) 118; E B197 (1982) 543.

2. F.A. Berends, R. Kleiss and S. Jadach, Nucl. Phys. B202 (1982) 63.

3. F.A. Berends, G. Burgers, W. Hollik and W.L. van Neerven, Phys. Lett. B203 (1988) 177.

4. G. Burgers in Polarization at LEP, CERN 88-06, P. 121, ed. by G. Alexander, G. Altarelli, A. Blondel, G. Coignet, E. Keil, D.E. Plane and D. Treille.

5. F.A. Berends, G. Burgers, C. Mana, M. Martinez and W.L. van Neerven, Nucl. Phys. B301 (1988) 583.

6. D.A. Ross and J.C. Taylor, Nucl. Phys. B51 (1973) 25;
   G. Passarino and M. Veltman, Nucl. Phys. B160 (1979) 151;
   A. Sirlin, Phys. Rev. D22 (1980) 2695.

7. D.Y. Bardin, P.C. Christova and O.M. Fedorenko, Nucl. Phys. B175 (1980) 435; B197 (1982) 1.

8. M. Böhm, W. Hollik and H. Spiesberger, Fortschr. Phys. 34 (1986) 687.

9. F. Jegerlehner, Z.Phys. C32 (1986) 195.

10. W. de Boer, 10th Warsaw Symposium on Elementary Particles, Kazimierz (1987).

11. T.H. Chang, K.J.F. Gaemers and W.L. van Neerven, Nucl. Phys. B202 (1982) 407.

12. W. Hollik, DESY 88-188 (1988).

13. F.A. Berends and A. Böhm in: High Energy Electron-Positron Physics, ed. by A. Ali and P. Söding, World Scientific, Singapore (1988).

14. P. Kalyniak, J.N. Ng and P. Zakarauskas, Phys.Rev. D29 (1984) 502.

15. P. Kalyniak, J.N. Ng and P. Zakarauskas, Phys. Rev. D30 (1984) 123.

16. F.A. Berends and R. Kleiss, Nucl. Phys. B260 (1985) 32.

17. E. Franco in Physics at LEP, CERN 86-02, p. 187, ed. by J. Ellis and R. Peccei.

18. W. Beenakker and W. Hollik, Z. Phys. C40 (1988) 141.

19. A.A. Akhundov, D.Y. Bardin and T. Riemann, Nucl.Phys. B276 (1986) 1.

20. D.C. Kennedy, B.W. Lynn and C.J.-C. Im, SLAC-PUB 4128 (1988).

21. W. Wetzel, Nucl.Phys. B227 (1983) 1; CERN 86-02 (1986), 40.

22. W. Beenakker, F.A. Berends and S. van der Marck, to be published.

23. D. Bardin, M. Bilenky, T. Riemann and M. Sachwitz, to be published; T. Riemann, these Proceedings.

24. W. Beenakker and W. Hollik, in ECFA Workshop on LEP 200, CERN 87-08, p. 185, ed. by A. Böhm and W. Hoogland.

25. D.Y. Bardin, A. Leike, T. Riemann and M. Sachwitz, Phys. Lett. B206 (1988) 539.

26. D.Y. Bardin, O.M. Fedorenko, T. Riemann, Dubna preprint E2-87-663; D.Y. Bardin, M.S. Bilenky, O.M. Fedorenko and T. Riemann, Dubna preprint E2-88-324.

27. S. Jadach, J.H. Kühn, R.G. Stuart and Z. Was, Z. Phys. C38 (1988) 609.

28. F.A. Berends, G. Burgers and W.L. van Neerven, Nucl. Phys. B297 (1988) 429; E Nucl. Phys. B304 (1988) 921.

29. R. Barbieri, J.A. Mignaco and E. Remiddi, Nuovo Cim. 11A (1972) 824, 865.

30. A. Devoto and D.W. Duke, Riv. Nuovo Com. 7, No. 6 (1984).

31. E.A. Kuraev and V.S. Fadin, Sov. J. Nucl. Phys. 41 (1985) 466.

32. G. Altarelli and G. Martinelli, Yellow Report CERN 86-02 (1986) 47.

33. O. Nicrosini and L. Trentadue, Phys. Lett. B196 (1987) 551.

34. V.S. Fadin and V.S. Khoze, Novosibirsk preprint 87-157.

35. R.N. Cahn, Phys. Rev. D36 (1987) 2666.

36. A. Borelli, M. Consoli, L. Maiani and R. Sisto, private communication and to be published.

37. F. Aversa and M. Greco, private communication and to be published.

38. F.A. Berends and R. Kleiss, Nucl. Phys. B177 (1981) 237.

39. D.R. Yennie, Phys. Rev. Lett. 34 (1975) 239; J.D. Jackson and D.L. Scharre, Nucl. Instr. 128 (1975) 13; M. Greco, G. Pancheri-Srivastava and Y. Srivastava, Nucl. Phys. B101 (1975)234; F.A. Berends and G.J. Komen, Nucl. Phys. B115 (1976) 114.

Radiative Corrections for e^+e^- Collisions   Editor: J.H. Kühn
© Springer-Verlag Berlin, Heidelberg 1989

Uncertainties from light quark loops

H. Burkhardt[1]

Geneva, 26 May 1989

1. Introduction

The Standard Model has three input parameters (not counting fermion masses, higgs mass and mixings) that determine the couplings and masses of the vector bosons. Popular renormalization schemes use experimentally well measured quantities like the fine structure constant $\alpha = 1 / 137.0604(11)$ and the Fermi constant $G_F = 1.166344(11)\cdot 10^{-5}$ GeV^{-2} determined from the muon lifetime. The third parameter used to be $\sin^2\theta_W$, determined from neutral−current processes. For LEP physics instead, M_Z will soon be measured to high precision and is therefore a good choice as third parameter. Using $\cos\theta_W = M_W / M_Z$ as definition of the weak mixing angle, the tree level expression for M_W becomes $M_W = A_0 / \sin\theta_W$ where

$$A_0 = \sqrt{(\pi\alpha / \sqrt{2}G_F)} = 37.281 \text{ GeV}.$$

The importance of radiative corrections in precision tests of the standard model has been stressed since many years [1].

The radiative corrections to the relations above can be summed up in a single value called Δr, so that $(1 - \Delta r)\cdot M_W^2 = A_0^2 / \sin^2\theta_W$. Comparison of low energy neutrino scattering data and the direct measurements of M_W, M_Z at the pbarp collider have established already today the existence of radiative corrections to the three sig-

---

[1] University of Siegen

ma level [2]. More precise measurements are expected soon from the e^+e^- colliders LEP, SLC. An interesting feature of these radiative corrections is, that they are sensitive through vacuum polarization diagrams to physics over a broad range of energies, including energy scales yet unreachable in direct studies. Due to the spontaneously broken gauge symmetry of the standard model, the self energies of the heavy vector bosons increase about quadratically with the mass of heavy fermions (non decoupling) [3]. This can lead to sizeable corrections and allows already with the current knowledge of Δr to place an upper limit on the top mass around 180 GeV with 90% confidence level. A similar limit holds for the maximum mass splitting of any new generation of fermions with standard model couplings.

Light fermion masses instead come in mainly through the photon self energy $\Pi_{\gamma\gamma}$. Since they represent known physics, their effect should be estimated with sufficient precision to be sensitive to new physics. This is no problem in the case of leptons. For the light quarks instead, the masses are not unambiguously defined and QCD corrections large. This would lead to a sizeable theoretical error and would spoil the precision expected from future measurements expected from LEP and SLC. The way out is to express the hadronic part of $\Pi_{\gamma\gamma}$ as dispersion integral using the experimentally measured cross section of e^+e^- annihilation into hadrons. A recent collection of articles on the whole sector of precision tests and radiative corrections in the standard model can be found in the summary on the LEP study on polarization [4]. Little relevant new experimental information has become available in the meanwhile.

The first part of this conference – contribution will briefly review the known results. The second part will discuss simple parametrizations of the hadronic corrections, that should be useful for the various computer programs that are currently prepared or improved for LEP/SLC physics.

2. Calculation using data, experimental errors

The real part of the renormalized hadronic vacuum polarization can be expressed as dispersion integral:

$$\mathrm{Re}\Pi_h(s) = \frac{s}{4\pi^2 \alpha} P \int_{4m_\pi^2}^{\infty} \frac{\sigma_h(s')}{(s'-s)} ds' = \frac{\alpha s}{3\pi} P \int_{4m_\pi^2}^{\infty} \frac{R(s')}{s'(s'-s)} ds'$$

R is the cross section for e^+e^- annihilation into hadrons through one photon exchange, normalized to the point like QED cross section for lepton pair production.

Figure 1 shows as solid line the parametrization of R used in the integration. Only some data points are shown with their statistical errors. As described in [4], resonances have been parametrized by Breit – Wigner forms using the leptonic widths from reference [5] (with the exception of the ρ where an effective leptonic widths of 6.5 ± 0.3 keV has been used) and the continuum through a set of straight lines. The whole integration and error calculation can be performed analytically for all space – and timelike values of four momentum transfer. For the error calculation, the continuum contribution has been subdivided into four broad regions. Rather conservative error estimates have been assigned to each of these regions.

The various hadronic contributions to $\Delta\alpha = -\Pi_{\gamma\gamma}$ at the mass of the Z are summarized in table 1. The uncertainty is dominated by the contribution from low energies (The result for W = 0 − 12 GeV alone is 1.64 ± 0.08). The parametrization for energies above W = 40 GeV follows second order QCD with $\alpha_S = 0.12$. The assumed relative error in R for energies above 12 GeV is 3 % and largely covers the experimental uncertainty in the knowledge of α_S.

Some still preliminary and unfortunately unpublished results from DM2 in the energy range W = 1.3 − 2.2 GeV and from Crystal Ball for W = 5.0 − 7.4 GeV

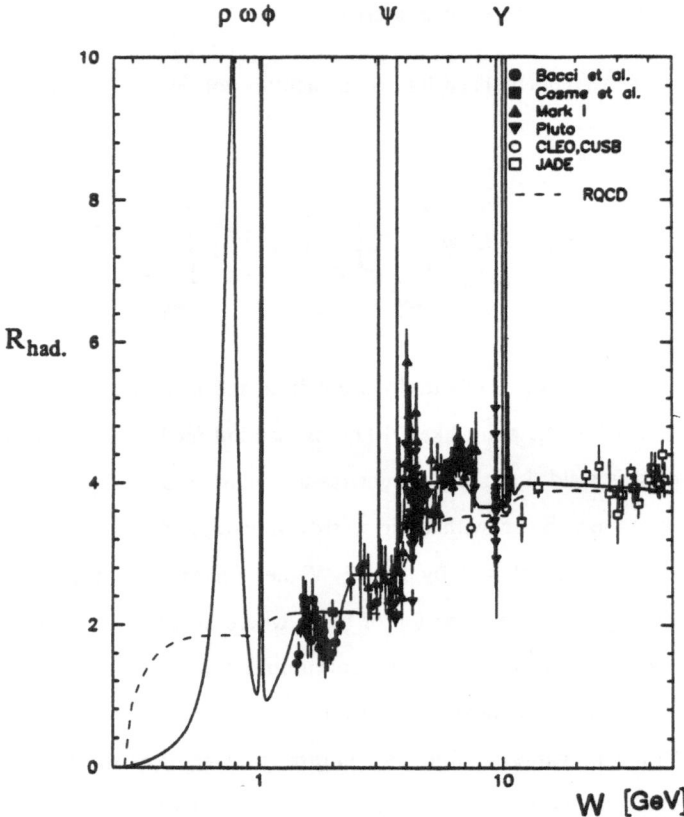

Figure 1: Parametrization of R including Resonances up to W = 50 GeV

have not been taken into account [6]. Both the DM2 and Crystal Ball results would probably help to reduce the overall relative error from currently 3 to maybe 2 %. The Crystal Ball results are at the same time expected to lie below the Mark II results and therefore to reduce also the central value (well within the present overall error).

Table 1:   Hadronic (udscb) contributions to $\Delta\alpha$ for $W = 92$ GeV

W range	rel.error	$\Delta\alpha$ in %
1.0 − 2.3 GeV	20 %	0.20 ± 0.039
2.3 − 9.0 GeV	10 %	0.72 ± 0.072
9.0 − 12.0 GeV	10 %	0.17 ± 0.017
12.0 − ∞	3 %	1.24 ± 0.037
all resonances	3 %	0.55 ± 0.018
total	3 %	2.88 ± 0.09

3. How to implement in practice

The photon self energy has to be known for some applications over a wide range of four momentum transfer. Bhabha scattering at the Z for example is dominated for large scattering angles by single Z exchange with momentum transfer $s = M_Z^2$. The same process is used for the Luminosity determination at small scattering angles (typically 50 mrad) where the t channel dominates ($t \cong -s{\cdot}\theta^2/4 = - (2.3\,\text{GeV})^2$, $\theta = 50$ mrad).

A simple and quite accurate parametrization for the hadronic (udscb) contribution to the photon self energy is of the form [7]:

$$- \operatorname{Re} \Pi_h (s) \cong - \operatorname{Re} \Pi_h (-t) \cong A + B \cdot \log (1 + C{\cdot}|s|)$$

(s,t in GeV²)

The values of A, B, C are given in table 2.

$\sqrt{|s|}$ [GeV]	A	B	C
0.0 — 0.3	0.00000	0.00835	1.000
0.3 — 3.0	0.00000	0.00238	3.927
3.0 — 100.0	0.00165	0.00299	1.000
100.0 — ∞	0.00221	0.00293	1.000

Table 2: Parametrization of Re Π_h

For the imaginary part we have Im Π_h (s) = $\alpha/3 \cdot$ R(s) where R(s) is the parametrization of the normalized hadronic cross section as used in the dispersion integral. Figure 2 shows the result of the full calculation of $-$ ReΠ_h and (as broken line) the parametrization for a broad range of time$-$ and spacelike momentum transfers. Within the plot resolution, the parametrization shows perfect overlap for spacelike momentum transfers. For timelike momentum transfers the parametrization should be only used for sufficiently large values of s far from resonances (some GeV above Upsilon threshold). It is strongly recommended to use the full calculation of $-$ ReΠ_h or the parametrization given here whenever the photon self energy is needed over a broad range of momentum transfers (Bhabha scattering). For standardization it would be desirable to use the same parametrization in all programs for LEP/SLC physics. (A compact FORTRAN program that calculates both the real and imaginary part of Π_h is available from the author and has been implemented in some generally used programs).

Some programs use the free field theory expression that is valid for leptons also in the case of quarks (f = u,d,s,c,b) :

$$\Pi_{\gamma\gamma h}(s) = \frac{\alpha}{3\pi}\sum_f N_c Q_f^2 P(s,m_f)$$

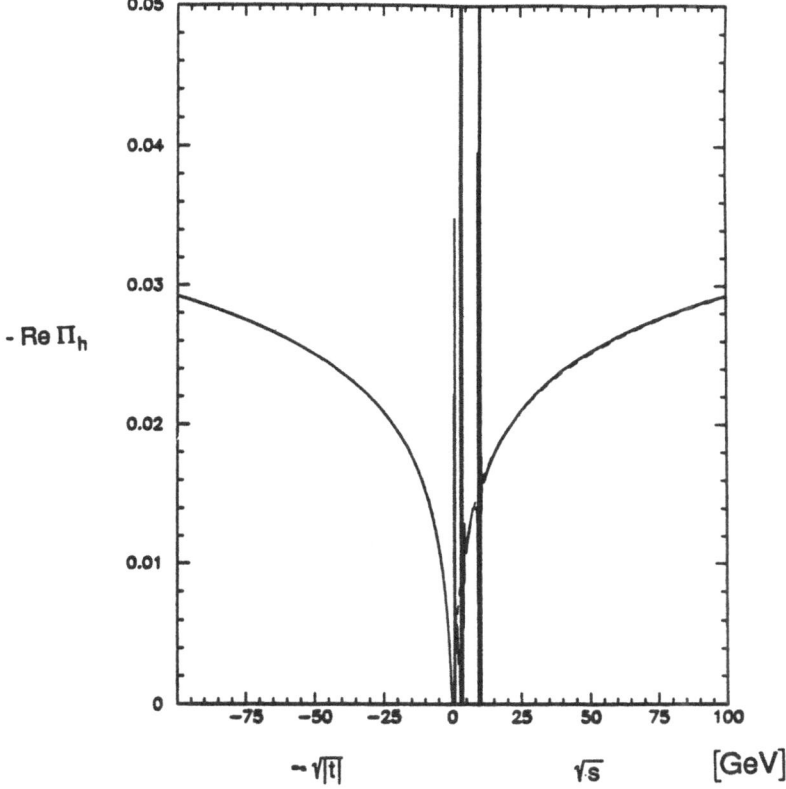

Figure 2: Result and Parametrization for time − and spacelike momentum transfers

The full expression for P(s,m$_f$) can be found in the contribution from Burgers, Hollik in reference [4].

In the limit s → 0 we get

$$-\Pi_{\gamma\gamma h}(s) = \frac{\alpha s}{5\pi}\sum_f \frac{Q_f^2}{m_f^2} = const\cdot s$$

which can be tuned to agree with the parametrization $0.00835 \log (1 + |s|) \cong$ $0.00835 \cdot s$ if the lightest quark masses are chosen to be not lighter than about 200 MeV.

For $s >> m_f^2$ we have $-\Pi_{\gamma\gamma h}(s) = \dfrac{\alpha}{\pi} \sum_f Q_f^2 \log s - \dfrac{\alpha}{\pi} \sum_f Q_f^2 (\dfrac{5}{3} + \log m_f^2)$

We see, that the factor in front of the logarithm has a constant value that cannot be adjusted through quark masses:

$$\frac{\alpha}{\pi} \sum_f Q_f^2 = \frac{11\alpha}{9\pi} = 0.00284$$

The value is a few percent lower than the $B \cong 0.003$ in the parametrization. This is due to QCD corrections $(1 + \alpha_S/\pi \cong = 1.031$ at Z energies) that have been neglected in the free field theory. If we fix the s.c,b masses at 0.5, 1.5 and 4.5 GeV, then we have to lower the u,d masses to about 30 MeV in order to get the right correction for energies around M_Z.

Apparently there is no satisfactory set of fixed quark masses, that allows to parametrize the hadronic contribution to $\Pi_{\gamma\gamma}$ over a broad region of momentum transfers. Moreover using fixed quark masses one has to worry about QCD corrections, that instead are automatically included in the dispersion relation approach.

References

[1]   W.J.Marciano, Phys.Rev.D20 (1979) 274
       A.Sirlin, Phys.Rev.D22 (1980) 971
[2]   U.Amaldi et al.,Phys.Rev.D36 (1987) 1385
[3]   W.F.L.Hollik DESY 88 − 188
[4]   G.Alexander et al. Ed., CERN 88 − 06 Vol I.
       in particular the article therein of
       H.Burkhardt, F.Jegerlehner, G.Penso, C.Verzegnassi, subm. to Z.Phys.C
[5]   Review of Particle Properties, Phys.Lett.B204(1988)
[6]   R.Baldini, private communication
       W. Lockman, private communication
[7]   H.Burkhardt, TASSO Note 192 (1981) and DESY F35−82−03 (1982).

Radiative Corrections for e⁺e⁻ Collisions   Editor: J.H. Kühn
© Springer-Verlag Berlin, Heidelberg 1989

Hadron Physics and Radiative Corrections

J.H. Kühn

Max-Planck-Institut für Physik und Astrophysik
– Werner-Heisenberg-Institut für Physik –
P.O.Box 42 12 12, Munich (Fed. Rep. Germany)

Abstract: The interplay between hadron physics, radiative corrections and precision tests of the Standard Model is studied. Examples considered in this review are: i) initial state radiation of hadrons, ii) the influence of QCD on the weak bosons' vacuum polarization functions, iii) hadronic corrctions to the forward backward asymmetry, iv) QCD corrections to the Z decay rate. It is shown that all these corrections are sufficiently well under control.

1. Introduction

Forthcoming experiments at electron positron colliders will probe the predictions of the standard model of electroweak interactions at a hitherto unrivalled level of precision. They will be sensitive to one of the most fundamental aspects of the theoretical formulation – quantum corrections as predicted within the perturbative treatment of a local quantum field theory.

Many of these experiments will – directly or indirectly – involve hadron physics:
i) Measurements requiring high statistics, like the determination of the left-right asymmetry A_{LR} will be performed with hadronic final states. The Z line shape will be explored with leptonic and hadronic final states. The restriction to leptonic final states is not necessarily a remedy of the problem of hadron physics: Around the Z peak hadronic and leptonic production rates are closely correlated through unitarity constraints.
ii) The forward backward asymmetry of quarks will be of interest for the determination of $\sin^2 \theta_W$ and could eventually even be sensitive to physics beyond the standard model.
iii) The influence of the hadronic vacuum polarization plays, furthermore, a crucial role for the evaluation of weak corrections to asymmetries and rates, involving quark and leptonic final states as well.

One thus has to assure that the final error on the quantities of interest is not dominated by "hadronic uncertainties", as is the case in neutrino nucleon scattering. There the systematic error in the present result [1] for the weak mixing angle $\sin^2 \theta_W = 0.233 \pm 0.003 \pm 0.005$ originates from the uncertainty in the choice of m_c — in other words from uncontrolled hadron physics.

In most of the calculations the hadronic system is substituted by free quarks. Occasionally perturbative QCD corrections are included. A notable exception is the hadronic part of Π_{AA}, the photon vacuum polarization, which must be evaluated on the basis of experimental data via dispersion relations. The other self energies Π_{ZZ}, Π_{ZA} and Π_{WW} are calculated using the parton model, often even ignoring QCD corrections.

A systematic analysis in this field has to proceed in two directions: The applicability and the limitations of the parton model have to be examined and further calculations have to be performed, either based on perturbative QCD or on dispersion techniques.

This review will be concerned with several topics which demonstrate the close relation between hadron physics and electroweak precision measurements and which are of relevance for e^+e^- reactions at high energies. Chapter 2 will deal with (real and virtual) initial state radiation of hadrons. The evaluation of hadronic corrections to the vacuum polarization from heavy quarks in chapter 3 will be based on a combination of perturbative QCD and dispersion relations. The rest of the paper is devoted to reactions with hadronic final states. QCD corrections to the forward backward asymmetry will be treated in chapter 4. Chapter 5 will be concerned with deviations from the parton model predictions for the total cross section $\sigma(e^+e^- \to hadrons)$ and for A_{LR} measured with hadronic final states. The emphasis will be on "non-universal" corrections to R which appear firstly in order $(\alpha_s^2)\,\alpha_s^3$ for the (axial) vector induced part of the rate and which originate from flavour singlet intermediate states. After a general discussion of the formalism and of vector current induced rates to $\mathcal{O}(\alpha_s^3)$ the results of our (two loop) calculation for the singlet part of axial current induced rates to $\mathcal{O}(\alpha_s^2)$ will be presented. For $b\bar{b}$ final states one obtains additional corrections with a top mass dependent coefficient between ≈ -1 and ≈ -5. The paper concludes with a discussion of the combined $\mathcal{O}(\alpha_s^2)$ and $\mathcal{O}(\alpha_s^3)$ corrections to the Z decay rate.

2. Hadronic initial state radiation*

Measurements of the total cross section for e^+e^- annihilation have reached a level of precision [3] where the influence of higher order radiative corrections is no longer negligible. Also the determination of the Z mass and width through the resonance line shape at future e^+e^- colliders will be influenced by radiative corrections and again the treatment to $\mathcal{O}(\alpha)$ is insufficient [4].

Radiative corrections are dominantly due to initial state radiation. To $\mathcal{O}(\alpha^2)$ purely photonic contributions as well as those from real and virtual leptonic and hadronic states are relevant. Photonic and leptonic corrections have been calculated in Refs. 5, 6, 7.

Hadronic corrections are known to contribute approximately 50% to the large logarithms that appear in the vacuum polarization $\Pi(q^2)$ for large q^2. A priori they might be as important in the high energy region for $\mathcal{O}(\alpha^2)$ vertex corrections as the aforementioned large leptonic terms. One might try to guess the characteristic scale for initial state radiation, which is presumably lower than s, thus reducing the relative importance of hadrons compared to electrons, or one might evaluate the lepton formulae with effective quark masses. However, a more rigorous treatment is at hand [2]. Just like $\Pi(q^2)$ also these $\mathcal{O}(\alpha^2)$ hadronic corrections are determined by the quantity $R(s) \equiv \sigma_{had}/\sigma_{point}$ measured in lower energy e^+e^- collisions, assuming that $R(s)$ approaches a constant value for large s.

---

*This chapter is largely based on [2]

In Ref. 2 an expression for the hadronic contribution to the virtual $\mathcal{O}(\alpha^2)$ corrections has been derived which is valid for arbitrary q^2. It becomes particularly simple in the large q^2 region. The information contained in $R(s)$ can be condensed in its asymptotic behaviour together with three moments which fix the coefficients of the $\ln^n q^2/m^2$ terms. A similar approach has been developed for real soft and hard hadron radiation which in the high energy region depends on the same moments. The formalism can be easily applied to the special case of lepton radiation and reproduces earlier results for virtual radiation [5, 6, 7] and the logarithmically enhanced terms from real radiation [5].

A special situation arises close to the Z peak. The Born cross section changes rapidly when the energy varies by $\Gamma_Z/2 \approx 1.3$ GeV. On the other hand, hadron production in e^+e^- annihilation has its threshold at $2m_\pi$, peaks at about 800 MeV and shows drastic variations up to 4 GeV. The "soft" approximation that can be justified for radiation of additional e^+e^- or $\mu^+\mu^-$ pairs is not applicable to this region. At that point one has to resort to numerical techniques. The numerical result will therefore also be compared to a simplified treatment based on ρ meson dominance which provides the bulk of real radiation. Finally the effect of hadronic and leptonic corrections on the Z line shape will be presented.

1. *Virtual Corrections*

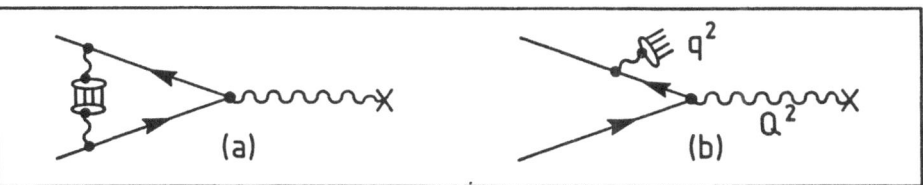

Fig. 1: Feynman diagram indicating the hadronic contribution to the $\mathcal{O}(\alpha^2)$ correction to the formfactor(a) and real hadron radiation(b).

The $\mathcal{O}(\alpha^2)$ correction to the electron form factor which originates from the leptonic and hadronic vacuum polarization $F_{vac}^{(4)}$ is shown in Fig. 1. It is obtained from the $\mathcal{O}(\alpha)$ result by replacing the photon propagator by the $\mathcal{O}(\alpha)$ renormalized vacuum polarization insertion [8]

$$\frac{g_{\alpha\beta}}{q^2 - \lambda^2 + i\epsilon} \rightarrow \frac{g_{\alpha\delta}}{q^2 - \lambda^2 + i\epsilon}(q^2 g^{\delta\epsilon} - q^\delta q^\epsilon)\left(\frac{\alpha}{\pi}\right)\Pi^{(2)}(q^2)\frac{g_{\epsilon\beta}}{q^2 - \lambda^2 + i\epsilon} \qquad (2.1)$$

where q denotes the momentum of the internal photon line and λ the photon mass introduced to regulate infrared singularities. $\Pi^{(2)}(q^2)$ is related to $R = \sigma_{had}/\sigma_{point}$ via

$$\Pi^{(2)}(q^2) = \frac{q^2}{3}\int_{4m_\pi^2}^\infty \frac{dq'^2}{q'^2}\frac{R(q'^2)}{q^2 - q'^2 + i\epsilon} \qquad (2.2)$$

where the threshold of σ_{had} is located at $2m_\pi$. $\Pi^{(2)}(q^2)$ vanishes at $q^2 = 0$ together with its first derivative so that no infrared divergencies appear and the limit $\lambda^2 \rightarrow 0$ is legitimate. The photon propagator is thus effectively replaced as follows:

$$\frac{g_{\alpha\beta}}{q^2 + i\epsilon} \rightarrow \frac{\alpha}{3\pi}\int_{4m_\pi^2}^\infty \frac{dq'^2}{q'^2}R(q'^2)\frac{g_{\alpha\beta}}{q^2 - q'^2 + i\epsilon} \qquad (2.3)$$

Exchanging the order of integration one finds

$$\left(\frac{\alpha}{\pi}\right)^2 F_{vac}^{(4)}(q^2) = \frac{\alpha}{3\pi} \int_{4m_\pi^2}^\infty \frac{dq'^2}{q'^2} R(q'^2) \frac{\alpha}{\pi} F^{(2)}(m_e^2, q'^2, q^2) \tag{2.4}$$

where $F^{(2)}(m_e^2, q'^2, q^2)$ denotes the vertex correction due to the exchange of a boson of mass $(q'^2)^{1/2}$, such that the rhs. can effectively be interpreted as originating from a superposition of vector masses of mass $(q^2)^{\frac{1}{2}}$. For the contribution from heavy leptons or hadrons we have $m_\pi \gg m_e$. Since we are furthermore only interested in the high energy limit $(q^2 \gg m_e^2)$, we may set $m_e = 0$ without encountering any mass singularities.

In this case the one loop vertex correction depends only on the ratio $u \equiv q^2/(q'^2 - i\varepsilon)$ and is particularly simple [9, 10]

$$F^{(2)}(0, q'^2, q^2) \equiv \rho(u) = -\frac{7}{8} - \frac{1}{2u} + \left(\frac{3}{4} + \frac{1}{2u}\right) \ln(-u) - \frac{1}{2}\left(1 + \frac{1}{u}\right)^2 (\zeta(2) - \text{Li}_2(1 + u)) \tag{2.5}$$

(Li$_2$ and Li$_3$ denote the di- and trilogarithms, $\zeta(2) = \pi^2/6$, and $\zeta(3) = 1.202\ldots$). $F_{vac}^{(4)}$ is then expressed in terms of R and ρ through the following equation

$$F_{vac}^{(4)}(q^2) = \frac{1}{3} \int_{4m_\pi^2}^\infty \frac{dq'^2}{q'^2} R(q'^2) \rho\left(\frac{q^2}{q'^2 - i\epsilon}\right) \tag{2.6}$$

which is the starting point of the subsequent discussion.

For a parametrization for $R(s)$ derived from data [11] eq. (6) can be evaluated numerically. One finds for the hadronic contribution to $F_{vac}^{(4)}(q^2)$:

$$F_{had}^{(4)} = -57.9 \quad \text{at} \quad \sqrt{q^2} = 93 \text{ GeV} \tag{2.7}$$

with an estimated error of about 5%. It is smaller than the corresponding contributions from electrons but larger than those from muons and τ's:

$$F_{electron}^{(4)} = -255.9 \quad F_{muon}^{(4)} = -29.5 \quad F_\tau^{(4)} = -2.2 \tag{2.8}$$

For practical applications it is, however, convenient to consider functions $R(s)$ which converge for large s towards an asymptotic value $R(\infty)$.

Defining moments R_n through*

$$R_n \equiv \int_0^1 \frac{dx}{x} \frac{\ln^n x}{n!} \left(R(4m_\pi^2/x) - R(\infty)\right) \tag{2.9}$$

$F_{vac}^{(4)}$ can be cast into the following form in the high energy limit

---

*It should be noted that $R(\infty)$ and R_0 also determine the large q^2 behaviour of the vacuum polarization, $\Pi^{(2)}(q^2) = (R(\infty) \ln q^2/4m^2 + R_0)/3$.

$$\mathrm{Re}F_{vac}^{(4)}(q^2) = R(\infty)\left[-\frac{1}{36}L^3 + \frac{1}{8}L^2 + (\frac{\zeta(2)}{6} - \frac{7}{24})L - \frac{\zeta(2)}{4} + \frac{5}{16}\right]$$
$$+ R_0\left[-\frac{1}{12}L^2 + \frac{1}{4}L + \frac{\zeta(2)}{6} - \frac{7}{24}\right] + R_1\left[-\frac{1}{6}L + \frac{1}{4}\right] - \frac{1}{6}R_2 \tag{2.10}$$

where $L \equiv \ln q^2/4m_\pi^2$. With the parametrization of Ref. 11 one finds for hadrons

$$R(\infty) = 4.0 \quad R_0 = -8.31 \quad R_1 = 13.1 \quad R_2 = -15.6 \tag{2.11}$$

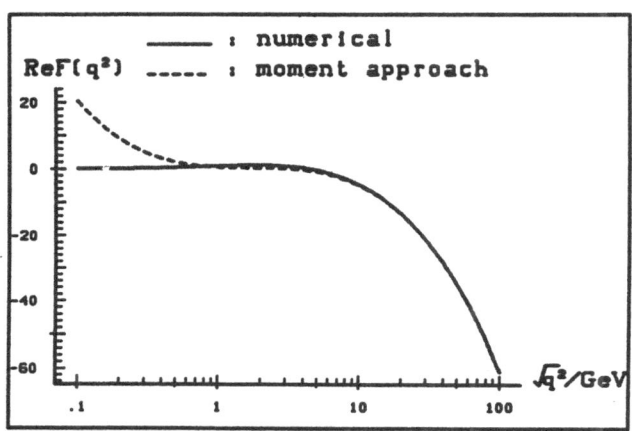

Fig. 2: $\mathcal{O}(\alpha^2)$ corrections to the electron form factor from hadrons (numerical integration and moment approach).

Eq. (10) leads to an excellent parametrization for $\sqrt{q^2} > 1$ GeV as may be seen from Fig. 2. For muons the moments can be calculated analytically so that the result of Refs. 5, 6 for the form factor contribution of muons is easily verified.

2. *Real Corrections*

Radiation of real hadrons can be calculated from

$$\frac{d^2\sigma}{dq^2dQ^2} = \sigma_{Born}(Q^2)\left(\frac{\alpha}{\pi}\right)^2 R(q^2)f(s,q^2,Q^2) \tag{2.12}$$

f denotes a kinematical factor, $\sigma_{Born}(Q^2)$ stands for the lowest order cross section for e^+e^-annihilation into any final state with invariant mass Q^2, and q^2 is the invariant mass of the radiated system (see Fig. 1b). Depending on the experimental setup, events with radiation of additional hadrons may be sorted out, e.g. in the case when Z decays into muon pairs and the soft hadrons are observed or when a natural cutoff on radiation is provided by running on a narrow resonance with width below $2m_\pi$ — such as a toponium resonance. In other circumstances these events will contribute to the total cross section, a situation characteristic for hadronic final states in the continuum.

The cross section for "soft" hadron radiation, which includes all events with radiated hadrons up to some energy $\Delta \ll \sqrt{s}$ is given by

$$\sigma_{soft}(s,\Delta) = \int_{4m^2}^{\Delta^2} dq^2 \int_{(\sqrt{s}-\Delta)^2}^{(\sqrt{s}-\sqrt{q^2})^2} dQ^2 \frac{d^2\sigma}{dq^2dQ^2} \tag{2.13}$$

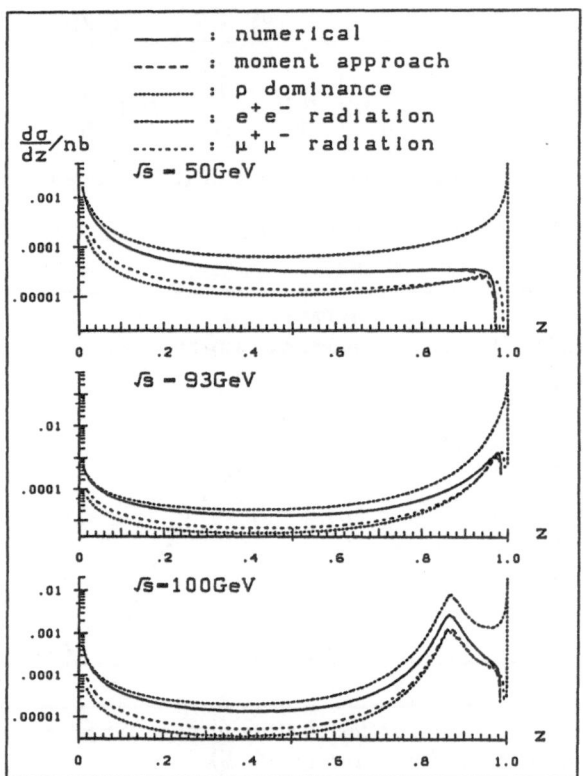

Fig. 3: *Differential cross section for hard radiation of hadrons, electrons and muon pairs as a function of $z = Q^2/s$. The hadronic radiation is evaluated through numerical integration, using the moment approach and ρ meson dominance. σ_{Born} stands for $\sigma(e^+e^- \to \mu^+\mu^-)$.*

If at the same time $R(\Delta^2)$ is sufficiently close to $R(\infty)$ and the variation of $\sigma_{Born}(Q^2)$ for Q^2 between $(\sqrt{s} - \Delta)^2$ and $(\sqrt{s} - 2m)^2$ is sufficiently small σ_{soft} may be expressed in terms of the moments R_n defined above ($\mathcal{L} \equiv \ln \Delta^2/m^2$),

$$\frac{\sigma_{soft}(s,\Delta)}{\sigma_{Born}(s)} = \left(\frac{\alpha}{\pi}\right)^2 \frac{4}{3}\left[R(\infty)\left(\frac{1}{24}\mathcal{L}^3 - \frac{\zeta(2)}{4}\mathcal{L} + \frac{\zeta(3)}{2}\right) + R_0\left(\frac{1}{8}\mathcal{L}^2 - \frac{\zeta(2)}{4}\right) + \frac{1}{4}R_1\mathcal{L} + \frac{1}{4}R_2\right] \quad (2.14)$$

For μ pairs this directly leads to the result given in Refs. 5, 7. When the virtual correction is added to the soft radiation the moment R_2 and the $\ln^3 s/4m^2$ term are cancelled ($L \equiv \ln s/4m^2$, $\ell \equiv \ln 2\Delta/\sqrt{s}$):

$$2\mathrm{Re}F_{vac}^{(4)}(s) + \frac{\sigma_{soft}(s,\Delta)}{\sigma_{Born}(s)} = \left(\frac{\alpha}{\pi}\right)^2\left\{\left[R(\infty)\left(\frac{L^2}{2} - \zeta(2)\right) + R_0 L + R_1\right]\left(\frac{2}{3}\ell + \frac{1}{2}\right) + \right.$$
$$\left.\left(R(\infty)L + R_0\right)\left(\frac{2}{3}\ell^2 - \frac{7}{12}\right) + R(\infty)\left(\frac{4}{9}\ell^3 + \frac{2}{3}\zeta(3) + \frac{5}{8}\right)\right\} \quad (2.15)$$

This result is consistent with the expectation that the maximal power of the logarithm should be n in n-th order perturbation theory [12].

Under similar assumptions on the behaviour of R and for $z \equiv Q^2/s$ chosen such that $2m/\sqrt{s} \ll \Delta/\sqrt{s} \le (1 - \sqrt{z})$ one may also calculate the cross section for hard radiation. The relevant formulas can be found in [2]

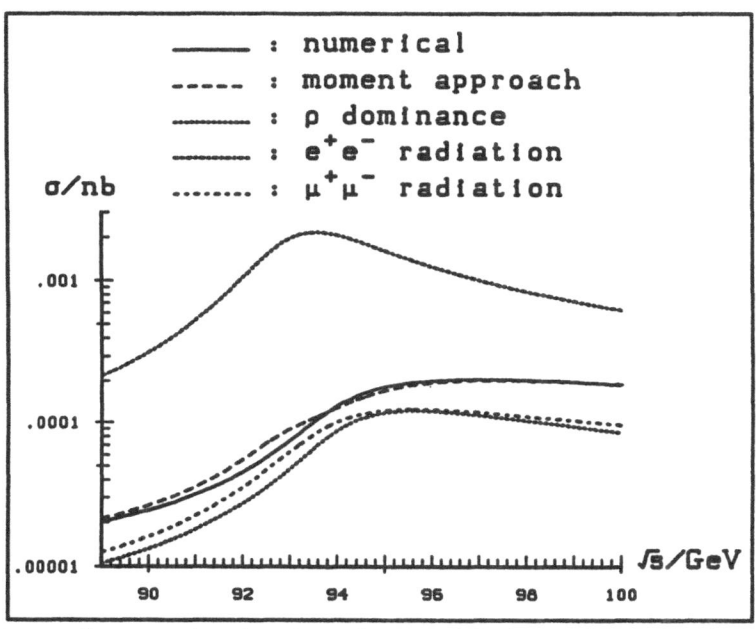

Fig. 4: Cross section for the radiation of hadron, electron and muon pairs, as a function of $\sqrt{s}$, integrated between $z_{min} = 0.5$ and 1.

All approximations discussed in the context of real radiation are strictly applicable to electron and muon pairs in the continuum and on the Z resonance and to hadron radiation in the continuum only. The discussion of real hadron radiation is more involved in the neighbourhood of the Z resonance, since $R(q^2)$ varies rapidly for $\sqrt{q^2}$ between $2m_\pi$ and 4 GeV. Within this energy range also σ_{Born} exhibits rapid variations and provides an effective cutoff on $\sqrt{q^2}$ which varies with $(M_Z - \sqrt{s})$ but is typically of order Γ_Z. In this case only the numerical evaluation of eq. (12) leads to an adequate result over the full range of z. It is, however, noteworthy that — as a consequence of the low effective cutoff — the hadronic system is dominantly a ρ meson for $z \ge 0.95$. This is evident from Fig. 3 where the differential distribution $d\sigma/dz$, evaluated with the complete hadronic contribution, is compared with the one for which only the ρ meson has been taken into account. This figure also shows radiation of electron and muon pairs. The cross section is fairly small in the region $0.1 \le z \le 0.6$ and events with small z will be quite distinct from the bulk of events close to $z = 1$. Hence the integrated cross section was evaluated with a cut at $z_{min} = 0.5$, and the result is practically independent from the precise choice of z_{min} as long as it is chosen between 0.1 and 0.6. In Fig. 4 the production cross section is shown separately for radiation of hadrons, electron and muons pairs. The close agreement between the numerical integration and the moment approach for the integrated cross section is remarkable.

The combin ed effect of hadron and lepton radiation on the line shape is at the edge of the planned experimental accuracy.

3. Hadronic corrections to Δr for large m_t

It is well known that a large splitting between top and bottom quark masses would lead to a significant decrease of Δr, the quantity which summarizes the influence of radiative corrections in the relation between m_W, m_Z, α and G_μ. For $m_t = 230$ GeV, a value consistent with present experimental information, the additional piece originating from a heavy top would amount to $\Delta r(tb) \approx -0.07$. In most calculations the hadronic interaction in the heavy quark sector is neglected and quarks are treated just like leptons. However, one expects the influence of hadronic interactions on these terms to be of order $\alpha_s/\pi \cdot 0.07 \approx 3 \cdot 10^{-3}$. They could therefore be larger than the uncertainty from the light hadrons' vacuum polarization [13] and comparable to the Higgs mass dependence, if m_H is varied e.g. from 10 to 100 GeV.

One may attempt to evaluate the influence of hadron physics on $\Delta r(tb)$ by calculating hadronic corrections to the various vacuum polarizations $\Pi_{AA}, \Pi_{ZA}, \Pi_{ZZ}$ and Π_{WW} entirely in the framework of perturbative QCD*. In this case the imaginary parts of the various $\Pi's$ are not described correctly in the threshold region and the quarkonium resonances $(t\bar{t})$ and $(t\bar{b})$ are not taken into account. Furthermore, the relation between the quark mass which appears in the calculation and the masses of the $(t, \bar{t})$- and T-mesons is not obvious.

In Ref. 17 another approach has been proposed. The imaginary part of the vacuum polarization is modelled in the region far above threshold using perturbative QCD, with the argument of the running coupling constant $\alpha_s(\mu^2)$ chosen alternatively as $\mu^2 = 4p_t^2$ or s. The resonance region has been evaluated with the resonance parameters (masses and couplings) for $(t\bar{t})$ and $(t\bar{b})$ bound states derived from a potential model [18]. The real part is then deduced via dispersion relations. The result of this approach is shown in Fig. 5.

Fig. 5a gives the hadronic correction to $\Delta r(tb)$ originating from the continuum region under three different assumptions on the behaviour of α_s: i) $\alpha_s = 0.12 = const.$, ii) $\mu^2 = s$ and iii) $\mu^2 = 4\vec{p}^2$, where $\vec{p}$ denotes the momentum of the top quark. The constituent quark mass $m_t = m_T - 400$ MeV is closely related to the mass of the top meson. General considerations suggest that the true answer should lie between the choices ii) and iii) (with slight preference for iii), such that the shaded band is indicative of present theoretical uncertainty.

For comparison also shown is the hadronic correction to $- \cos^2 \theta_W / \sin^2 \theta_W \, \Delta\rho(tb)$ which coincides for large m_t with $\Delta r(tb)$ but differs significantly below $m_t \approx 100$ GeV. Fig. 5b gives the contribution of $(t\bar{t})$ and $(t\bar{b})$ resonances to Δr and $- \cos^2 \theta_W / \sin^2 \theta_W \, \Delta\rho$, which has to be added.

---

*For attempts in this direction see [14, 15, 16].

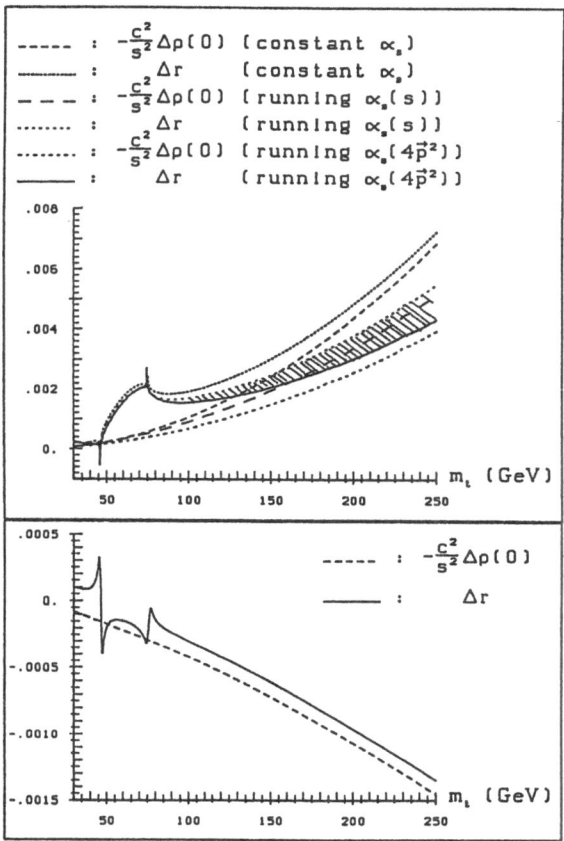

Fig. 5: QCD corrections for the tb−doublet contribution to Δr and to the leading term $\propto \Delta \rho(0)$. a) Continuum contribution for different choices of α_s. b) Resonance contribution.

The sum of these two pieces amounts to a change of $\Delta r(tb)$ by about $2 \cdot 10^{-3}$ for m_t between 70 and 200 GeV, well consistent with the order of magnitude estimate indicated above. It corresponds to a shift of the W mass (for fixed m_Z, G_μ, α) of about 25 MeV. The corresponding changes in the prediction for the left-right asymmetry A_{LR}, which are also studied in Ref. 17 are significantly smaller.

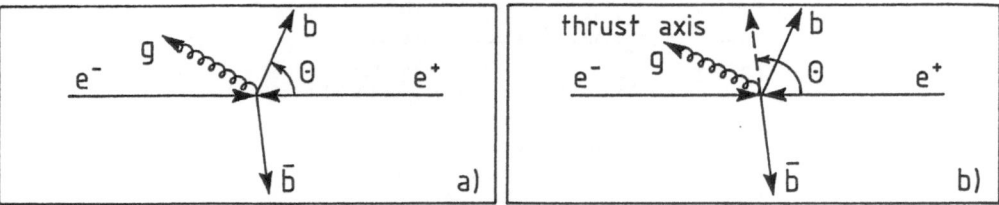

Fig. 6: Two different definitions of the scattering angle: a) through the quark direction, b) through to the thrust axis.

4. Forward-backward asymmetries with quark final states

The measurement of forward backward asymmetries with quark final states will constitute an important tool to test the validity of the Standard Model and to determine the relative strengths of the vector-axial-vector couplings of the various fermion species, in particular of the electron. All formulas below apply to light (u, d, s) and heavy (c, b) quarks equally well. However, in practice tagging of charmed and bottom quarks will be easier. Thus the influence of QCD and of mass corrections has to be studied. In Born approximation and on top of the Z the asymmetry is given by

$$A_{FB}^0 = \frac{3}{4} \frac{\sigma_F^0}{\sigma_V^0 + \sigma_A^0} \tag{4.1}$$

with

$$\sigma_F^0 = 2v_e a_e 2v_q a_q \beta; \qquad \sigma_V^0 = (v_e^2 + a_e^2)v_q^2(3 - \beta^2)/2; \qquad \sigma_A^0 = (v_e^2 + a_e^2)a_q^2\beta^3 \tag{4.2}$$

Once QCD corrections are included a careful definition of the scattering angle is mandatory. One may, for example, choose the direction of the b quark (Fig. 6a), or alternatively, the direction of the thrust axis (Fig. 6b). QCD corrections to A_{FB} for arbitrary quark mass, based on the first definition have been calculated in Ref. 19. For small $\mu^2 = 4m_q^2/s$ and on top of the Z the corrections to the various terms appearing in eq. (1) can be cast into the form

$$\sigma_i = \sigma_i^0 \left(1 + c_i(\frac{\alpha_s}{\pi})\right) \tag{4.3}$$

with

$$c_F = 0 + \frac{2\pi}{4}\mu + \frac{2\mu^2}{3}\left[6 + \frac{\pi^2}{8} + \frac{1}{8}(\log\frac{\mu^2}{4})^2 - \frac{3}{2}\log\frac{\mu^2}{4} - \frac{5}{2}\log 2\right] + \dots$$

$$c_V = 1 + 3\mu^2 + \dots \tag{4.4}$$

$$c_A = 1 + 3\mu^2(\log(4/\mu^2) - \frac{1}{2}) + \dots$$

The dominant effect is thus a reduction of the asymmetry by about 4% (for $\mu^2 = 0$) through the factor $(1 - \frac{\alpha_s}{\pi})$, and mass corrections lower the coefficient of $\frac{\alpha_s}{\pi}$ from 1 to about 0.9.

Similar results are obtained if the thrust axis is chosen as reference direction. Defining two-jet-events through the requirement that the jet invariant masses are less than $\sqrt{y}E_{cm}$ and that x_b (the momentum fraction of the b quark) is larger than a certain cutoff ε, one finds corrections to the two jet forward backward asymmetry which depends on y and ε:

$$A_{FB}(2\text{jet}) = A_{FB}^0(1 - k(y,\varepsilon)\frac{\alpha_s}{\pi}) \qquad (4.5)$$

The coefficient k is in general fairly small, typically below 0.5 for a reasonable range of the cutoff parameters [20]. We note that the impact of uncertainties from higher order QCD corrections on the experimental determination of the weak couplings is fairly small: b quark asymmetries on top of the Z amount to about 0.12 for $M_Z \approx 92$ GeV. Taking the uncertainty from higher order QCD corrections as 1/4 of the leading $\mathcal{O}(\alpha_s)$ term this results in a "systematic" theoretical error in the prediction for A_{FB} of about 10^{-3} — well below the anticipated experimental error. B-meson asymmetries are therefore suited for precision test.

5. Non-universal QCD corrections

1. *Influence of QCD corrections on A_{LR} and the Z line shape*
In leading order of the electroweak coupling the total cross section for e^+e^- into hadrons can be cast into the following form

$$\sigma_{\{^R_L\}} = \frac{4\pi\alpha^2}{3s}\left[\frac{(v_e \mp a_e)^2}{y^2}\left|\frac{s}{s - M_Z^2 + iM_Z\Gamma_Z}\right|^2 (r^{(V,V)} + r^{(A,A)})\right.$$

$$+ 2Q_e\frac{(v_e \mp a_e)}{y}\text{Re}\left(\frac{s}{s - M_Z^2 + iM_Z\Gamma_Z}\right)r^{(em,V)} \qquad (5.1)$$

$$\left. + Q_e^2 r^{(em,em)}\right]$$

with

$$Q_e = -1 \quad v_e = 2I_e^3 - 4Q_e\sin^2\theta_W \quad a_e = 2I_e^3 \quad y = 4\sin^2\theta_W\cos^2\theta_W \qquad (5.2)$$

R and L denote the electron beam polarization (positrons are assumed as unpolarized). The functions $r^{(i,j)}$ are defined as the transverse parts of $\sum_n\langle 0|J_\mu^i|hk\rangle\langle h|J_\nu^j|0\rangle$ and are the natural generalizations of the familiar $R \equiv \sigma_{had}/\sigma_{point} = r^{(em,em)}$ to the currents of interest in the high energy region. In the massless parton model they are given by

$$r^{(V,V)} = 3\sum_q\frac{v_q^2}{y^2} \qquad r^{(A,A)} = 3\sum_q\frac{a_q^2}{y^2} \qquad r^{(em,em)} = 3\sum_q Q_q^2 \qquad r^{(em,V)} = 3\sum_q Q_q\frac{v_q}{y}$$

$$(5.3)$$

The left right asymmetry is evidently independent of the final state as long as the leading $|Z|^2$ term is considered. In the very moment when $Z - \gamma$ interference and the pure QED contribution are included, the question arises to which level of accuracy the predictions of the parton model for the line shape, the normalization and for A_{LR} can be maintained. The normalization of $r^{(i,j)}$ and thus of the cross section is obviously affected by QCD corrections. If these corrections were different for the different i,j and for different quark final states then the parton model predictions would be invalidated not only for the overall normalization, but also for the line shape and in particular for A_{LR} in a rather complicated and perhaps even uncontrolable way. A closely related problem arises if one considers the second term in eq. (2): If $r^{(em,V)}$ is real, $Z - \gamma$ interference vanishes on top of the Z resonance. If $r^{(em,V)}$ were complex, a non-negligible final state dependence of A_{LR} could arise at the peak.

In this chapter these questions will therefore be adressed in a rather general form. It will be investigated to which extent QCD corrections for $r^{(i,j)}$ are given by a universal factor and the leading non universal terms will be identified and calculated. It will be convenient to write these corrections in the form

$$r^{(i,j)} = r^{(i,j)}_{Eorn}\left(1 + c_1^{(i,j)}\left(\frac{\alpha_s}{\pi}\right) + c_2^{(i,j)}\left(\frac{\alpha_s}{\pi}\right)^2 + c_3^{(i,j)}\left(\frac{\alpha_s}{\pi}\right)^3 + \dots\right) \qquad (5.4)$$

For clarity and completeness we stress that $r^{(ij)}$ contains in addition to the charge factors mass corrections which read

$$r^{(i,j)}_{Born} = r^{(i,j)}_{Born}\Big|_{\mu^2=0} \times \begin{cases} \beta(3-\beta^2)/2 & \text{for } i,j = em \text{ or } V \\ \beta^3 & \text{for } i = j = A \end{cases} \qquad (5.5)$$

with $\beta \equiv \sqrt{1-\mu^2}$ and $\mu^2 \equiv 4m_q^2/s$.

2. QCD corrections to $\mathcal{O}(\alpha_s)$ including mass terms

$\mathcal{O}(\alpha_s)$ corrections including non vanishing quark masses have been calculated in Refs. 14, 19.

The leading terms for small μ^2 are given by

$$c_1^{(i,j)} = \begin{cases} 1 + 3\mu^2 + \dots & \text{for } i,j = em \text{ or } V \\ 1 + 3\mu^2(\log(4/\mu^2) - \frac{1}{2}) + \dots & \text{for } i = j = A \end{cases} \qquad (5.6)$$

The coefficient $c_1^{(AA)}$ exceeds unity by $\approx +0.17$ for b quarks so that the mass correction in 1st order QCD is as important as the entire 2nd order of the Z width.

$c_2^{(ij)}$ and all higher corrections have been calculated for massless quarks only. In the remainder of this chapter which will be concerned with two or three loop amplitudes quarks in the final state will be considered as massless, thus implicitly assuming $m_Z < 2m_t$.

3. Vector currents up to $\mathcal{O}(\alpha_s^3)$

QCD corrections are identical to $\mathcal{O}(\alpha_s^2)$ for all vector current induced cross sections,

$$c_2^{(i,j)} = 1.986 - 0.115 n_f \quad \text{for } i,j = em \text{ or } V \qquad (5.7)$$

since the quark line that couples to the external current forms one closed loop in all Feynman diagrams under consideration.

As noted in Ref. 21 a new diagram with three gluons as intermediate states appears in third order. It contributes specifically for the flavour singlet configurations. The resulting terms can be considered separately and are proportional to the square of the singlet content of the respective current. It is thus convenient to split c_3 into singlet (S) and non-singlet (NS) parts

$$c_3^{(i,j)}(NS) = 64.861 \quad \text{for } i,j = em \text{ or } V$$

$$c_3^{(em,em)}(S) = -\frac{(\Sigma Q_q)^2}{3\Sigma Q_q^2}1.679$$

$$c_3^{(V,V)}(S) = -\frac{(\Sigma v_q)^2}{3\Sigma v_q^2}1.679 \tag{5.8}$$

$$c_3^{(V,em)}(S) = -\frac{(\Sigma v_q)(\Sigma Q_q)}{3\Sigma v_q Q_q}1.679$$

Numerically one finds

$$c_3^{(em,em)}(S) = -0.051$$

$$c_3^{(V,V)}(S) = -0.549 \tag{5.9}$$

$$c_3^{(V,em)}(S) = +0.202$$

for 5 massless quarks and $\sin^2\theta_W = 0.23$.*

The masses of real and virtual quarks are assumed to be zero in this calculation, i.e. the top quark contribution has been ignored. In principle also diagrams involving virtual top quarks should be included. However, power counting – combined with the requirements of gauge invariance – demonstrates that these contributions are suppressed by $(m_Z/2m_t)^2$ (consistent with the Appelquist Carazzone decoupling theorem) and can thus be neglected for large m_t. For $2m_t/M_Z = \mathcal{O}(1)$ no full calculation is at hand. Taking the opposite extreme $2m_t/M_Z \ll 1$ to estimate the potential influence of such terms, one finds again coefficients smaller than one

$$c_3^{(em,em)}(S) = -0.336$$

$$c_3^{(V,V)}(S) = -0.251 \tag{5.10}$$

$$c_3^{(em,V)}(S) = +0.351$$

which are completely negligible compared to $c_3(NS)$. Therefore one expects that the singlet contribution has no practical influence on c_3 also for arbitrary m_t between these limiting cases.**

4. *Phase of* $r^{(em,V)}$

Closely related arguments can be applied to the discussion of the phase of $r^{(em,V)}$ [23]. Whereas the other functions $r^{(i,j)}$ are evidently positive, neither the sign nor the phase of $r^{(em,V)}$ are fixed a priori. However, it will be shown in the following that the QCD

---

*As a trivial consequence one observes that the QCD corrections to the W decay rate are simply given by the non-singlet corrections.

**This qualitative argument is supported by the smallness of vector current induced part of the $Z \to ggg$ decay rate calculated in Ref. 22.

induced phase is extremely small. For this purpose it will first be shown that $r^{(em,V)}$ is real in a fictitious $SU(n)$ invariant theory. $SU(n)$ breaking due to the large top mass will then introduce small corrections which are subsequently estimated.

Let us first consider a fictitious model with n mass degenerate quarks (applicable for $n_f = 5$ to the case $m_q^2 \ll s \ll m_t^2$ with $q \neq t$). One may then expand J^{em} and J^V in terms of $SU(n)$ currents j^i ($i = 1, \dots n^2 - 1$; for J^{em} and J^V only the diagonal components are relevant) and the singlet current j^s with real coefficients a

$$J^{em} = \sum_i a_i^{em} j^i + a_s^{em} j^s$$
$$J^V = \sum_i a_i^V j^i + a_s^V j^s \tag{5.11}$$

One thus obtains

$$r^{(em,V)} \propto \sum_i a_i^{em} a_i^V \sum_h \langle 0|j^i|0\rangle\langle h|j^i|0\rangle + \sum_i a_s^{em} a_s^V \sum_h \langle 0|j^s|h\rangle\langle h|j^i|0\rangle$$
$$+ \sum_i a_i^{em} a_s^V \sum_h \langle 0|j^i|h\rangle\langle h|j^s|0\rangle + \sum_i a_s^{em} a_i^V \sum_n \langle 0|j^s|h\rangle\langle h|j^i|0\rangle \tag{5.12}$$

The last two terms vanish as a consequence of $SU(n)$ invariance, the first two terms are evidently real.

Even if $4m_t^2 > s$, virtual top contributions lead to a violation of $SU(n)$ invariance which leads to an a priori undetermined phase of $\sum_h \langle 0|j^s|h\rangle\langle h|j^i|0\rangle$. Such terms are induced by diagrams of the following type

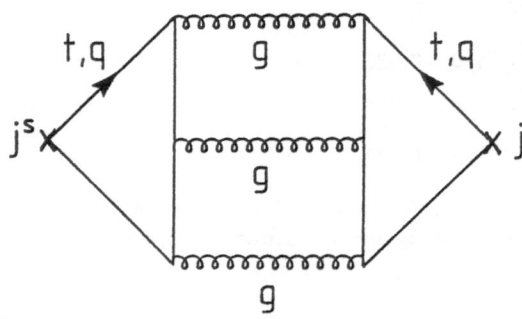

Fig. 7: *Feynman diagram whose absorbtive part contributes to the non-vanishing phase of $r^{em,V}$ if the top contribution is included in the loops.*

and may appear in third or higher order. The induced phase vanishes in the limit of $4m_t^2/s \to 0$, as discussed above and in the limit of $m_t^2/s \to \infty$ as a consequence of decoupling. For $4m_t^2/s \approx 1$ it is expected to be of order $(\alpha_s/\pi)^3$ with a non-enhanced coefficient. This leads to a tiny $Z - \gamma$ interference of order

$$\mathcal{O}\left(\frac{\Gamma_Z}{M_Z}(\frac{\alpha_s}{\pi})^3\right) \approx \mathcal{O}(10^{-6}) \tag{5.13}$$

5. *Corrections for axial currents*

Flavour non-singlet corrections to $r^{(AA)}$ in the massless limit are the same as those for the vector current. In one and two loop

$$c_1^{(AA)} = 1$$
$$c_2^{(AA)} = 1.986 - 0.115 n_f$$

(5.14)

Whereas induced flavour singlet contributions to $r^{(i,j)}$ arise for vector currents only in $\mathcal{O}(\alpha_s^3)$, they are present already in second order* for $r^{(AA)}$ as is evident from Fig. 8.

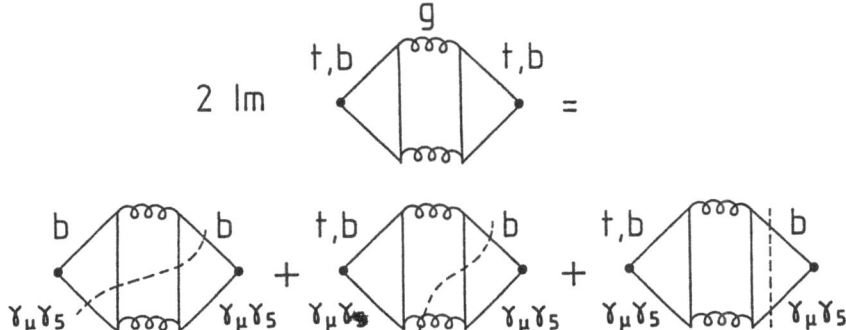

Fig. 8: *Feynman diagrams pertinent to the $\mathcal{O}(\alpha_s^2)$ corrections calculated in eq. (15). Permutations are omitted.*

The triangle anomaly originating from the t and b quark loops individually is compensated in the sum ($a_t = -a_b$!). Mass dependent terms, however, survive this cancellation. In the limit of $m_t^2/s \to \infty$ the result increases logarithmically – a signal of the break down of anomaly cancellation if top is removed from the theory.

There is increasingly strong experimental evidence that $2m_t \gtrsim M_Z$. Therefore one expects that the coefficient of this $\mathcal{O}(\alpha_s^2)$ correction is sizeable in practice. The relevant contributions originate from two-, three- and four-particle intermediate states as shown in Fig. 8. (Two on-shell gluons are forbidden by Yang's theorem.) Each term is individually infrared and ultraviolet finite and can be interpreted as contribution to the two-, three- and four-jet cross section. A lengthy calculation [25] leads to the following correction terms which appear for Γ_A in addition to those discussed above:

$$\delta\Gamma_A(b\bar{b}) = \frac{1}{3}(\frac{\alpha_s}{\pi})^2 I((M_Z/2m_t)^2)\Gamma_A^{QPM}$$

(5.15)

with

___

*For vector currents this part is absent as a consequence of Furry's theorem [24].

$$I(a) = I_2(a) + I_3(a) + I_4$$

$$= +6l + 2A^2 - \frac{15}{4}$$

$$+ \sqrt{\frac{1}{a} - 1} \left\{ 2\,\mathrm{Cl}_2(2A) + 2A(2l - 3) - \frac{1}{a}\left[\mathrm{Cl}_2(2A) + 2A(l - 1)\right] \right\} \qquad (5.16)$$

$$- \frac{1}{a}(4A^2 + 1) - \frac{1}{a^2}\left(\mathrm{Cl}_3(2A) - \zeta(3) + A\,\mathrm{Cl}_2(2A)\right)$$

$$\approx -9.250 + 1.037a + 0.632a^2 + 6l$$

where $A = \arcsin\sqrt{a}$, $l = \ln 2\sqrt{a}$, $\zeta(3) = 1.202...$ and Cl_i denotes Clausen's function of i-th order [26]. $2m_t \geq M_Z$ and $m_b = 0$ is assumed. I_2, I_3 and I_4 characterize the two, three and four parton configurations. The function $I_4 = \frac{\pi^2}{3} - \frac{15}{4}$ is independent of m_t, I_3 approaches a constant for large m_t and I_2 increases logarithmically. I_2, I_3 and I_4 are shown in fig. 2a, the overall correction $I/3$ in fig. 2b.

The magnitude of the additional terms is well comparable to the $\mathcal{O}(\alpha_s^2)$ part of the nonsinglet corrections (eq. (14)) and varies between -0.18 and -0.81% for m_t between $M_Z/2$ and 250 GeV and $\alpha_s/\pi = 0.04$.

It should be mentioned that also final states with light quarks (u, d, c, s) are individually affected by these considerations – however, the individual changes compensate in the total rate. Specifically, the effect on two and three parton configurations is opposite equal for u- and d-type quarks respectively and can be read directly from eq. (15). Additional terms in the cross section with the four partons $u\bar{u}\,u\bar{u}$ and $d\bar{d}\,d\bar{d}$ respectively are equal in sign and magnitude. They are compensated by a corresponding term from the "mixed" configuration $u\bar{u}\,d\bar{d}$ which appears with a relative factor -2.

The logarithmically increasing term in I_2 can be deduced directly from Refs. 27, 28. Consider a kinematical situation where the masses of both U and D quarks in the loop are larger than $\sqrt{s}/2$ (as considered in Ref. 29) and where the quarks in the final state are massless. Then an effective flavour singlet piece is induced in the axial current

$$J_A = \frac{1}{2}(u\bar{u} - d\bar{d}) + \frac{1}{2}(u\bar{u} + d\bar{d})(\frac{\alpha_s}{\pi})^2 \frac{1}{6}\left(I_2(s/(2m_U)^2) - I_2(s/(2m_D)^2)\right) \qquad (5.17)$$

and

$$\frac{1}{6}\left(I_2(s/(2m_U)^2) - I_2(s/(2m_D)^2)\right) \xrightarrow[m_Q \to \infty]{} \ln\frac{m_U}{m_D} \qquad (5.18)$$

in accordance with Refs. 27, 28, 29.

Acknowledgements: I would like to thank Bernd Kniehl, Maria Krawczyk and Robin Stuart for pleasant collaborations on various topics treated in this review.

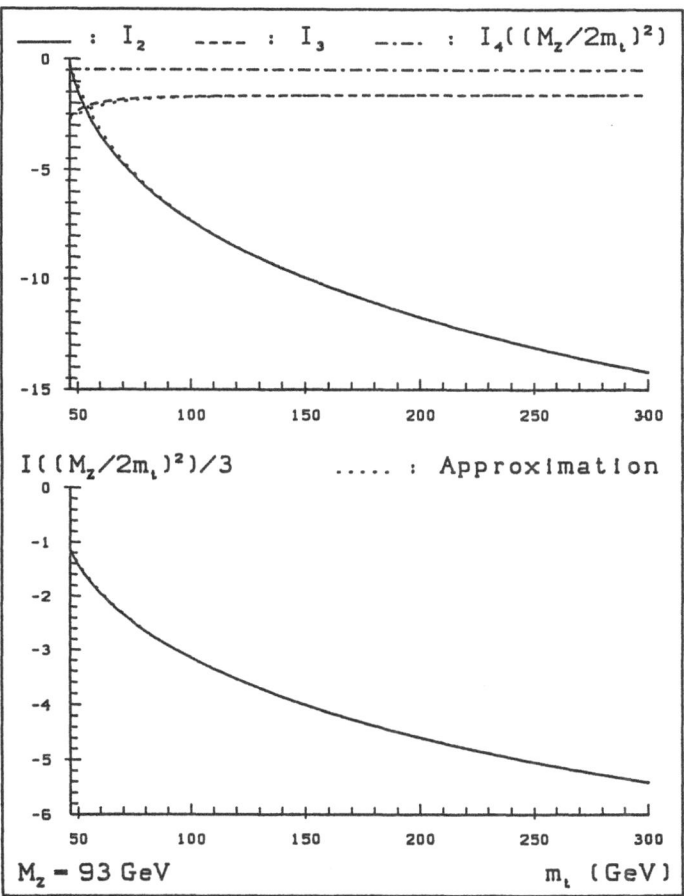

Fig. 9: The functions I_2, I_3, I_4 which determine the corrections for two, three and four parton final states. b) The function $I/3$ which determines the overall corrections.

References

[1] U. Amaldi at al., *Phys. Rev.* D **36** (1987) 1385.

[2] B.A. Kniehl, M. Krawczyk, J.H. Kühn and R.G. Stuart, *Phys. Lett.* **209** B (1988) 337.

[3] W. deBoer, these proceedings.

[4] G. Altarelli et al., in "Physics at LEP", CERN Report 86-02 (1986).

[5] E.A. Kuraev and V.S. Fadin, Sov. J. Nucl. Phys. **41** (1985) 466.

[6] G.J.H. Burgers, *Phys. Lett.* **164** B (1985) 167.

[7] F.A. Berends, G.J.H. Burgers and W.L. van Neerven, *Phys. Lett.* **185 B** (1987) 395; *Nucl. Phys.* **B 297** (1988) 429; Erratum, *Nucl. Phys.* **B 304** (1988) 92.

[8] R. Barbieri, J.A.Mignaco and E. Remiddi, *Nuovo Cimento* **11 A** (1972) 824,865.

[9] B. Grzadkowski, J.H. Kühn, P. Krawczyk and R.G. Stuart, *Nucl. Phys.* **B 281** (1987) 18.

[10] M. Böhm, H. Spiesberger and W. Hollik, *Fortsch. Phys.* **34** (1986) 687.

[11] H. Burkhardt, private communication and Tasso Note No.192.

[12] V.V. Sudakov, *Soviet. Phys. JETP* **3** (1956) 65.

[13] W. Burkhardt, these proceedings.

[14] T.H. Chang, K.J.F. Gaemers and W.L. van Neerven, *Nucl. Phys.* **B 202** (1982) 407.

[15] A. Djouadi, Preprint PM/87-53;
A. Djouadi and C. Verzegnassi, *Phys. Lett.* **195 B** (1987) 265.

[16] F. Jegerlehner, in "Proc. XI Int. School of TP", Szczyrk, Poland, World Scientific(1988).

[17] B.A. Kniehl, J.H. Kühn and R.G. Stuart,
Phys. Lett. **214 B** (1988) 621.

[18] J.L. Richardson, *Phys. Lett.* **82 B** (1979) 272.

[19] J. Jersak, E. Laermann and P.M. Zerwas, *Phys. Lett.* **98 B** (1981) 363.

[20] A. Djouadi, J.H. Kühn and P.M. Zerwas, in preparation.

[21] S.G. Gorishny, A.L. Kataev and S.A. Larin, *Phys. Lett.* **212 B** (1988) 238.

[22] J. van der Bij, *Nucl. Phys.* **B 313** (1989) 237.

[23] S. Jadach, J.H. Kühn, R.G. Stuart and Z. Was, *Z. Phys.* **C 38** (1988) 609.

[24] W.H. Furry, *Phys. Rev.* **51** (1937) 125.

[25] B.A. Kniehl and J.H. Kühn, MPI-PAE/PTh 12/89, *Phys. Lett.* **B** in print.

[26] R. Lewin, Dilogarithms and Associated Functions, MacMillan, London

[27] S. Adler, *Phys. Rev.* **177** (1968) 2426.

[28] J. Collins, F. Wilczek and A. Zee, *Phys. Rev.* **D 18** (1978) 242.

[29] Y. Kizukuri et al., *Phys. Rev.* **D 23** (1981) 2095.

Radiative Corrections for e⁺e⁻ Collisions Editor: J.H. Kühn
© Springer-Verlag Berlin, Heidelberg 1989

Electroweak Monte Carlos for LEP: a status report

Ronald Kleiss
CERN-TH

1. Introduction

In this talk I shall try to give an idea of the activities of the study group on Monte Carlo software which is a part of the current workshop on physics at LEP. The study of the existing software is not trivial in the sense that the requirements on programs to predict the physics we shall encounter at LEP with high accuracy are quite stringent: I shall argue that a precision of about 0.1% is necessary. Also most of the programs are in a state of flux: new versions are brought out regularly, support for some programs has stopped, and so on. Therefore, I restrict myself to more global remarks and a few numerical comparisons. I shall start by discussing the accuracy requirements on the Monte Carlo and other software, and the processes for which we have this accuracy in mind. I will briefly describe the two types of theoretical input that go into a typical Monte Carlo prorgam to simulate $e^+e^- \to \mu^+\mu^-$. This is followed by a brief discussion of a few technical points which are relevant to an appreciation of the differences between programs. Then I shall describe our results so far on a comparison between various programs. Finally I shall also make some comments on Bhabha scattering.

2. Precisions, processes and predictions

LEP is a machine for precision physics. An error of about 50 MeV on the Z mass seems experimentally realistic: this amounts to about 0.05%. Another typical number is the expected error on $\sin^2\theta_W$ from e.g. the forward-backward asymmetry: this amounts to about 0.0012 i.e. about 0.5% (These numbers I took from [1]). Hence we can say tentatively that for those processes which will be used to determine the standard model parameters, the experimental error will typically be something like 0.3%. Our theoretical prediction should be of course better: assuming these errors add in quadrature, we see that about 0.1% should be a desirable precision of the theoretical predictions (note that a precision of better than 0.05% will be useless since the Z mass itself, that fundamental parameter, will not be known better than that anyway). One observation follows immediately: a Monte Carlo simulation, in order to be good to 0.1%, will need about 10^6 events. At a speed of about 100 events per IBM 168 second this amounts to roughly 3 hours of CPU time – just feasible. Therefore, a good Monte Carlo program should have at least this speed.

Of course, a 0.1% precision is not needed in the prediction for every process: it would be useless to try to predict, say, an arbitrary supersymmetric scenario to this level! One can distinguish three broad classes of processes:

- **high precision**: These are the processes which will be used to determine the standard model parameters: $e^+e^- \to \mu^+\mu^-, e^+e^-$, hadrons, $\tau^+\tau^-, \nu\bar{\nu}$. If the Higgs is sufficiently light then also $e^+e^- \to H f\bar{f}$ will be in this class, for which as we saw an accuracy of 0.1-0.3% is desirable.

- **medium precision**: Here we find processes that by themselves are not so interesting, but will give background events in the precision measurements: two-photon processes like $e^+e^- \to e^+e^-\mu^+\mu^-, e^+e^-e^+e^-, e^+e^-$hadrons, and so on, and $e^+e^- \to$ photons. We can probably be content with, say, 1-3% for this class.

- **low precision**: this is all the exotic stuff, like new generations, supersymmetric scenarios, compositeness, additional vector bosons, and what not. It is clearly not useful to go all-out in predictions for these possibilities: at this moment I believe that we can be content with about 10-30% accuracy. Of course, if one of these scenarios is actually realized at LEP, we will need more precise predictions, and moreover all our radiative correction calculations will have become obsolete: in the following, I will ignore this possibility, and assume the minimal standard model to hold.

If we restrict ourselves to the high-precision class, we see that there is in fact one archetypical process: $e^+e^- \to \mu^+\mu^-$. This contains the full set of electroweak effects; moreover, by changing the fermion type from muons into quarks and so on we obtain in principle all the other processes except $e^+e^- \to e^+e^-$, where the t channel photon exchange introduces an additional complication. I do not say that a good understanding of muon pair production is sufficient to do LEP physics: but if we do *not* understand muon pair production to 0.1%, we may as well forget about the other processes. In fact most of the theoretical effort so far has gone into software for $e^+e^- \to f\bar{f}$: the Bhabha scattering process has been sadly neglected, as we shall see.

Before starting to discuss Monte Carlo software, I would like to stress that of course Monte Carlo is not the only way to make predictions. In fact a number of calculations for inclusive quantities like the total cross section exist. Since the expressions are in any case complicated most of these results are in the form of computer code as well: I shall refer to these as semianalytical results. Let us consider typical aspects of these two classes of predictions:

- **semianalytical predictions**: as I said these are usually only available for *inclusive* quantities, where an integration of the whole or a part of phase space as been performed. Examples are the total cross section $\sigma(s)$ and the forward-backward asymmetry A_{FB}, as well as the Z width Γ_Z. Often these results cannnot be directly compared with experiment, since for instance the total cross section is not measured, only a restricted angular range. On the other hand, semianalytical results are often obtained independently by several groups, and can hence be considered reliable. Also they have no statistical Monte Carlo error, and the software typically runs fast enough so that one can play around with for instance the top quark mass M_t and the Higgs mass M_H.

- **Monte Carlo predictions**: here completely specified events are generated, so they are very *exclusive*. For comparison with experiment they are very good since all kinds of cuts and acceptances can easily be implemented [1]. However, there are setbacks. It is usually not feasible to play around with, say, M_t and M_H, because each generated event sample costs a lot of time. Also the magic limit of 0.1% may be hard to reach: not only are large numbers of events required, but by that time also the quality of the random number generator may come into play. Also, many Monte Carlo algorithms are subtle, and it is not always clear that one is in fact generating the distributions one thinks to be generating. Another important consideration is that a Monte Carlo treats everything as probabilities rather than quantum mechanical amplitudes: therefore, all interference effects (photon-Z interference, initial-final-state radiation interference, and so on) have to be put in by hand, and *all cross sections have to be positive eveywhere in phase space*. This is in contrast to the semianalytical case, where one is just adding numbers and doesn't care whether they are positive or negative.

In my opinion, the main usefulness of semianalytical results is that *they provide checks on the Monte Carlo programs*. In fact we shall encounter examples of this in our discussion of the line shape.

3. Weak ingredients

It is of course unpractical to discuss all theoretical information on the higher-order radiative corrections that goes into each semianalytical or Monte Carlo program. However some general remarks can be made. First of all, there are the 'purely weak' effects. Here all the subtleties of the renormalization procedure and the choice of input parameters come into play. By now it can be said that basically all groups agree to these effects (as can be seen in the contribution of F. Berends to this workshop [2]). The Z line shape including these effects can be assumed to be known reliably to about 0.2%. There is still some disagreement on the effects of a heavy t quark (more than, say, 200 GeV) in higher order in the result for Δr, but apparently it is a matter of time before the various groups will agree. Note that a similar situation occurred some time ago with the effect of light quark loops: by now, everybody agrees on a common number [3]. I will not go into a detailed discussion of the numbers that come out: that is done by F. Berends. The only important point for us is the fact that the *total* cross section can be approximated very well by a quite simple modification of the Born cross section:

---

[1] A subtle but important point which is often overlooked is that a good Monte Carlo also gives an idea of the statistical fluctuations that can be expected: this is important for measurements of small event samples

typically one has formulae like

$$\sigma(s) \sim \frac{\tilde{\rho} G_F^2 s}{(s - M_Z^2)^2 + s^2 \Gamma_Z^2 / M_Z^2} \quad .$$ (1)

For more details I refer to [2]. This is of course not totally surprising, but it has a nice consequence for Monte Carlo programs: we have two options to proceed, leading to fundamentally different programs.

- **stand-alone Monte Carlo** : in this approach one just computes all the radiative corrections as given by the standard model. The input parameters are the parameters of the electroweak Lagrangian: M_Z, α, $\alpha_s(M_Z^2)$, G_F, M_H, M_t, and so on. There is no need for a modified Born expression or so, and in principle this can give you the answer to any accuracy within the limits posed by the given order of perturbation theory you are working in. A setback is that in this approach one is confined to precisely the standard model – very few deviations from it are possible.

- **QED dresser:** this is a Monte Carlo that bases itself on the use of a form of the modified Born like the one in (1), and proceeds to apply the necessary QED corrections to this ansatz. Now the input parameters are things like widths and branching ratios, rather than masses of possibly unobservable particles like t or H: from the point of view of an experimentalist this is more attractive. The accuracy is fundamentally limited by the quality of the modified Born approximation: since we hear [2] that this is good to about 0.2% as well (at least in the resonance region) one may feel completely justified in doing so. On the other hand, to obtain the standard model predictions, one has to get, say, Γ_Z from somewhere else: typically, from a stand-alone program. A nice feature is that many deviations from the standard model can easily be described, like an anomalous value for the Z width, or the existence of an additional vector boson, and so on.

In fact the modified Born approximation provides something of a translation between these two approaches, each of which is realized in some existing Monte Carlo programs. This is good: it provides us with an idea of the usefulness of the QED dressing approach (it should be mentioned that the quality of the modified Born approximation has so far only been established for the total cross section σ, and not for the angular distribution $d\sigma/d\Omega$: work is in progress on this [4]).

4. QED ingredients

In addition to the purely weak corrections, we have to add QED corrections due to virtual and real photons. In fact, it is the bremsstrahlung which makes Monte Carlo simulation so essential for e^+e^- phenomenology: without it one could never get an accurate idea of the effects of experimental cuts etcetera. Let us first consider some orders of magnitude. In the resonance region, initial-state radiation amounts to a correction of about -40% in first order, with higher orders contributing some additional 10%. Final-state radiation gives only the famous small number $3\alpha/4\pi \sim 0.17\%$ (which however is still relevant if we stick to a precision goal of 0.1%): however, when resonably tight experimental cuts are applied this may change into about -10%. Initial-final interference is a truly small number at the Z peak and for loose cuts [5], but again if either tight cuts are applied or one moves away from the resonance the effect may become a few percent. We conclude that none of these effects can be safely neglected for precision predictions: we have to deal with the full QED cross section, including complicated kinematics and so on. For some very restricted classes of problems maybe one could neglect some QED contributions from e.g. the final state, but my personal view is that in real life we shall have to take every effect into account.

Let me now turn to a potential difficulty. As is well known, the bremsstrahlung correction is formally infinite, which divergence has to be cancelled by the inclusion of virtual photons. For Monte Carlo programs this leads to a nasty problem which in the jargon is known as the k_0-problem, or the positivity problem. It arises precisely from the infrared cancellation, as follows. Consider the case of one-loop QED corrections to some cross section σ_0. In a very simplified and abbreviated notation, the various QED cross sections can be written as

$$
\begin{aligned}
\sigma^{\text{virtual}} &\sim \sigma_0(1 + \beta \log(m_\gamma/E) + \cdots) \ , \\
\sigma^{\text{soft}} &\sim \sigma_0 \beta \log(E_0/m_\gamma) + \cdots \ , \\
\sigma^{\text{hard}} &\sim \sigma_0 \beta \log(E/E_0) + \cdots \ ,
\end{aligned}
\tag{2}
$$

where $\beta = 2\alpha/\pi(\log(s/m_e^2) - 1)$ and m_γ is a small regulatory photon mass: E is the beam energy. Note that I have left out all nonleading terms and so on: the above notation is more or less symbolic. I have split up the real bremsstrahlung cross section into a soft photon part and a hard photon part; hard photons have an energy larger than E_0 and can in principle be observed. The parameter E_0 is in fact artificial and arbitrary. In the generation of a Monte Carlo event one will have to decide at some point whether the event will contain a hard (i.e. observable) photon or not. This means that we have to add the virtual and soft photon cross section. The two alternatives (photon or no photon) are then given by the cross sections

$$
\begin{aligned}
\sigma^{0\gamma} &\sim \sigma_0(1 + \beta \log(k_0) + \cdots) \ , \\
\sigma^{1\gamma} &\sim \sigma_0 \beta \log(1/k_0) + \cdots \ ,
\end{aligned}
\tag{3}
$$

where $k_0 = E_0/E$. Reasonably, the *total* cross section $\sigma^{0\gamma} + \sigma^{1\gamma}$ does not depend on k_0. However we have to keep in mind that in a simulation events coming from $\sigma^{0\gamma}$ will have zero photon energy. Consequently there is a 'gap' in the photon spectrum between 0 and k_0. If an experiment is sensitive to this gap then the result will in fact show some k_0 dependence. Therefore, k_0 should be as small as possible. To get an idea, note that for $k_0 = 0.01$ we are neglecting photons with an energy of 500 MeV, which may not be realistic. Even experiments that do not see such photons could notice the effect: a 500 MeV photon can produce an acollinearity in the μ pair of 0.5 degrees. It is generally agreed that k_0 had better not be larger than about 0.001. However, as k_0 goes to zero, the cross section $\sigma^{0\gamma}$ will become negative! In that case one cannot talk any more of probabilities and the whole idea of Monte Carlo becomes invalid. In a semianalytical calculation we may just take the limit $k_0 \rightarrow 0$ but for Monte Carlo this is not allowed. Including in β also the effects of final-state radiation and the interference, we get

$$\beta = 2\frac{\alpha}{\pi}\left(\log(\frac{s}{m_e^2}) + \log(\frac{s}{m_\mu^2}) - 2 - 4\log\tan(\theta/2)\right) \ , \tag{4}$$

where θ is the scattering angle. From this (still simplified) formula we can derive a lower limit on the allowed k_0 values, which now depend on the angle as well: at the Z peak, the limit at 5,90,and 175 degrees is respectively 0.0001,0.002, and 0.01. We see that if the Monte Carlo is to be good for all angles, then we have to choose k_0 at least 0.01: but I just argued that this was too large! Note that this k_0-problem exists irrespectively of the magnitude of the first order correction itself. Indeed, going to second order we have the following cross sections for production of zero, one or two observable photons:

$$\begin{aligned}
\sigma^{0\gamma} &\sim \sigma_0(1 + \beta\log(k_0) + \beta^2\log^2(k_0)/2 + \cdots) \ , \\
\sigma^{1\gamma} &\sim \sigma_0(1 + \beta\log(k_0))\beta\log(1/k_0) + \cdots \ , \\
\sigma^{2\gamma} &\sim \sigma_0\beta^2\log(1/k_0)^2/2 + \cdots \ ,
\end{aligned} \tag{5}$$

which is seen to lead to precisely the same lower limit on the value of k_0. We must conclude that whereas a semianalytical result may always be improved by going to one more order in perturbation theory, the effect of a nonzero k_0 in a Monte Carlo result cannot be made smaller by going to any finite higher order: the event topologies cannot be improved.

It has been realized by several authors (for instance,see the discussion in [6]) that the only way out of the k_0-problem is to go to fully infinite order in α: the so-called exponentiation procedure. To get an idea of this, consider the two-loop result for $\sigma^{0\gamma}$. We can recognize the expansion of an exponential, which in fact was be proven to be valid in all orders in the classic paper by Yennie, Frautschi and Suura [7]. Resumming this exponential, we obtain the so-called exponentiated cross section

$$\sigma^{0\gamma} \sim \sigma_0 k_0^\beta + \cdots \ , \tag{6}$$

which now is finite and positive for all k_0. However, $\sigma^{0\gamma}$ is an inclusive quantity, and one has to find more exclusive quantities like the bremsstrahlung spectrum by some trick, like differentiating $\sigma^{0\gamma}$ with respect to E_0, upon which the well-known exponentiated spectrum arises which goes not as E_0^{-1} but as $\beta E_0^{\beta-1}$. Some additional complications are the following. Firstly, exponentiation is only rigorous in the limit of vanishing bremsstrahlung energy: extending it to finite energies (involving the so-called $\mathcal{R}$-pocedure described in [6]) involves some arbitrariness. This is also reflected in the fact that only the leading terms in the exponentiated expressions are unambiguous, and a number of exponentiation schemes exist, all differing in some nonleading terms. Fortunately, as mentioined in [2], around the Z all these schemes agree very well numerically (at LEP200 the discrepancies will probably be bigger). Secondly, the kinematics of an unlimited number of photons is difficult to handle. In practice, all photons softer than, say, 10^{-5} or so are neglected: a remnant of the old k_0, but now much more harmless. Also, it has not been settled completely how the initial-final interference should be handled. As an illustration, consider the angle-dependent β described above: due to its occurrence in the exponent, exponentiation and integration over the scattering angle do not commute with each other: the result depends on what you do first, exponentiating or integrating, and which one is correct? At the Z peak these ambiguities are numerically unimportant, but it is my opinion that we will have to resolve them sooner or later, certainly before LEP200 starts up.

The exponentiation principle is implemented in existing Monte Carlo programs in a number of different ways which I shall now describe.

- **the YFS approach**: this is an interesting and beatiful algorithm described in [6]. It aims at implementing the Yennie-Frautschi-Suura approach of [7] as completely as possible, and in my opinion does a good job of it. The only problem is in the so-called $\mathcal{R}$-procedure introduced by the authors, which is in fact ambiguous and arbitrary. Again, for LEP100 this is probably not so important. Since the YFS exponentiation allows for an infinite number of soft photons, this algorithm can in principle give events with arbitrarily high photon multiplicity: in practice, multiplicities go up to 4 or 5 photons harder than 100 MeV, at $\sqrt{s} = 100 GeV$.

  the MUSTRAAL approach: this is much more simple-minded. One just takes a one-loop correction program like the old MUSTRAAL [8], and in it modifies the photon spectrum in accordance with the exponentiation prescription. This means that there will be more events without photons, fewer events with visible but soft photons, and about the same number of events with hard photons. In the absence of anything better, it is of course about the only thing one can do, and the total cross section may not come out too bad. Since algorithms of this kind still generate at most one photon, the event topologies are of course somewhat poor.

- **the structure function approach**: this approach looks a lot like what is done in QCD calculations. QED structure functions (the emission of a photon by either the electron or the positron) were first discussed by a number of authors, of which the most useful reference seems to be [9]. This was applied in [10] to perform a semianalytical calculation of the Z line shape. Later on, a similar computation was performed in [11], and an extension into a Monte Carlo algorithm decribed in [12]. In mainly consists of taking an Altarelli-Parisi-like kernel decribing radiation of one bremsstrahlung photon, and iterating it a number of times, keeping correct track of momentum conservation etcetera. Hence, such an algorithm is capable of generating any number of bremsstrahlung photons. This type of algorithm is quite intuitive and transparent, and moreover lends itself to direct comparison with semianalytical calculations.

- **the DYMU2 approach**: this is a clever modification of the structure function approach, described most completely in [13]. It consists of modifying the QED structure function such that *two* applications of the structure function, one corresponding to radiation from the incoming electron, and one to radiation from the incoming positron, give a result for the total cross section which is precisely equal to the exact result of [14]. In other words, the complete bremsstrahlung effect is mimicked by two photons, and the total cross section should come out right. One more structure function for final-state radiation gives yet another photon, so that 3 bremsstrahlung photons can be generated.

It should be said that so far most emphasis has been on initial-state radiation and to a lesser extent on final-state radiation: interference is treated less well, or not at all, in most programs. I think this will improve in the near future. Before we turn to actual programs for $e^+e^- \rightarrow \mu^+\mu^-$ and their performance, I want to discuss one subtle problem in the structure function approach that has escaped attention so far.

5. A problem with structure functions?

Let us consider the general formula on which the structure function approach is based. For simplicity taking only up to two photons from the initial state, one writes for the differential cross section the following expression, in analogy with QCD:

$$\frac{d\sigma(s)}{d\Omega} \sim \int dx_1 dx_2 D(x_1) D(x_2) \frac{d\sigma_0(s')}{d\Omega} \ , \tag{7}$$

which is again supposed to be understood as a symbolic expression. The $x_{1,2}$ stand for the photon energy (and angular) variables, and D denotes a structure function. The cross section after QED corrections is $\sigma(s)$. Note that the 'undressed' cross section σ_0 must be evaluated at a reduced energy due to the energy lost to the bremsstrahlung photons. The structure function strategy is now as follows. First one generates the bremsstrahlung variables, according to some appropriate algorithm. Then the centre-of-mass frame (CMS) of the muons is determined. One boosts to this frame, and there generates the muon variables (in this case, the scattering angle) using the Born prediction (at the appropriate energy s'). Finally, one boosts back to the lab frame. A little appreciated subtlety here is the following. After hard photon emission, not only is the CMS energy reduced from $\sqrt{s}$ to $\sqrt{s'}$, but also the electron and positron beams will have a different direction, due to the boost: in general they will no longer be back-to-back. How, then, to define a muon scattering angle? The result will depend on which of the two beam directions we choose to orient ourselves to. In other words, we not only have to determine the *effective energy*, s', but also an *effective beam direction*. Can this be done, and what is the effective beam direction? Elsewhere, I treat this problem in some detail [15]: here, I shall just quote the result. Obviously, if no bremsstrahlung is emitted, the effective beams correspond to the actual beams. If one photon is emitted, effective beams can always be found, but they depend on the bremsstrahlung helicity. If the photon helicity is equal to the electron helicity, then the effective beam direction in the CMS is that of the electron; in the other case it is equal to the positron helicity. For this case, then, the structure function approach can be made exact by inclusion of an extra step: after generating the photon energy and angles, one also chooses its helicity (the two alternatives having unequal but well-defined probabilities) and from that determines the effective beam direction: then, a muon scattering angle can be sensibly defined. When two photons are emitted, the situation is more complicated. If the helicities of the two photons are equal, precisely the same holds as for the one-photon case: effective beam directions can be defined, depending on the photon helicities. However, when the photon helicities are opposite, it can be proven rigorously that *no effective beam direction exists*, i.e. the muon angular distribution has a form that can not be written using $d\sigma_0/d\Omega$, no matter what one tries to use for a beam direction. This means that the structure function approach will break down at the two-loop level in a way that cannot be repaired by modifying the structure function, or introducing a K factor or so. Note that for the line shape this problem does not occur since there we are only dealing with the total cross section, integrated over the muon variables, so no reference to a beam direction has to be made there. For more details I refer to [15]. Fortunately, once at least one of the two photons becomes soft or collinear with the beams, an effective beam can again be defined. Therefore, the problem will only be apparent in the contributions to the cross section arising from hard acollinear photons: this means that the two-loop leading and subleading effects will be treated correctly, but the terms proportional to $(\alpha/\pi)^2$, without large logs, will be generated incorrectly by a structure function Monte Carlo program. Hence I do not expect numerically noticeable effects due to this problem at LEP100: however at higher energies, where much more hard bremsstrahlung will be present, it may become important.

6. Programs and results for muon pair production

I shall now describe the results of a comparison between various Monte Carlo programs for $e^+e^- \rightarrow \mu^+\mu^-$. Let me first list the programs that we have studied, with some of their relevant features.

1. ZBATCH, by G.Burgers, with important contributions by W. Hollik. This is a semianalytical stand-alone program which computes the total cross section, the only cut allowed being on the $\mu^+\mu^-$ invariant mass. The most useful reference for it is [16]. To date its result for the line shape is considered the most reliable and accurate one: as I have indicated before, it serves as a benchmark or reference point, against which all our other results on the total cross section are checked. I would like to stress that without a program like this the task of comparing the Monte Carlo programs would be very much more difficult!

2. EXPOSTAR, by D.C. Kennedy et al.[17]. Again, this is a semianalytical stand-alone program to compute the line shape, just like ZBATCH. It mainly differs from it in that a different scheme is used for the computation of the weak corrections (the so-called 'star'-scheme), and that the QED effects are implemented using structure functions: however, the photon transverse momentum is neglected, and therefore I consider it more useful as a semianalytical program than as a Monte Carlo, although in fact it generates muon momenta.

3. **MUONMC**. In fact this is an adaptation of the program **BABAMC** by F.A.Berends, R.Kleiss and W.Hollik [18], written for Bhabha scattering. It is stand-alone and computes the full weak corrections, but the QED corrections are only taken into account to first order. As we shall see this is unacceptable, and we mainly include this program for ilustration. A similar program, called **BREM5**, by B.W.Lynn et al. (unpublished) is also based on this code: again, since it is only $\mathcal{O}(\alpha)$ in the photonic corrections, it will appear to be no good for precision tests.

4. **MMGE**, by J.P. Alexander et al.[19]. This is a QED dresser, based on the old **MUSTRAAL** code, with ad-hoc exponentiation of the photon spectrum.

5. **KORALZ**, written by a group of which S.Jadach and Z.Wąs are the principal members [20]. This program has at the moment the greatest potential to be a truly all-round Monte Carlo for LEP: it is stand-alone, and has YFS exponentiation built in. Moreover it also (in fact, principally) does τ pair production and decay. If any Monte Carlo is going to be the 'ultimate Monte Carlo', then it may well be this one. Unfortunately, at this moment the code is very large (about 12,000 lines) and apparently not yet coherent enough; also, it is slower than we would like. Still, it has to be taken very seriously.

6. **MOE**, by G. Bonvicini and L.Trentadue [12]. This is a QED dresser working with QED structure functions. It is fast and not too large – however, at this moment it does not yet generate muon momenta, only bremsstrahlung momenta. This makes it more useful for neutrino counting than for $\mu^{+}\mu^{-}$ physics. It is expected that this will be improved in the near future.

7. **DYMU2**, the program referred in section 4. It is written by two experimentalists, J.-E. Campagne and R. Zitoun [13]; it is a quite compact and fast QED dresser.

8. **CALASY**, by S.Jadach and Z. Wąs. This semianalytical program computes only the forward-backward asymmetry as defined in [5]: it should be considered as an attempt to provide a benchmark number for the asymmetry of the same quality as **ZBATCH** does for the total cross section

How does one proceed to compare all these codes with each other? We have restricted ourselves to simple quantities, which can be otained with reasonably small statistical error: the total cross section σ and the forward-backward asymmetry A_{FB}. To begin we have tried to give precisely the same input parameters to the various programs, corresponding to $M_Z = 92$ GeV, $M_H = 100$ GeV, $M_t = 60$ GeV, and $\alpha_s(M_Z^2) = 0.12$. Where this is not directly possible, as in the case of QED dressers, we have picked the corresponding values for, say, Γ_Z and so on. We have then computed σ and A_{FB}, using only a cut on the $\mu\mu$ mass of 0.2 of $\sqrt{s}$. This is much too loose to be realistic experimentally, but remember that at this moment our goal is to compare the Monte Carlo programs with each other, not with experiment. A loose cut on $M_{\mu\mu}$ should be sufficient to create a situation where all Monte Carlo programs should ideally agree.

The following plots were prepared by H. Burkhardt of ALEPH. In fig.1 we present the actual agreement on the total cross section σ, for a range of $\sqrt{s}$ values. We have chosen to normalize all results to the result of **ZBATCH**. One immediately sees that the

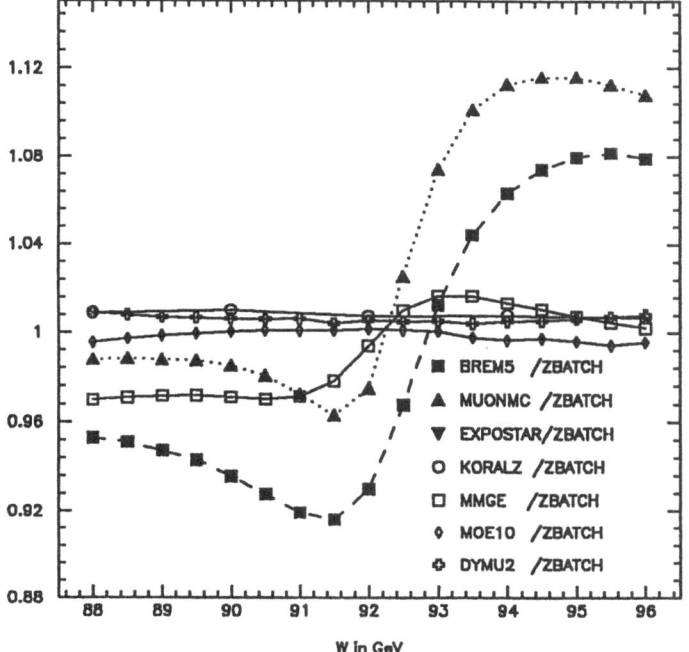

Figure 1: Cross sections

normalized to ZBATCH

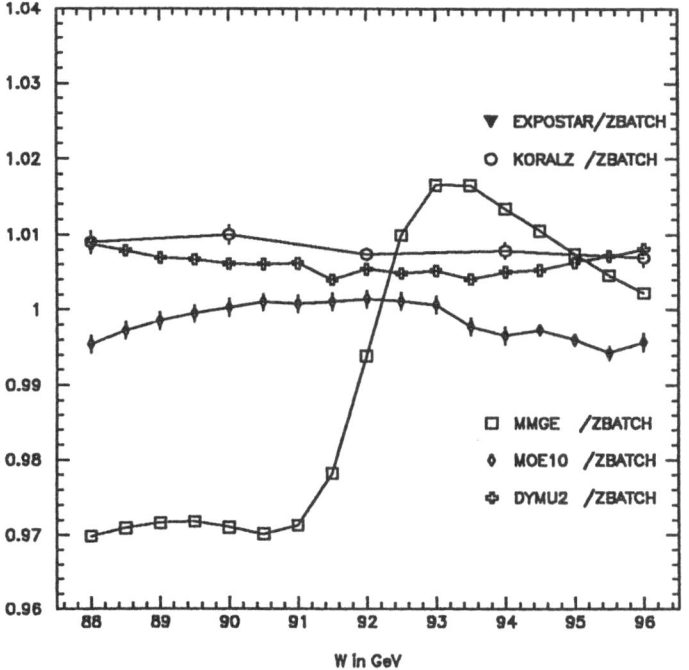

Figure 2: Cross sections

normalized to ZBATCH,

finer vertical scale

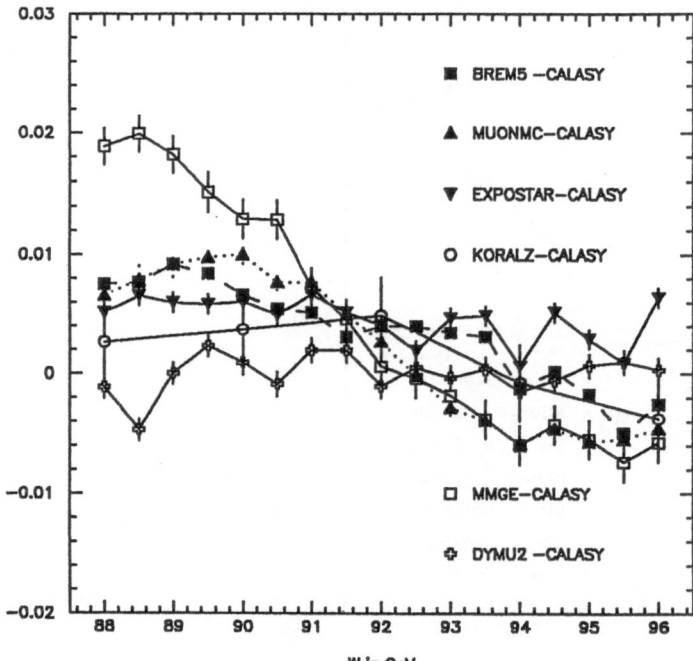

Figure 3: A_{FB} minus the

result given by CALASY

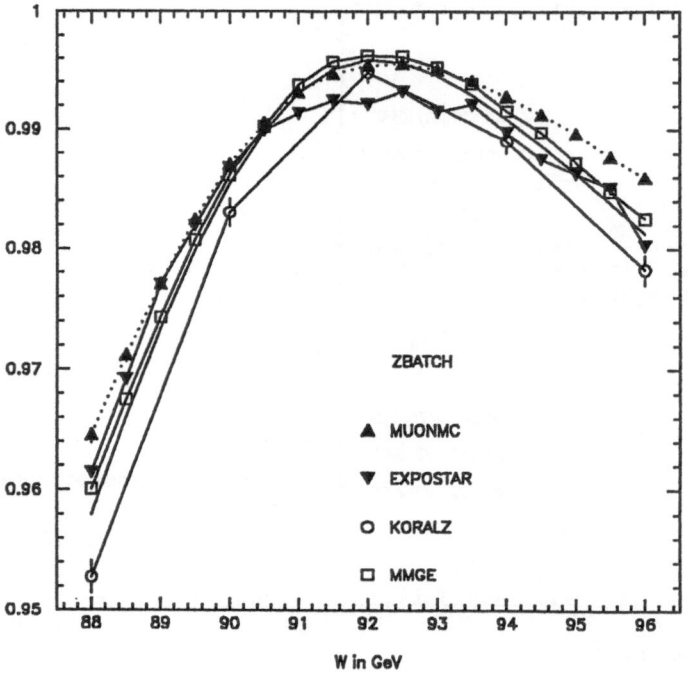

Figure 4: Suppression

of the cross section

by the $M_{\mu\mu}$ cut

$\mathcal{O}(\alpha)$ results of MUONMC and BREM5 are off by an unacceptable amount: the radiative tail of the Z is overestimated by some 10%. Don't use them!

Forgetting about the non-exponentiated programs and expanding the vertical axis, we obtain figure 2. The next worst agreement with ZBATCH is now seen to come from MMGE. This is not really surprising since the exponentiation procedure is not very sophisticated. Moreover, the original code contains bugs (mainly repaired by W.de Boer) and seems to be no longer supported by the authors. It is not really recommended. The other semianalytical result, that of EXPOSTAR agrees very well with that of ZBATCH – again, after some debugging. As I have said I do not take the program seriously as a Monte Carlo but as a check on ZBATCH it is very nice. Unfortunately the authors do not seem to support it any longer. We are now left with the only three candidates for a really good Monte Carlo: KORALZ, MOE, and DYMU2. Of these only KORALZ is standalone. It agrees not completely with ZBATCH, which at this moment is attributed to the electroweak library: replacing it with the library of Hollik that also goes into ZBATCH, the agreement becomes very good. I think the authors will have to make a commitment to one of these two libraries, and drop the other one (or debug it). Apart from some criticism on the size and speed of the program, it looks fine. MOE is the least finished of these three programs, since no muons are generated as yet. The agreement on the line shape is not too surprising, since the weak parameters have to be put in by hand. Still, here we see a check on the structure function approach to the bremsstrahlung exponentiation problem. The same holds more or less for DYMU2: any agreement on the line shape should not impress us too deeply. In fact, somewhat surprisingly, we see a more or less constant difference with ZBATCH. The authors claim to undertand this and it will presumably be repaired shortly.

Let us now turn to the A_{FB}. In fig.3 we show the values for A_{FB} from which we have subtracted the result of CALASY. The agreement in this case seems to be worse than for σ: worse, in fact, than the forseeable measurement error on A_{FB}. It is not easy to draw any definite conclusion here since there is no agreed-upon result of the quality of ZBATCH around. Also, one sees that the KORALZ prediction is a bit hampered by large statistical errors, due to the low speed of the program.

Finally, fig.4 presents the effects of a cut in $M_{\mu\mu}$ on the cross section. As we see, different programs are somewhat differently influenced by these cuts. It seems clear that if we would make even more cuts the agreement between programs would deteriorate further. A tentative conclusion is that at present there are 3 programs that deserve serious consideration: ZBATCH, KORALZ, and DYMU2. Of these, only KORALZ is really all-round.

7. Programs for Bhabha scattering

Let us now turn to Bhabha scattering. On the whole, the situation is much more confused and worse than for muon pairs. We have at our disposal the following programs:

1. **BABAMC**, by F.A. Berends, W. Hollik and R. Kleiss [18]. As stated before, it is only first-order in the QED corrections and will therefore be definitively wrong at large angles. However, it is one of the only two Bhabha Monte Carlo programs so far that contain the complete 1-loop electroweak corrections. The other one is the package **SPRING/BASES** by Tobimatsu and Shimizu [21]: this we have not studied so far.

2. **BHLUMI**, by S. Jadach and B.F.L. Ward [22] This is the Bhabha-scattering version of the YFS2 algorithm which is used by the same authors in their **KORALZ**. The prorgam is only supposed to be used at relatively small angles, and contains only QED and no weak effects.

3. **HOWLEEG** by W. de Boer. This is an adapted version of **OLDBAB**, a low-energy Bhabha scattering program of F.A. Berends and R. Kleiss [23]: it contains ad-hoc exponentiation following the **MUSTRAAL** approach. Again, since no Z contribution is taken into account, only small-angle scattering can be simulated.

4. **TEEGG** by D.Karlen [24]. This program was originally written to study the Bhabha background to single-photon production, used for neutrino counting. Consequently, no lower limit on the e^+e^- scattering angles is implemented, and so practically all events have zero or very small scattering angle. It contains QED up to second order. We have not included it in our studies so far.

As we see there is only a small overlap between the various programs: if we want to compare **BABAMC**, **BHLUMI** and **HOWLEEG** we will have to go to small angles, where all weak effects can essentially be neglected. In fact we can switch off the weak sector in **BABAMC** by replacing $M_Z \rightarrow A\, M_Z$ and $G_F \rightarrow A^{-2}G_F$, where A is some big number like 10 or 100. As a consequence, we loose all information on the performance of the program in the weak sector, but also the number of parameters decreases: in fact, only α remains. Also, since the Z peak has gone the dependence of all corrections on s will be weak, even in the Z region.

As far as the angular acceptance is concerned, there are two possibilities:

1. **symmetric**: both electron and positron must come out into the same angular window, for instance between 10 and 100 mrad from the beam.

2. **asymmetric**: either the positron or the electron must come out inside a somewhat smaller region, say, 20 to 90 mrad, or 11 to 99 mrad: then, the *other* one is again accepted from 10 to 100 mrad.

The motivation for an asymmetric setup is twofold: on the one hand there are uncertainties in the precise value of the measured scattering angle, due to uncertainties in the longitudinal position of the interaction vertex – this setup is supposed to repair this a bit since back-to-back pairs that do not really come from the centre of the detector can still be accepted even close to the edge of the acceptance. On the other hand, it has been claimed that the first-order correction is large and negative in the symmetric case [25] and that by enlarging the acceptance a bit one might decrease the size of the first-order correction to bring it under control. In fact this last argument is fallacious as I hope to show. We are at this moment still working on the results for these two setups.

Preliminarily, it seems that for the symmetric setup the first-order correction as given by BABAMC is quite small, and this is corroborated by HOWLEEG. Not surprisingly, the effect of exponentiation in HOWLEEG is tiny in this case. In contrast, BHLUMI disagrees with BABAMC/HOWLEEG by a few percent, and shows a big effect of exponentiation. As the authors themselves agree, there may still be problems with the overall normalization in BHLUMI, but the event topologies should be much better. Indeed, distributions of the e^+e^- acollinearity and so on seem to be the same for all three programs. In the asymmetric setup, the first-order corrections are (somewhat surprisingly) large and positive. Exponentiation here has some 1% effect. Again BHLUMI disagrees from the other two. This is as far as our comparison has gone. I expect that the agreement between BHLUMI and the rest will improve in the near future, but let me stress again that even in that case we can only compute the small-angle Bhabha cross section reliably. At large angles, *there exists no good Monte Carlo for large-angle Bhabha scattering*. Some hard work is urgently needed here!

Before finishing, I would like to discuss for one minute the order of magnitude of the corrections in small-angle Bhabha scattering. From the fact that the angular range is very small (10-100 mrad) we can conclude that a Bhabha pair, to be accepted, needs a very small acollinearity angle (90 mrad). In large-angle scattering this is a strict cut: only photons with an energy less than 0.09 of the beam energy can still be emitted isotropically under such cuts. Taking this as an upper limit on the photon energy, we indeed find a large negative correction to the cross section. From the talk of P. Rankin at this workshop I hear that the SLAC small Angle Monitors (SAM), which go from 50 to 160 mrad, and the MiniSAM which goes from 15 to 25 mrad, would result in a correction of some -30%. In that case, higher orders would be absolutely essential. However, in so arguing we should not forget the effect of hard bremsstrahlung. Typically, hard photons are mainly emitted parallel to the incoming and outgoing fermions. When collinear to the outgoing fermions, such photons will *not* generate any acollinearity, and the upper limit on their energy is essentially the beam energy E. For inital-state radiation, there is a nice formula, which reads as follows. Let us denote the scattering angle of the electron and positron by θ_1 and θ_2, respectively. Assuming for simplicity that θ_1 is the largest of the two, and *assuming that the photon is emitted parallel to the beams*, the energy of the photon is given by

$$E_{\text{photon}} = E\left(1 - \frac{\tan(\theta_2/2)}{\tan(\theta_1/2)}\right) \ , \tag{8}$$

which is exact for massless electrons. It can then easily be shown that in fact the upper limit on the photon energy (again, for photons that go along the beam!) is about 0.4 for the MiniSAM and 0.69 for the SAM. When so much hard radiation is allowed, one should not be surprised that the first-order correction is very small! We can now also understand the fact that in the asymmetric case the corrections are large. From the above formula, we see that the photon is emitted in the direction of the fermion with the largest scattering angle. Now in the asymmetric case, we have a correction from those events where both particles are inside the tighter cut (as we have seen, this correction is very small), plus the contribution from the events where one of them is at smaller angle. From our kinematical considerations we see that those events are a kind of 'spill-over' from events that originally (i.e. without bremsstrahlung) would have come out at the smaller angle, and hence be rejected, but in which one fermion has radiated a photon, got a larger scattering angle, and ended up inside the tighter cut. Since the angular distribution is so peaked in Bhabha scattering, there are many of these 'spill-over' events – a large correction. How to treat this large correction? It is clear that we will need higher order effects here. However, the largeness of the correction is in this case due to *hard* photon effects rather than soft photon effects. Therefore I expect the YFS exponentiation to be more or less useless here – we need a real two-loop Monte Carlo .

Conclusions

As far as muoun pair production is concerned, things start to look under reasonable control: we have more or less accurate predictions for the line shape, and two programs (KORALZ and DYMU2) with the potential to really be reliable and accurate enough to be useful. At this moment, I think that for realistic cuts a precision of about 0.5% in the predictions is attainable. More work is obviously needed to go down to 0.1%, in particular for the asymmetries. For A_{FB} some more semianalytical results are urgently needed. For hadrons, more or less the same remarks hold, except that there does not seem to be any Monte Carlo version for hadrons yet! It should not be too difficult to make versions of both KORALZ and DYMU2 that do hadronic decays. Bhabha scattering is not in good shape: we have no program that is good in the central detector. Some diligent work is urgently needed also here!

Acknowledgements

I am grateful to all mebers of the study group for their incterest and their hard work; in particular I would like to mention the efforts of H. Burkhardt, E. Locci, G. Bonneaud, A. Blondel, J. Ludwig, B. Gary, D. Schaile, V. Schegelsky, D. Karlen, M. Dam and F. Bianchi. I have also benefited much from discussions with F. Berends, S. Jadach. B. Ward, Z. Wąs, W. Hollik, F. Jegerlehner and many other colleagues. Finally I would like to thank the organizers of this workshop for the excellent and stimulating environment they have created.

References

[1] A.Blondel, in the CERN Yellow Report on Polarization at LEP, CERN 88-06, p.1.

[2] F.A.Berends, these proccedings, and references quoted therein.

[3] II. Burkhardt et al., in the CERN Yellow Report on Polarization at LEP, CERN 88-06, p.145, and II. Burkhardt, these proceedings.

[4] T. Riemann, these proceedings.

[5] S.Jadach and Z.Wąs, CERN-TII.5127/88

[6] S.Jadach and B.F.L.Ward, Univ.of Krakow preprint TPJU-15/88 and SLAC-PUB-4834.

[7] D.R. Yennie, S.C. Frautschi and II.Suura, Annals of Physics 13(1961)379.

[8] F.A. Berends, S. Jadach and R. Kleiss, Comp. Phys. Comm. 29(1983)185.

[9] E.A. Kuraev and V.S. Fadin, Sov. Journ. Nucl. Phys. 41(1985)466.

[10] G. Altarelli and G. Martinelli, in the CERN Yellow Report "Physics at LEP", CERN 86-02, p.47.

[11] O. Nicrosini and L. Trentadue, Phys. Lett. 196B(1987)551.

[12] G. Bonvicini and L. Trentadue, preprint UM-IIE-88-36.

[13] J.-E. Campagne. and R. Zitoun, University of Paris preprint LPNIIE-88.08, and J.-E. Campagne, Ph.D. Thesis, University of Paris VI/VII, 1989.

[14] F.A. Berends, G.J.II. Burgers and W.L. van Neerven, Nucl. Phys. 297(1988)249.

[15] R. Kleiss, to be published.

[16] G. Burgers, in the CERN Yellow Report on Polarization at LEP, CERN 88-06, p.121, and references quoted therein.

[17] D.C. Kennedy, B.W.Lynn, C.J.-C. Im and R.G. Stuart, SLAC-PUB-4128.

[18] F.A. Berends, W. IIollik and R. Kleiss, Nucl. Phys .B304(1988)712.

[19] J.P. Alexander, G. Bonvicini, P.S. Drell, and R. Frey, Phys. Rev. D37(1988)56.

[20] S. Jadach, Z. Wąs, R.G. Stuart and B.F.L. Ward, unpublished (can be obtained via the authors).

[21] Tobimatsu and Shimizu, Progr. Theor. Phys. 74(1985)567, 75(1986)905, 76(1986)334.

[22] S. Jadach and B.F.L. Ward, unpublished (can be obtained via the authors).

[23] F.A. Berends and R. Kleiss, Nucl. Phys. B228(1983)537.

[24] D. Karlen, Nucl. Phys. B289(1987)23, and Ph.D. thesis, Stanford University, SLAC-0325.

[25] P. Rankin, these proceedings.

Radiative Corrections for e⁺e⁻ Collisions   Editor: J.H. Kühn
© Springer-Verlag Berlin, Heidelberg 1989

The Exclusive Exponentiation in the Monte Carlo -
The Case of The Initial State Bremsstrahlung[†]

Stanisław Jadach

Institute of Physics, Jagellonian University

30-059 Kraków, ul. Reymonta 4, Poland

and

CERN, Geneva, Switzerland

B. F. L. Ward

Department of Physics and Astronomy,

The University of Tennessee, Knoxville, Tennessee 37996-1200, U.S.A.

and

CERN, Geneva, Switzerland

Abstract. The structure of the QED infrared singularities is reviewed and the various types of the Monte Carlo approaches to deal with them are briefly described. It is shown how to exponentiate the lowest first and second order QED calculation following the Yennie-Frautschi-Suura prescription and how to implement the corresponding multiphoton distributions in the Monte Carlo.

1. Introduction

The problem of the exponentiation of the infrared divergences in the framework of the conventional Quantum Electrodynamics is discussed in most complete way in the classical paper of Yennie, Frautschi and Suura [1] and the work presented in this talk relies heavily on the result obtained there. The casual reader of the above and other papers on the exponentiation, see also ref. [2], may think however that these papers deal only with the inclusive integrated cross section. We shall spend, therefore, the first part of the talk on demonstrating what are the exclusive multiphoton distributions and we shall show that in fact they are well known in the framework of ref. [1]. For pedagogical reasons we shall first consider simplified one-dimensional examples. The corresponding Monte Carlo generation algorithms will be discussed briefly. Then, we shall show what are the main ingredients in the full scale *second order* exclusive exponentiation, i.e. what are the corresponding multiphoton distributions. Particular emphasis will be put on explaining what is the practical difference between the lowest, first and second order exponentiation.

---

† Presented by S. Jadach

The talk will be concluded with some limited discussion on the second order Monte Carlo program (for the fermion production process in e^+e^- annihilation) and the corresponding numerical results. Let us note finally that more detailed information on the presented results and ideas can be found in refs. [3,4,5].

2. Infrared divergences - the Toy Model

One way of characterizing various Monte Carlo programs for QED calculations is to examine how they treat the virtual and real infrared divergences. In the following we shall describe in a simplified "pedestrian" way what is the structure of the infrared divergent poles in the QED distribution. We shall start with the first and second order (single and double bremsstrahlung) cases and then shall go to the infinite order case (exponentiation). The corresponding Monte Carlo algorithms will be briefly discussed.

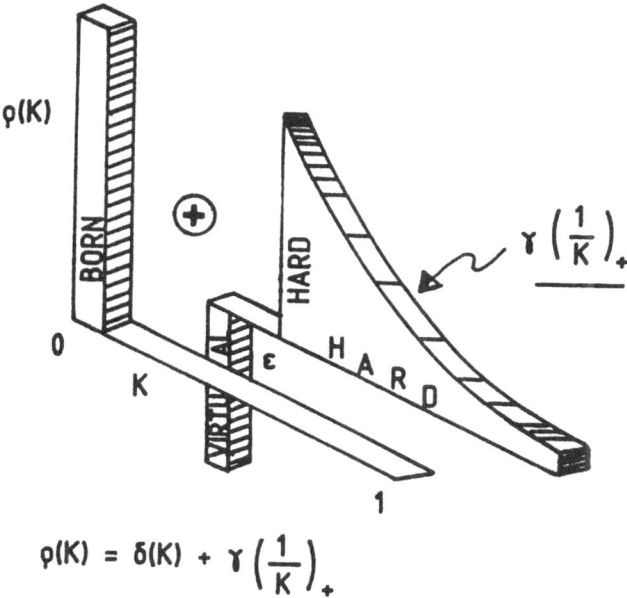

$$\varrho(K) = \delta(K) + \gamma \left(\frac{1}{K} \right)_+$$

Let us start with the well known $O(\alpha)$ case. The main purpose of this will be the introduction of the notation which will be used in a more complicated second and infinite order cases. In the following discussion we shall limit ourselves to the leading infrared divergent term. The nonleading finite contributions will be included in the discussion later on. Let us consider initial state bremsstrahlung in e^+e^- annihilation. In such a simplified picture (leading infrared divergences only) the total cross section can be written as follows

$$\sigma(s) = \int\limits_0^1 dK \rho_1(K) \sigma_{\text{Born}}(s(1-K)) \qquad (2.1)$$

where

$$\rho_1(K) = \delta(K) + \gamma\left(\frac{1}{K}\right)_+ = \delta(K)(1 + \gamma\ln\varepsilon) + \gamma\frac{1}{K}\theta(K-\varepsilon) \qquad (2.2)$$

is the photon energy $K = E^\gamma/E^{beam}$ distribution and the "effective" perturbative expansion parameter is $\gamma \simeq 2\frac{\alpha}{\pi}\int_{m_e^2/s}^{1}\frac{d\theta}{\theta} \simeq 2\frac{\alpha}{\pi}\ln\frac{m_e^2}{s} \simeq 0.10$. We keep track of the possible strong dependence of the Born cross section σ_{Born} on the center of mass energy $\sqrt{s}$. If this dependence is neglected then the integrated cross section in (2.1) will be equal to $\sigma_{\text{Born}}(s)$ due to $\int dK\left(\frac{1}{K}\right)_+ = 0$. The parameter ε was introduced here as an explicit regulator in the function $\left(\frac{1}{K}\right)_+$ i.e. it is the regulator of the infrared singularity. See also Fig. 1 for a pictorial representation. The negative contribution $\gamma\ln\varepsilon$ in the cross section below ε threshold

$$\sigma(K < \varepsilon) = \sigma_{\text{Born}}(s)(1 + \gamma\ln\varepsilon) \qquad (2.3)$$

is normally obtained by combining the virtual photon (vertex) contribution and the real soft photon contribution - both regularized with fictitious photon mass. The photon mass drops out and ε remains as an effective infrared cut-off/regulator. In order to simplify the following discussion we denote the sum of these two contributions (vertex + virtual-soft) as the virtual one i.e. the term $\gamma\ln\varepsilon$ we shall call the virtual contribution.

In the Monte Carlo of the class of ref. [6] one uses the finite value of $\varepsilon \simeq 10^{-2}$ and the distribution $\rho(K)\sigma_{\text{Born}}(s(1-K))$ is generated at the very beginning of the program. If $K < \varepsilon$ then the photon multiplicity is assigned zero, $n = 0$, otherwise, for $K > \varepsilon$, the photon multiplicity $n = 1$ is assigned. In the rest of the program the angular variables of the photon and fermions are generated. In order to keep the probability of $n = 0$ positive one cannot put ε too low. This restriction is relevant for the Monte Carlo - in the analytical calculation of the integrated cross section (for loose experimental cut-offs) the limit $\varepsilon \to 0$ is always understood.

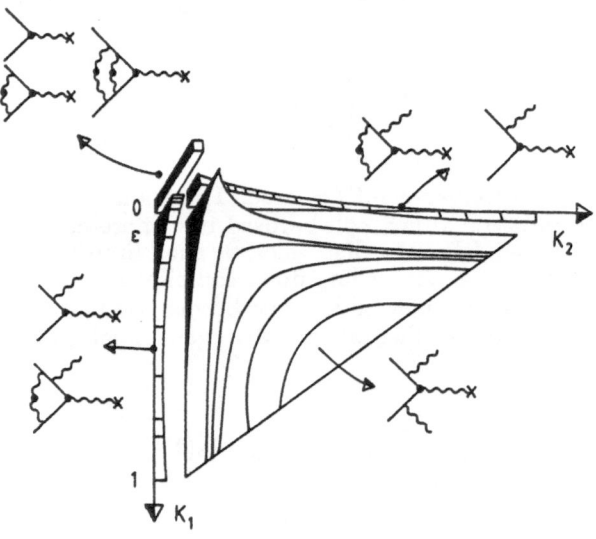

Let us come now to the *second order* cross section

$$\sigma = \int\limits_{K_1+K_2<1} dK_1 dK_2 \; \sigma_{\text{Born}}(s(1-K_1-K_2))\rho(K_1,K_2) \qquad (2.4)$$

where the double photon energy distribution (the leading infrared contribution as previously) can be written as follows

$$\rho_2(K_1,K_2) = \delta(K_1)\delta(K_2)\left(1 + \gamma \ln \varepsilon + \frac{1}{2}\gamma^2 \ln^2 \varepsilon\right)$$

$$+ \frac{1}{2}\delta(K_1)\frac{\theta(K_2-\varepsilon)}{K_2}\gamma(1 + \gamma \ln \varepsilon) + \frac{1}{2}\delta(K_2)\frac{\theta(K_1-\varepsilon)}{K_1}\gamma(1 + \gamma \ln \varepsilon)$$

$$+ \frac{1}{2}\gamma^2 \frac{\theta(K_1-\varepsilon)}{K_1}\frac{\theta(K_2-\varepsilon)}{K_2},$$

$$(2.5)$$

see Fig. 2 for a graphical representation. The above structure of the infrared divergences can be obtained by means of the explicit $O(\alpha^2)$ calculation, see ref. [7]. On the other hand the double photon energy distribution can be obtained by expanding the simple generating functional and keeping the terms up to $O(\gamma^2)$

$$1 \equiv \exp\left[\int\limits_0^1 dK \left(\frac{1}{K}\right)_+\right] \xrightarrow{O(\gamma^2)} \int\limits_0^1 dK_1 dK_2 \; \rho_2(K_1,K_2). \qquad (2.6)$$

Of course, the same procedure provides $\rho_1(K)$ (by retaining terms of $O(\gamma)$) and the higher order multiphoton distributions. Before we go to the higher order case let us discuss a hypothetical *second order* Monte Carlo generator. In such a Monte Carlo in the beginning of the generation procedure one would generate the two dimensional distribution $\rho_2(K_1,K_2)\sigma_{\text{Born}}(s(1-K_1-K_2))$ and assign the photon multiplicity $n = 0,1,2$ according to the number of the photons which go under the ε threshold. If σ_{Born} is strongly dependent on the energy $\sqrt{s}$ then it is more reasonable to rearrange the integral in the following way

$$\sigma = \int\limits_0^1 dK \sigma_{\text{Born}}(s(1-K))\rho_2(K)$$

$$\rho_2(K) = \int dK_1 dK_2 \rho(K_1,K_2)\delta(K-K_1-K_2) \qquad (2.7)$$

$$= \delta(K)\left(1 + \gamma \ln \varepsilon + \frac{1}{2}\gamma^2 \ln^2 \varepsilon\right) + \frac{\theta(K-\varepsilon)}{K}\gamma(1 + \gamma \ln \varepsilon)$$

$$+ \frac{1}{2!}\gamma^2 \int dK_1 dK_2 \delta(K-K_1-K_2)\frac{\theta(K_1-\varepsilon)}{K_1}\frac{\theta(K_2-\varepsilon)}{K_2}$$

In this case   in the corresponding Monte Carlo one would first generate the one-dimensional distribution $\rho(K)\sigma_{\text{Born}}(s(1-K))$, assign $n = 0$ for $K < \varepsilon$ and $n = 1,2$ for $K > \varepsilon$. The total energy K should be redistributed among two photon according to $\rho(K_1,K_2)$.

Let us note that in the above simple example one is already able to see the interesting phenomenon of the *competition of the soft photons for the total energy*. In spite of the fact that $\int_0^1 \int_0^1 dK_1 dK_2 \; \rho_2(K_1,K_2) \equiv 1$ the constrained distribution integrates

$$\int\limits_{0}^{1} dK \rho_2(K) = 1 - \gamma^2 \frac{\pi^2}{12} \tag{2.8}$$

to a value smaller than one. The inspection of the $\rho_2(K)$ distribution reveals that much of the negative contribution is located in the range $\varepsilon < K < 2\varepsilon$. This is related to the fact that the variable K in this range cannot be obtained by summing the two energies $K_1, K_2 > \varepsilon$. The similar phenomenon shows up for an infinite number of soft photons, see later this Section.

Let us now go to the case of the infinite order with an arbitrary number of photons in the game. Neglecting the $O(\gamma^3)$ and higher terms the second order formula can be rewritten as follows

$$
\rho_2(K) = e^{\gamma \ln \varepsilon} \Big\{ \delta(K) + \int dK_1 \delta(K - K_1) \frac{\theta(K_1 - \varepsilon)}{K_1} \\
+ \frac{1}{2} \int dK_1 dK_2 \delta(K - K_1 - K_2) \frac{\theta(K_1 - \varepsilon)}{K_1} \frac{\theta(K_2 - \varepsilon)}{K_2} \Big\}. \tag{2.9}
$$

The above expression suggests quite strongly the infinite order generalization of the following type

$$
\rho(K) = e^{\gamma \ln \varepsilon} \Big\{ \delta(K) + \sum_{n=1}^{\infty} \frac{1}{n!} \prod_{i=1}^{n} \frac{dK_i}{K_i} \theta(K_i - \varepsilon) \delta\Big(K - \sum_{i=1}^{n} K_i\Big) \Big\} \tag{2.10}
$$

Let us note immediately that the above formula is well suited for the Monte Carlo calculation since for a given photon multiplicity *the differential cross section is always positive*. There is no problem (except of computing time) to set the value of ε arbitrarily low. The Monte Carlo exercise based on this distribution was done in ref. [8]. In reality, the exclusive distribution (2.10) is not the result, however, of a pure guesswork based on the second order explicit calculation but it results from the careful analysis of the infrared virtual and real singularities in all orders of the perturbative QED which was done in the work of Yennie Frautschi and Suura [1]. In fact the result of this reference coincides exactly with formula (2.10). For readers aquainted with [1] the formula (2.10) may look unfamiliar, however. Let us elaborate on this rather important point. The main difference between (2.10) and the notation of ref. [1] and also other papers on the exponentiation, (see for instance ref. [2]) is that here we introduced the explicit regulator for the infrared singularity ε and we avoided the use of the Mellin transform which is unacceptable for the Monte Carlo. We can however translate (2.10) to a notation typical for classical papers on exponentiation rather easily:

$$
\rho(K) = e^{\gamma \ln \varepsilon} \int \frac{dx}{2\pi} e^{iKx} \Big\{ 1 + \int dK_1 \gamma \frac{\theta(K_1 - \varepsilon)}{K_1} e^{-ixK_1}
$$

$$
+ \frac{1}{2!} \int dK_1 \gamma \frac{\theta(K_1 - \varepsilon)}{K_1} e^{-ixK_1} \int dK_2 \gamma \frac{\theta(K_2 - \varepsilon)}{K_2} e^{-ixK_2} + ... \Big\} \tag{2.11}
$$

$$
= \int \frac{dx}{2\pi} e^{iKx} \exp\Big\{ \gamma \ln \varepsilon + \gamma \int dK' \frac{\theta(K' - \varepsilon)}{K'} e^{-ixK'} \Big\}
$$

The explicit dependence on the ε regulator may be removed by noticing that $\gamma \ln \varepsilon = -\gamma \int_{\varepsilon}^{1} \frac{dK'}{K'}$ and $\int_{0}^{\varepsilon} (e^{-iK'x} - 1) \to 0$ for $\varepsilon \to 0$ and finally we arrive at a compact, manifestly infrared finite and regulator free formula

$$\rho(K) = \int \frac{dx}{2\pi} e^{iKx} \exp\left\{\gamma \int\limits_0^\infty \frac{dK'}{K'}\left(e^{-xK'}-1\right)\right\} = \gamma K^{\gamma-1}\frac{e^{\gamma C}}{\Gamma(1+\gamma)}, \qquad (2.12)$$

which is precisely what can be found in ref. [1] ($C = 0.57721566...$). The above exercise can be done in the opposite direction i.e. the formula (2.10) can be obtained from (2.12) by introducing ε regularization, expanding the exponent and performing the x–integration. *The main lesson from this exercise is that although at first sight the formula (2.12) looks as an expression for the total cross section it contains, however, the full information on the differential cross sections with an arbitrary number of photons, as shown in eq. (2.10).* In other words the inclusive formula (2.12) is a *generating functional* for the exclusive formula (2.10). The two are completely equivalent numerically* and algebraically.

3. Full scale exclusive exponentiation

Let us now discuss a realistic case of second order exponentiated initial state bremsstrahlung for fermion pair production in electron positron annihilation. The two main differences to the previous Section are the restoration of the full phase space and the introduction of the nonleading (in the infrared limit) contributions. The QED multi-differential distribution can be encoded again either in the Mellin type formula analogous to (2.12) or in a manifestly exclusive manner as in eq. (2.10). The Mellin-type generating functional [1] reads

$$\begin{aligned}
\sigma = \int & \frac{d^4x}{(2\pi)^4} \int \frac{d^3q_1}{q_1^0}\frac{d^3q_2}{q_2^0} e^{ix(p_1+p_2-q_1-q_2)} \\
& \exp\left(2\alpha \mathrm{Re}B(p_1,p_2) + \int \frac{d^3k}{k^0}\tilde{S}(p_1,p_2,k)e^{-ixk}\right) \\
& \left\{\tilde{\beta}_0(p_i,q_j) + \int \frac{d^3k_1}{k_1^0}e^{-ixk_1}\tilde{\beta}_1(p_i,q_j,k_1)\right. \\
& \left.+\frac{1}{2!}\int \frac{d^3k_1}{k_1^0}\int \frac{d^3k_2}{k_2^0}e^{-ixk_1-ixk_2}\tilde{\beta}_2(p_i,q_j,k_1,k_2)\right\}
\end{aligned} \qquad (3.1)$$

where

$$\begin{aligned}
2\alpha B(p_1,p_2) &= \frac{i\alpha}{4\pi^2}\int \frac{d^4k}{k^2-m_\gamma^2}\left(\frac{2p_1-k}{k^2-2p_1k}+\frac{2p_2+k}{k^2+2p_2k}\right)^2, \\
\tilde{S}(p_1,p_2,k) &= -\frac{\alpha}{4\pi^2}\left(\frac{p_1}{p_1k}-\frac{p_2}{p_2k}\right)^2
\end{aligned} \qquad (3.2)$$

and the exclusive Monte-Carlo-friendly formulation reads

---

* In ref. [8] it was checked that the Monte Carlo calculation based on (2.10) gives precisely the same numerical answer as the analytical result shown in the right hand side of eq. (2.12).

$$\sigma = \sum_{n=0}^{\infty} \frac{1}{n!} \int \frac{d^3q_1}{q_1^0} \frac{d^3q_2}{q_2^0} \left(\prod_{i=1}^{n} \frac{d^3k_i}{k_i^0} \tilde{S}(p_1,p_2,k_i) \theta\left(\frac{2k_i^0}{\sqrt{s}} - \varepsilon\right) \right)$$

$$\delta^4\left(p_1 + p_2 - q_1 - q_2 - \sum_{i=1}^{n} k_i\right)$$

$$\exp\left(2\alpha \mathrm{Re}B(p_1,p_2) + \int \frac{d^3k}{k^0} \tilde{S}(p_1,p_2,k)\theta\left(\varepsilon - \frac{2k^0}{\sqrt{s}}\right)\right)$$

$$\left\{ \tilde{\beta}_0(p_1,p_2,q_1,q_2) + \sum_{l=1}^{n} \frac{\tilde{\beta}_1(p_1,p_2,q_1,q_2,k_l)}{\tilde{S}(k_l)} + \sum_{\substack{l,j=1 \\ l \neq j}}^{n} \frac{\tilde{\beta}_2(p_1,p_2,q_1,q_2,k_l,k_j)}{\tilde{S}(k_l)\tilde{S}(k_j)} \right\}.$$

$$(3.3)$$

The above formulas coincide with the one dimensional simplified distributions of the previous Section for $\tilde{\beta}_1 = \tilde{\beta}_2 \equiv 0$. To see it more clearly let us show how to recover the characteristic Sudakov formfactor $e^{\gamma \ln \varepsilon}$ in the above equations. Introducing the conventional infrared regulator $\varepsilon << 1$ we find

$$\exp\left(2\alpha \mathrm{Re}B + \int \frac{d^3k}{k^0} \tilde{S}e^{-ixk}\right) = \exp\left(R(\varepsilon) + \int\limits_{2k^0/\sqrt{s}>\varepsilon} \frac{d^3k}{k^0} \tilde{S}\, e^{-ixk}\right) \qquad (3.4)$$

where

$$R(\varepsilon) = 2\alpha \mathrm{Re}B + \int\limits_{2k^0/\sqrt{s}<\varepsilon} \frac{d^3k}{k^0} \tilde{S} = 2\frac{\alpha}{\pi}\left(\ln \frac{s}{m_e^2} - 1\right)\ln \varepsilon + \frac{\alpha}{\pi}\left(\frac{1}{2}\ln \frac{s}{m_e^2} - 1 + \frac{1}{3}\pi^2\right) \simeq \gamma \ln \varepsilon$$

is present in the exponent in eq. (3.3). The exclusive sum over the photon multiplicity in (3.3) is obtained by means of expanding $\exp\left[\int \frac{d^3k}{k^0}\tilde{S}e^{-ixk}\theta(2k^0/\sqrt{s} - \varepsilon)\right]$ and integrating over d^4x. The integration over photon angles leads to multi-photon energy distributions precisely as those in eq. (2.10). The function $\tilde{\beta}_0(p_1,p_2,q_1,q_2)$ is up to a normalization factor the Born differential cross section $d\sigma_{\mathrm{Born}}/d\Omega$ at the reduced center of the mass energy $s' = (p_1 + p_2 - \sum k_i)^2$. The K variable is replaced here by $v = 1 - s'/s$.

The above case of $\tilde{\beta}_1 = \tilde{\beta}_2 = 0$ and the example discussed in the previous Section represents *the lowest order exponentiation*. As can be read from the eqs. (3.1) and (3.3) in this case the differential cross section for production of n photons is

$$d\sigma_n \sim \prod_{i=1}^{n} \frac{d^3k}{k_i^0} \tilde{S}(k_i) \exp\left(2\alpha \mathrm{Re}B(p_1,p_2)\right)|\mathcal{M}_{\mathrm{Born}}|^2, \qquad (3.5)$$

that is, up to a normalization constant, we have[†]

[†] In Ref. [1] it was proven in arbitrary order of the perturbative QED that the most singular term in the differential cross section with n real photons looks like that in eq. (3.5) and that the residue coefficient $\tilde{\beta}_0$ is always the same, *independently* of n.

$$\tilde{\beta}_0 = |\mathcal{M}_{\text{Born}}|^2.$$

The factor $\exp\left(2\alpha\text{Re}B\right)$ steams from the sum over an infinite number of virtual photons and the distribution of the n real photons is good as long as $k_i^0 << \sqrt{s}/2$, i.e., there is no single hard photon.

Let us note an important point about the definition of $\tilde{\beta}_0$: the Born amplitude $\mathcal{M}_{\text{Born}}$ is, strictly speaking, defined within the two body phase space $p_1 + p_2 = q_1 + q_2$. The $\tilde{\beta}_0$ in eqs. (3.1), (3.3) and (3.5) are defined for momenta which do not obey this equation. It means that the definition of $\tilde{\beta}_0$ must *necessarily* embody a mapping procedure (so called reduction procedure, see ref. [3]) $p_i, q_i \to \mathcal{R}p_i, \mathcal{R}q_i$ such that $\mathcal{R}p_1 + \mathcal{R}p_2 = \mathcal{R}q_1 + \mathcal{R}q_2$. This transformation should tend to identity when the sum of photon energies goes to zero.

The principal question: *what is the first and second order exponentiation* should be restated as follows: *what is the proper multi-distribution in the case when in addition to many soft real photons there are one or two hard photons?* The answer was found in the Yennie-Frautschi-Suura work [1] and is already built in the eqs. (3.1) and (3.3). To see it more clearly let us look into the n-photon differential cross section in eq. (3.3) assuming that $\tilde{\beta}_2 = 0$ and $\tilde{\beta}_1 \neq 0$ (first order exponentiation)*

$$d\sigma_n \sim \tilde{\beta}_0 \prod_{i=1}^{n} \tilde{S}(k_i) + \sum_{l=1}^{n} \tilde{\beta}_1(k_l) \prod_{i \neq l} \tilde{S}(k_i). \tag{3.6}$$

If all photons are soft ($k_i^0 << \sqrt{s}/2$) then the second sum is negligible because $\tilde{\beta}_1(k)$ is not singular in the infrared limit $k^0 \to 0$. If one photon, say $l = L$, is hard $k_L^0 \sim \sqrt{s}/2$ then in the second sum only the term $l = L$ is non-negligible and we can rewrite (3.6) as follows

$$\tilde{S}(k_1)...\tilde{S}(k_{L-1})\left(\tilde{\beta}_0(k_L)\tilde{S}(k_L) + \tilde{\beta}_1(k_L)\right)...\tilde{S}(k_n) = \tilde{S}(k_1)...\tilde{S}(k_{L-1})\rho_1(k_L)...\tilde{S}(k_n),$$

where the $\rho_1(k)$ is the conventional single bremsstrahlung matrix element calculated from the Feynman rules. The above implies the definition of the $\tilde{\beta}_1(k)$, see refs. [3,5]

$$\tilde{\beta}_1(k) = \rho_1(k) - \tilde{S}(k)\tilde{\beta}_0. \tag{3.7}$$

Again, as it the case of $\tilde{\beta}_0$, the definition of $\tilde{\beta}_1$ must *necessarily* include the reduction procedure which eliminates from the four-momentum balance all photons but one. It should be also clarified that in the above definition one should use the lowest order $\tilde{\beta}_0^{(0)} \sim |\mathcal{M}_{\text{Born}}|^2$ while in the first term of the eq. (3.6) one uses rather

$$\tilde{\beta}_0^{(1)} = |e^{-2\alpha B}\mathcal{M}(\mathcal{O}(\alpha))|^2 = (1 - 2\alpha\text{Re}B)(1 + 2\text{Re}F_1)|\mathcal{M}_{\text{Born}}|^2$$

$$= (1 - 2\alpha\text{Re}B + 2\text{Re}F_1)|\mathcal{M}_{\text{Born}}|^2 = \left(1 + \frac{\alpha}{\pi}\left(\ln\frac{s}{m_e^2} - 1\right)\right)|\mathcal{M}_{\text{Born}}|^2$$

where $\mathcal{M}(\mathcal{O}(\alpha)) = (1 + F_1)\mathcal{M}_{\text{Born}}$ is again the first order (one loop) result obtained directly from the Feynman rules.

---

* In fact, as is shown in ref. [1], the differential cross section for n real photons can be expanded exactly (without any approximation) into n terms which include $\tilde{\beta}_0, \tilde{\beta}_1, ...\tilde{\beta}_n$. Here, we keep the two most singular terms. The point to be noted is that the residue coefficients $\tilde{\beta}_0$ and $\tilde{\beta}_1$ are precisely the same regardless the value of the photon multiplicity n.

What happens if among n photons there is not one but rather *two hard photons* In this case the distributions with $\tilde{\beta}_1$ only are not sufficiently good. One has to include $\tilde{\beta}_2$, update virtual corrections in $\tilde{\beta}_0$ and $\tilde{\beta}_1$ one order higher and generally this will be called the second order exponentiation. We refer the reader to ref. [3] for more details.

The recepy for the Yennie-Frautschi-Suura exponentiation can be summarized as follows:

1. Calculate in the traditional way, from Feynman diagrams, the second order QED matrix element with zero, one and two real photons, including the virtual corrections regularized by a photon mass.[‡] Subtract the ultraviolet divergences.

2. Eliminate the virtual infrared contribution by factoring out the $\exp(2\alpha B)$ term. (The photon mass will disappear.)

3. Calculate the $\tilde{\beta}_2$, $\tilde{\beta}_1$ by extracting in a recursive way the singular factors $\tilde{S}(k)$, as in eq. (3.7) (do not forget to include the reduction procedure in the definition).

4. Construct the differential cross section for n real photons using $\tilde{\beta}_{0,1,2}$ according to eq. (3.3) and integrate over the phase space.

5. Check that you do not use the calculation for the region of the phase space which includes three or more hard photons (or evaluate that such a contribution is negligible).

4. The second order Monte Carlo

The Monte Carlo program for the initial state bremsstrahlung in the electron-positron annihilation with the second order exclusive exponentiation is described in a detail in ref. [3]. The Monte Carlo algorithm depicted in Fig. 3 consists of the following steps:

1. Choose $v = 1 - s'/s$ according to a distribution which is roughly $\gamma v^{\gamma-1}\sigma_{\text{Born}}(s(1-v))$.

2. If $v < \varepsilon$ then choose the photon multiplicity $n = 0$, otherwise generate $n - 1$ photons according to Poisson distribution with the average $\gamma \ln \frac{1}{\varepsilon}$.

3. Generate the photon four-momenta according to the density

$$\frac{d^3 k}{k^0} \tilde{S}(k).$$

4. Construct the four-momenta of all photons and fermions.

[‡] One should use, in principle, the exact $O(\alpha^2)$ matrix elements as an input in the exponentiation procedure. In ref. [3] one uses for this purpose the leading and next-to-leading logarithmic approximations because (a) the relevant exact matrix elements are not available in the literature and (b) because such approximations are good enough to reach the precision necessary for LEP/SLC experiments.

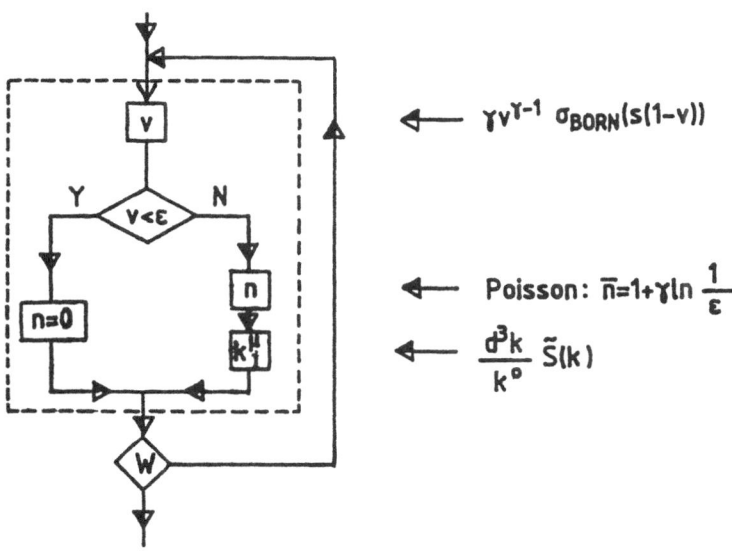

$$\longleftarrow \quad \gamma v^{\gamma-1}\, \sigma_{BORN}(s(1-v))$$

$$\longleftarrow \quad \text{Poisson: } \bar{n}=1+\gamma\ln\frac{1}{\varepsilon}$$

$$\longleftarrow \quad \frac{d^3k}{k^0}\,\tilde{S}(k)$$

5. Reject events according to the weight which adds the effects due to $\tilde{\beta}_1$ and $\tilde{\beta}_2$.

Let us finally present the example of a numerical results from the YFS2 Monte Carlo of ref. [3]. In Table 1 below we show the values of the integrated cross sections in the R-units. The following input parameters were used: $M_Z = 92GeV$, $\Gamma_Z = 2.45346$, $\sin^2\theta_W = 0.228818$. Upper limit on the photon phase space was $s'/s > 0.2$ where $\sqrt{s'}$ is an effective mass of the muon pair. The cross section σ_E is the best available non-Monte-Carlo result from ref. [7]; we show their "exponentiated" second order result, with the omission of the production of additional fermion pairs. No electroweak corrections were included. The two calculations agree to within 0.2%.

Table 1

$\sqrt{s}$ [GeV]	σ Monte Carlo	$(\sigma_E - \sigma)/\sigma \cdot 10^3$
40	1.1026 ± 0.0006	$-0.3 \pm .5$
90	33.65 ± 0.01	0.6 ± 0.4
91	73.33 ± 0.02	0.6 ± 0.3
92	131.53 ± 0.04	0.8 ± 0.3
93	93.38 ± 0.03	0.7 ± 0.3
94	58.38 ± 0.02	1.4 ± 0.4
100	13.21 ± 0.01	0.6 ± 0.7

REFERENCES

1. D. R. Yennie, S. C. Frautschi and H. Suura, Annals of Phys. **13** (1961) 379.

2. M. Greco, G. Pancheri and Y. N. Srivastava, Nucl. Phys. **B101** (1975) 234; Phys. Lett. **56B** (1975) 367.

3. "YFS2 - the second order Monte Carlo for fermion pair production at LEP/SLC with the initial state radiation of two hard and multiple soft photons", Jagellonian University preprint, TPJU-15/88 (SLAC-PUB-4834, UTHEP-88-0901) December 1988, to be submitted to Comp. Phys. Commun.

4. S. Jadach and B. F. L. Ward, SLAC-PUB-4543 (1988), Phys. Rev. **D38** (1988) 2897.

5. B. F. L. Ward, Phys. Rev. **D36** (1987) 939; Acta Phys. Pol. **B19** (1988) 465.

6. F. A. Berends, R. Kleiss and S. Jadach, Comp. Phys. Commun. **29** (1983) 185.

7. F. A. Berends and W. L. Van Neerven and G. J. H. Burgers, Nucl. Phys. **B297** (1988) 249.

8. S. Jadach, "Yennie-Frautschi-Suura soft photons in the Monte Carlo generators", preprint of Max-Plack-Institut, München, MPI-PAE/PTh 6/87 (1987).

Radiative Corrections for e^+e^- Collisions   Editor: J.H. Kühn
© Springer-Verlag Berlin, Heidelberg 1989

THE A_{pol} AS TEST OF THE STANDARD MODEL AT LEP*

Stanisław Jadach

Institute of Physics, Jagellonian University

30-059 Kraków, ul. Reymonta 4, Poland

and

Zbigniew Wąs †

Max Planck Institut für Physik und Astrophysik

D-8000 München 40, Föhringer Ring 6, FRG.

Abstract. In this paper we discuss the measurement of the τ polarization at LEP and its application for precision tests of the Standard Model. The observables which can be constructed for this purpose and different classes of radiative corrections which affect these quantities are carefully reviewed. We discuss the expected statistical and systematic errors for A_{pol} and compare them with the size of electroweak corrections.

Finally, going beyond the standard model, we argue that the size of the experimental errors, for theoretically interesting quantities deteriorates significantly if the Z-fermion couplings are non-universal.

1. Introduction

The experiments at SLC and LEP will offer an opportunity for precision tests of the electroweak sector of the Standard Model (SM) [1]. The general principle of these tests is the following: using, for example, α, G_μ and M_Z as an input one calculates the values of some precisely measurable quantities and compares them with experimental results. If the accuracy of the experiment is good enough to match the size of $O(\alpha)$ electroweak corrections then the agreement constitutes positive experimental evidence for the validity of the SM beyond the tree level. On the other hand, any sizable discrepancy would indicate new physics. In addition to the left-right asymmetry A_{LR}, which offers the best sensitivity but requires polarized beams, and the forward-backward asymmetry A_{FB}, the τ spin polarization asymmetry A_{pol} in the reaction $e^+e^- \to \tau^+\tau^-$ is of particular interest.

In a sense the above three observables are complementary, they all measure couplings of leptons to the Z. From the experimental point of view they do it, however, in a different way and thus will be differently influenced by systematic errors. On the other hand, they are testing slightly different aspects of the model. For instance, they involve coupling constants of leptons (μ, e, τ) from various families and therefore check lepton universality.

---

* Presented by Z. Wąs

† Permanent adress: Institute of Nuclear Physics, Cracow ul. Kawiory 26a, Poland

In this paper we concentrate on the τ polarization asymmetry A_{pol}, which can be measured in τ pair production and the subsequent decay process. This observable was considered [2,3] as an important data point in precision tests of the SM model. It remained, however, in the shadow of the forward-backward asymmetry A_{FB} for muons which is experimentally much easier to measure. It appears, however, that although A_{FB} can be measured more easily and with a smaller statistical error it is less sensitive to the $\tau - Z$ vector coupling constant and thus to $\sin^2 \theta_W$. It is also subject to large QED corrections [4] and other problems resulting from its rapid dependence on the center of mass energy (CMS) $\sqrt{s}$ across the Z resonance. The τ polarization asymmetry A_{pol} benefits, similar to A_{LR}, from the same property of the linear dependence on the vector coupling constant to the Z. It varies less strongly with $\sqrt{s}$ than A_{FB}. It is not prone to effects from the initial/final state bremsstrahlung interference, similar to A_{LR} [5] and A_{FB} [6] (to some extent).

The clear disadvantage of A_{pol} is that it has to be measured using τ decay distributions and since this can be done only for some decay modes a substantial loss in statistics occurs. Systematic uncertainties arising from a misidentification of the τ decay modes may also degrade the value of A_{pol} as a promising observable. In the following we shall try to answer the question as to how important are the potentially most dangerous experimental uncertainties and also how pure QED effects (hard bremsstrahlung) influence the τ polarization measurement. We shall also show how the quality of the measurement will deteriorate if we go beyond the SM and relax some of its constraints.

The layout of the paper is the following: In Section 2 we collect the basic facts on the τ polarization asymmetry A_{pol}. We introduce the notation and in particular we define the experimental measurables related to A_{pol}. Then we show the dependence of A_{pol} on the center of mass energy and the scattering angle. We concentrate mainly on the one decay mode $\tau \to \pi \nu$ which will be used most extensively as τ polarimeter. In the discussion we shall use Born differential distributions both for production and for decay processes, sometimes limiting ourselves to Z exchange only. The dependence on the $Z-e$, $Z-\tau$ and $W-\tau$ couplings will be shown explicitly. We construct then two π energy asymmetries which respectively measure Z couplings to leptons (universality must be assumed) and Z couplings to τ.

In Section 3 we concentrate on QED effects. We show how various types of the bremsstrahlung, i.e. photon emission from incoming beams, outgoing τ's and decay products, influence the A_{pol} measurement. The numerical results are presented. We also present estimates of the experimental and statistical errors in the measurements of A_{pol} and how their size depends on the assumption of lepton universality.

Section 4 contains the summary and concluding remarks. We compare statistical and systematic errors of A_{pol} with the expected errors for the measurement of A_{FB}. Finally we will check how the size of the experimental errors matches the weak effects entering through top and Higgs mass dependence.

2. Basic facts on the τ polarization asymmetry

The Born differential distribution for τ pair production $e^+e^- \to \tau^+\tau^-$ including the dependence on the τ^- longitudinal polarization p reads (see eg. [7]):

$$\frac{d\sigma_{\text{Born}}}{d\cos\theta}(s, \cos\theta; p) = (1 + \cos^2\theta)F_0(s) + 2\cos\theta F_1(s)$$
$$+ p[(1 + \cos^2\theta)F_2(s) + 2\cos\theta F_3(s)], \tag{2.1}$$

with the four formfactors defined as follows

$$F_0(s) = \frac{\pi\alpha^2}{2s}(q_e^2 q_\tau^2 + 2\text{Re}\chi(s)q_e q_\tau v_e v_\tau + |\chi(s)|^2(v_e^2 + a_e^2)(v_\tau^2 + a_\tau^2)),$$

$$F_1(s) = \frac{\pi\alpha^2}{2s}(\quad 2\text{Re}\chi(s)q_e q_\tau a_e a_\tau + |\chi(s)|^2 2v_e a_e 2v_\tau a_\tau),$$

$$F_2(s) = \frac{\pi\alpha^2}{2s}(\quad 2\text{Re}\chi(s)q_e q_\tau v_e a_\tau + |\chi(s)|^2(v_e^2 + a_e^2) 2v_\tau a_\tau), \tag{2.2}$$

$$F_3(s) = \frac{\pi\alpha^2}{2s}(\quad 2\text{Re}\chi(s)q_e q_\tau a_e v_\tau + |\chi(s)|^2 2v_e a_e (v_\tau^2 + a_\tau^2)),$$

and

$$\chi(s) = \frac{s}{s - M_Z^2 + is\Gamma_Z/M_Z}.$$

The $q_e, v_e, a_e, q_\tau, v_\tau, a_\tau$ are the Z coupling constants to the electron and τ respectively.

The above formfactors are related directly to the Born cross section σ_{Born}, the forward backward asymmetry A_{FB}, the τ polarization asymmetry A_{pol}, the combined polarization forward backward asymmetry $A_{\text{pol}}^{\text{FB}}$

$$\sigma_{\text{Born}}(s) = \frac{8}{3}F_0(s),$$

$$A_{\text{FB}}(s) = \frac{1}{\sigma_{\text{Born}}}\left\{\sigma(\cos\theta > 0) - \sigma(\cos\theta < 0)\right\} = \frac{3}{4}\frac{F_1(s)}{F_0(s)} \simeq \frac{3}{4}\frac{2v_e a_e}{v_e^2 + a_e^2}\frac{2v_\tau a_\tau}{v_\tau^2 + a_\tau^2},$$

$$A_{\text{pol}}(s) = -\frac{1}{\sigma_{\text{Born}}}\left\{\sigma(p = +1) - \sigma(p = -1)\right\} = -\frac{F_2(s)}{F_0(s)} \simeq -\frac{2v_\tau a_\tau}{v_\tau^2 + a_\tau^2},$$

$$A_{\text{pol}}^{\text{FB}}(s) = -\frac{1}{\sigma_{\text{Born}}}\left\{\sigma(\cos\theta > 0, p = +1) - \sigma(\cos\theta > 0, p = -1)\right.$$

$$\left. - \sigma(\cos\theta < 0, p = +1) + \sigma(\cos\theta < 0, p = -1)\right\}$$

$$= -\frac{3}{4}\frac{F_3(s)}{F_0(s)} \simeq -\frac{3}{4}\frac{2v_e a_e}{v_e^2 + a_e^2},$$

$$\tag{2.3}$$

and the angular dependence of the τ polarization asymmetry

$$A_{\text{pol}}^{\text{Born}}(s, \cos\theta) = -\frac{d\sigma^{\text{Born}}(\cos\theta, p = +1) - d\sigma^{\text{Born}}(\cos\theta, p = -1)}{2d\sigma^{\text{Born}}(\cos\theta, p = 0)}$$

(2.4)

$$= -\frac{(1 + \cos^2\theta)F_2(s) + 2\cos\theta\, F_3(s)}{(1 + \cos^2\theta)F_0(s) + 2\cos\theta\, F_1(s)}.$$

We have also indicated the approximate relation to the coupling constants at the Z peak (γ exchange neglected). The polarization p of the τ^- can be measured by looking into the energy distribution of a given decay product,

$$h(u) = \frac{1}{N}\frac{dN}{du}, \qquad u = \frac{2E_{\text{dec.prod.}}}{\sqrt{s}}, \qquad \int_0^1 h(u)du = 1.$$

(2.5)

Here we shall limit ourselves to $\tau \to \pi\nu$ decay mode, which is expected to be best suited for the measurement of A_{pol}. The decay distribution reads

$$h_\pi(p; u) = h_0^\pi(u) + p h_1^\pi(u) \simeq 1 - p\omega(2u - 1),$$

$$\int_0^1 h_0^\pi(u)du \equiv 1, \qquad \int_0^1 h_1^\pi(u)du \equiv 0.$$

(2.6)

The parameter

$$\omega = \frac{2g_V g_A}{g_V^2 + g_A^2},$$

(2.7)

includes the coupling constants of $W^\pm$ boson to τ. The standard $V - A$ hypothesis corresponds to $\omega = -1$.

What is actually measured in the experiment is the double differential distribution in $\cos\theta$ and u which is obtained by appropriatly combined eq. (2.1) and eq. (2.6)

$$\frac{d\sigma^{\text{Born}}}{d\cos\theta du}(s, \cos\theta, u) = \quad (1 + \cos^2\theta)F_0(s)h_0(u) + 2\cos\theta F_1(s)h_0(u)$$
$$(1 + \cos^2\theta)F_2(s)h_1(u) + 2\cos\theta F_3(s)h_1(u).$$

(2.8)

On the other hand the main aim of the experiments is to measure quantities

$$\mathcal{A}_{i,\ (i=e,\mu,\tau)} = 2v_i a_i/(v_i^2 + a_i^2).$$

(2.9)

In eq. (2.9) we will use the index $i = l$ when universality of the Z couplings to leptons is asumed.

Let us now turn to the discussion of the phenomenological consequences of eqs. (2.1), (2.6), (2.8).

At the top of the Z resonance eq. (2.4) can be simplified:

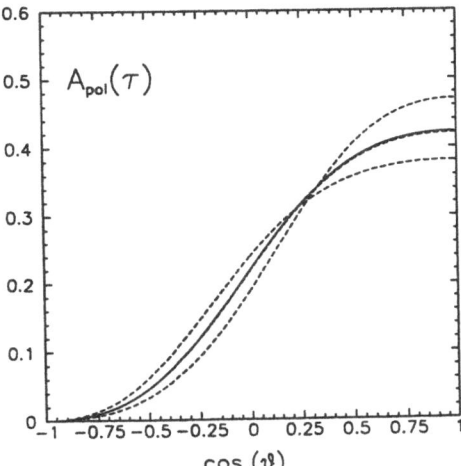

Figure 1. A_{pol} *as a function of* $\cos\theta$. *The solid line was obtained with eq.* (2.10) *whereas three dashed lines were obtained with eq.* (2.4), $\sqrt{s} = M_Z, M_Z \pm 2GeV$ *($M_Z = 93$ GeV, $\sin^2\theta_W = 0.222$, $\Gamma = 2.557 GeV$).*

$$A_{pol}(\cos\theta) = \frac{\mathcal{A}_\tau + \frac{2\cos\theta}{1+\cos^2\theta}\mathcal{A}_e}{1 + \frac{2\cos\theta}{1+\cos^2\theta}\mathcal{A}_e\mathcal{A}_\tau}. \qquad (2.10)$$

As we can see from Fig. 1 the τ's produced in the forward directions are strongly polarized, up to twice the mean polarization, whereas polarization is very small in the backward regions and approaches zero for $\cos\theta = -1$. The mean of the τ polarization is determined by couplings of Z to τ whereas the asymmetry in the polarization is due to the couplings of Z to electron. In Fig. 1 we see also that the τ polarization given by the approximate eq. (2.10) differs only very little from the predictions of eq. (2.4) and that the τ polarization changes only very little with the CMS energy. This mild dependence on the CMS energy is also the reason that τ polarization is rather mildly affected (see Fig. 2) by the potentially big initial state bremsstrahlung correction.

In Fig. 3 we show the two-dimensional u, $\cos\theta$ distribution of the $\tau \to \pi$ decay. As we can see the slope of the π energy spectrum can be used, at every angle θ separately to measure angular τ polarization distribution. For the purpose of the further discussion we limit ourself to the more global asymmetries $A_\pi^{(1,2)}$

$$A_\pi^{(1,2)} \equiv \frac{N^{(1,2)}(u_\pi > u_0) - N^{(1,2)}(u_\pi < u_0)}{N^{(1,2)}(u_\pi > u_0) + N^{(1,2)}(u_\pi < u_0)} \simeq \frac{1}{2}A_{pol}^{(1,2)} \qquad (2.11)$$

where u_0 denotes the mean energy of the π and $N^{(1,2)}(u_\pi > u_0)$, $N^{(1,2)}(u_\pi < u_0)$ denote respectively the number of the detected pions faster and slower than the mean energy. In the case 1) we take into account all π's accessible for the detection. In the case 2) we exclude backward region ($\cos\theta < -0.35$) where the τ sample is nearly

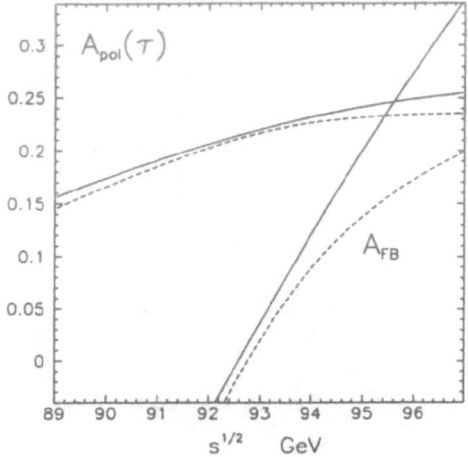

Figure 2. *The influence of the initial state bremmstrahlung on A_{pol} and on the forward-backward asymmetry A_{FB}. The steep solid and dashed lines represent A_{FB} and the flat solid and dashed lines depict A_{pol}. The solid lines represent the Born approximations whereas dashed ones include initial state bremsstrahlung corrections according to the approximation described in [8] ($M_Z = 93$ GeV, $\sin^2 \theta_W = 0.222$, $\Gamma = 2.557 GeV$).*

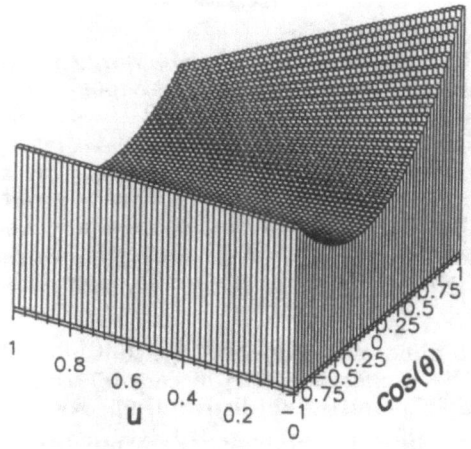

Figure 3. *The 2-dimensional distribution, described by the eq.(2.8) of the $\tau \to \pi$ decay at $\sqrt{s} = M_Z$ ($M_Z = 93$ GeV, $\sin^2 \theta_W = 0.222$, $\Gamma = 2.557GeV$).*

unpolarized. The mean τ polarization averaged over the region (1) $A_{pol}^{(1)}$ depends practically only on $\mathcal{A}_\tau$ whereas in the case (2), $A_{pol}^{(2)}$ depends on the combination of $\mathcal{A}_\tau$ and $\mathcal{A}_e$.

3. QED bremsstrahlung effects and experimental errors

The basics of QED corrections to A_{pol} close to the Z are as follows: 1) initial state bremsstrahlung effect is negligible, 2) the direct influence of final state bremsstrahlung on A_{pol} is even smaller than the one of the initial state bremsstrahlung; the indirect effect on the π energy distribution for measuring A_{pol} is however sizable.

The direct influence of the initial state photon emission is negligible because A_{pol} (like A_{LR}) depends weakly on $\sqrt{s}$ (contrary to A_{FB}) and the smearing of the center of the mass energy close to the top of the Z resonance is strongly cut off by the Z line shape. The final state bremsstrahlung does not change in practice the polarization of the τ. The helicity non-conservation induced by photon emission is of order $\frac{\alpha}{\pi} A_{pol} \simeq 10^{-4}$, but since A_{pol} is measured from the slope of the energy distribution of a decay product, the softening of its energy spectrum due to the emission of a photon from the τ prior to its decay and from the τ decay product itself influences the experimental estimate of A_{pol} quite significantly.

The first numerical estimate of the above QED effects was obtained in ref. [10,3] (see also [9]) with help of the $O(\alpha)$ QED Monte Carlo calculation. The ultimate result for the size of QED effects will always require a Monte Carlo calculation because QED effects are inherently dependent on the experimental cut-offs. Monte Carlo is also necessary for any reliable estimate of the systematical errors.

The Monte Carlo program [11] which could in principle be used to perform such a high precision study already exists. However this study requires good knowledge of the detector properties. This will have probably to wait until after first data samples are analyzed, and some experience is gained how detectors work in practice.

Let us recall the numerical results of [9] on the effects of the QED bremsstrahlung. In [9] $M_Z = 93 GeV$, $\Gamma_Z = 2.557 GeV$ and $\sin^2 \theta_W = 0.222$ were used and we will use these values also in this note. In the lowest possible level of approximation when only Z exchange is included we obtain from (2.10) that the mean τ polarization is given by the following eq.

$$A_{pol} = \mathcal{A}_\tau \qquad (3.1)$$

The direct QED corrections which modify the relation (2.10) between A_{pol} and the couplings of the τ lepton to the Z originate from pure s-channel photon exchange and from bremsstrahlung in the initial and final states. As can be seen from Table 1, these direct QED corrections, which modify relation (2.10) are small (compared to $A_{pol} \simeq 0.2$), they do not exceed 0.007. The reason is as follows: The biggest correction due to the emission of hard photons in the initial state is strongly reduced because on the top of the Z the photon spectrum is damped by the resonance shape. In Fig. 4 the corresponding distributions are shown. Contrary to A_{FB}, the τ polarization asymmetry A_{pol} depends weakly on the CMS energy in the region of the Z resonance and is thus not vulnerable to the emission of soft photons.

γ contribution to Born	Initial state brem.	Final state brem.
-0.0012	-0.0049	-0.0003

Table 1 *Corrections to eq. (2.10) from: purely photonic contribution to the Born approximation, from initial state bremsstrahlung and from final state bremsstrahlung calculated up to $O(\alpha)$. All corrections are calculated for $\sqrt{s} = M_Z$.*

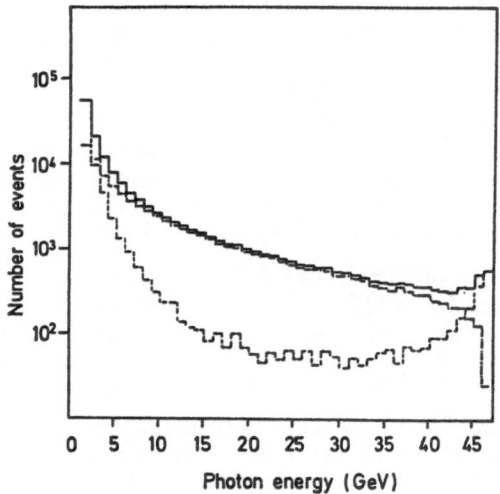

Figure 4. *The $O(\alpha)$ photon energy spectrum for $\sqrt{s} = M_Z$. The solid line includes all photons, the dashed initial state bremsstrahlung, and the dotted line final state bremsstrahlung only. No cuts are applied.*

Final state bremsstrahlung is practically unable to modify the τ polarization directly. The main correction originates from the helicity-flip [10] contribution which is of $O(0.0003)$. As in the case of the asymmetry A_{LR} [12] there is a rather strong cancellation between the corrections of boxes and initial-final state bremsstrahlung interference. This class of corrections is completely negligible even for relatively strong cut-offs.

QED bremsstrahlung distorts the relation (2.6) between the slope of the π energy spectrum and A_{pol}. In fact this indirect effect of QED on the measurement of the τ polarization is significantly larger than the direct ones previously discussed. The emission of the hard photon in the initial or final state reduces the invariant mass of

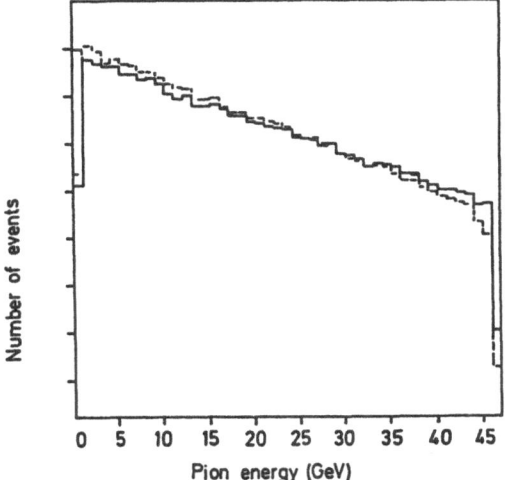

Figure 5. *The π energy spectrum with (dashed line) and without (solid line) QED $O(\alpha)$ corrections. No cuts applied.*

the $\tau^{\pm}$ pair and thus thedistributions of their decay products are deformed[*]. Figure 5 presents the deformation of the energy spectrum in the $\tau \to \pi\nu$ decay mode, from hard photon emission to $O(\alpha)$. The energy of the photons emitted in the final state is usually larger than the energy of photons emitted in the initial state (Fig. 4). The contribution of the photons emitted in the final state produces then the bulk, 75%, of the correction. The relative variations of the asymmetry in the π energy are given in Table 2.

To eliminate partially this variation in the measurement of the τ polarization we may take into account only those events where the $\pi's$ have an energy larger than $E_{min} = 4.3 GeV$ and smaller than $E_{max} = 42.5 GeV$. The application of the cut-offs reduces the correction on A_{pol} to 0.013.

We may summarize the discussion of QED corrections by saying that the indirect corrections in contrast to the direct ones are sizable and also depend on experimental cut-offs.

Let us now turn to the question of statistical and systematic errors in the measurement of A_{pol} with help of $A_{\pi}^{(1,2)}$. We present a very brief discussion, which leads however, to practically the same results as more detailed in [9]. It is expected [14] that every collaboration will collect about $3 \cdot 10^6$ Z events per year of LEP running. This means that after including branching ratios for $Z \to \tau^+\tau^-$ and $\tau\pm \to \pi\nu$ we may expect about $2 \cdot 10^4$ $\pi's$ for the A_{pol} measurement. On the other hand the contamination of the π sample with the events $\tau \to \rho\nu$, $\rho \to \pi\gamma\gamma$ is expected to be the main source of systematic error in the measurement of A_{pol}. According to the estimates in [9] the error is expected to be $\Delta A_{\pi} = 0.006$.

---

[*] We exclude effects due to the bremsstrahung in τ decay from our discussion. This class of correction is, however, small. See [13] and references therein for more details.

138

A/ no cuts	Initial state brem.	Final state brems.
$\Delta A_\pi / A_\pi$	3.4%	8.1%
B/ cut on E_π (u)	Initial state brem.	Final state brem.
$\Delta A_\pi / A_\pi$	0.1%	6.1%

Table 2 *Relative variation of the pion energy asymmetry $\Delta A_\pi / A_\pi$ due to the QED $O(\alpha)$ initial state bremsstrahlung and final state bremsstrahlung. A) no cuts on π energy, B) π energy in the range $4.3 - 42.5 GeV$. Samples of 10^6 π were used and all results have 1% statistical error.*

Let us now present how this numbers translate on the errors of A_l and A_r. First recall the definitions (2.11) of π energy asymmetries $A_\pi^{(1,2)}$. In the case 1) we have according to eqs. (2.10)

$$A_r \simeq 2 A_\pi^{(1)} \tag{3.2}$$

This gives a 0.014^* statistical error and 0.012 systematic in the measurement of A_r. In the case (2) we exclude nearly 50% of events but on the other hand $A_\pi^{(2)}$ is much more sensitive to A_l

$$A_l \simeq A_\pi^{(2)}. \tag{3.3}$$

There is a significant improvement on the size of both, statistical and systematic errors. We get a statistical error of 0.01 and a systematic error of 0.006 in the measurement of A_l. With the help of the lowest order formula this numbers can be translated into the error on $\sin^2 \theta_W$. We get respectively 0.0012 for the statistical and 0.0008 for the systematic error.

---

* As it was presented in [15] (see also [7]) the statistical error can be further decreased up to even 30% if one includes all τ decay modes for the analysis of A_{pol} and also exploits spin correlations of τ^+ and τ^- decay products. This ambitious program requires however much better, than it is at present, knowledge of the systematic errors for all τ decay modes.

4. Summary and conclusions

Let us now compare expected statistical and systematic errors in the measurement of A_{pol} and forward backward asymmetry A_{FB} for muons. From the same sample of $3 \cdot 10^6 \, Z$ we get 10^5 muon pairs. This brings the statistical error in the measurement of A_{FB} down to 0.003 which transmits to 0.0011 error on $\sin^2 \theta_W$. The systematic error in the measurement of A_{FB} is limited by the uncertainty in the knowledge of the difference between the CMS energy and the Z mass. If we recall [2] that the Z mass itself will be measured with precision of about $30 MeV$, we get in this way a systematic error on A_{FB} not smaller than 0.003 and respectively a systematic error on $\sin^2 \theta_W$, $\Delta \sin^2 \theta_W = 0.0011$. Even for the rather small value of $\sin^2 \theta_W = 0.222$, used in this paper, we find that the measurement of A_{pol} is competitive with A_{FB}. The closer we go with $\sin^2 \theta_W$ to 0.25 the bigger the error in A_{FB} measurement of $\sin^2 \theta_W$. This is due to the quadratic dependence of the forward-backward asymmetry on $\mathcal{A}_l$ and thus on $\sin^2 \theta_W$.

Finally let us turn to the most interesting point of the electroweak effects. We have practically excluded this topic from our considerations because within the SM, predictions for integrated A_{pol} are the same as for A_{LR}. The point is that the 0.01 error in measurement of $\mathcal{A}_l$ matches in size the variation due to the top mass increase from 50 to $120 GeV$ see eg. [16].

We can summarize this talk as follows. The measurement of τ polarization asymmetry A_{pol} is expected to be a very good, possibly even the best precision test of the SM at LEP. Effects of the QED corrections are well understood. The detailed study of the interplay of QED corrections and experimental conditions is not yet completed.

The assumption of universality of Z couplings to leptons play an important role in minimizing both statistical and systematic errors.

REFERENCES

1. S. L. Glashow, Nucl. Phys. **22** (1961) 579;
   S. Weinberg, Phys. Rev. Lett. **19** (1967) 1264; Phys. Rev. **D5** (1972) 1412;
   A. Salam, in *Elementary Particle Theory*, ed. N. Svartholm, Stockholm, 1968,
   p 361.

2. G. Altarelli et al., in Physics at LEP, CERN 86-02 (1986) Vol. 1 p. 1.

3. F. Dydak et al. ECFA Workshop on LEP 200, CERN 87-08 (1987) Vol. 1, p. 157.

4. R. Kleiss, Hard Bremsstrahlung in $e^+e^- \to \mu^+\mu-$ Matching Theory and Experiment, in Physics at LEP 86-02 Vol 1 p. 153.

5. S. Jadach, J. H. Kühn, R. G. Stuart, Z. Wąs, Z. Phys. **C38** (1988) 609; J. H. Kühn, R. G. Stuart, Phys. Lett. **B200** (1988) 360.

6. S. Jadach, Z. Wąs, Phys. Lett. **B219** (1989) 103.

7. J. H. Kühn, F. Wagner, Nucl. Phys. **B236** (1984) 16.

8. S. Jadach, Z. Wąs, First and higher order noninterference QED radiative corrections to the charge asymmetry at the Z resonance, Munich preprint MPI-PAE/PTh 33/89.

9. F. Boillot and Z. Wąs, "Uncertainties in the τ polarization measurement in LEP/SLC", MPI-München, (1988) preprint MPI-PAE/Exp 196., to appear Z. Phys.

10. Z. Wąs, Acta Physica Polonica **B18** (1987) 12.

11. S. Jadach, Z. Wąs, R. G. S. Stuart and B. F. L. Ward, "KORALZ the Monte Carlo program for τ and μ pair production processes at LEP/SLC", unpublished, may be obtained form Z. Wąs: WASM @ CERNVM.

12. J. H. Kühn, S. Jadach, R. G. Stuart and Z. Wąs, Z. Phys. **C38** (1988) 609, J. H. Kühn, R. G. Stuart, Phys. Lett. **B200** (1988) 360.

13. G. Altarelli et al., in Physics at LEP, CERN 89 in preparation

14. LEP design report, CERN LEP / 84-01, vol. II.

15. C. A. Nelson, Tests for New Physics from τ spin correlation functions for $Z \to \tau^+\tau^- \to A^+B^-X$, Preprint, New York 1989, SUNY BING 1/30/89.

16. W. Hollik, DESY report, DESY-88-188, December 1988, and references therein.

Part II

Electroweak Corrections

Radiative Corrections for e⁺e⁻ Collisions   Editor: J.H. Kühn
© Springer-Verlag Berlin, Heidelberg 1989

The Z boson width: higher order effects and influence on the Z line shape

WOLFGANG HOLLIK

CERN, Geneva, Switzerland

June 2, 1989

Abstract

The width of the Z boson has two faces: the one which determines, in the interplay with bremsstrahlung, the peak maximum of the Z cross section, and the other one which is a prediction of the Standard Model allowing a confrontation with experimental data as a test of the theory. We give an overview how the total and partial Z widths are calculated including the $O(\alpha^2)$ contributions. The dependence on the Higgs and top mass is discussed as well as the effect of the these presently unknown parameters on the Z line shape. The relation of the peak maximum to the physical Z mass turns out to be independent of the Higgs and top mass within 2 MeV which allows to measure M_Z as an independent input parameter. Some observable effects in Z decays resulting from non-standard charged Higgs bosons are also presented.

1. Introduction

One of the basic measurements at the e^+e^- colliders LEP and SLC will be the determination of the shape of the Z resonance. This will provide us with two of the most interesting and important electroweak parameters: the mass and the width of the neutral vector boson. For precision tests of the Standard Model and for searches for signals of possible new physics it is indispensible to know the predictions of the Standard Model with high accuracy, including higher order corrections. The QED corrections [1], in particular real and virtual photonic corrections in the initial state, constitute the largest part of the radiative corrections and lead to a distortion of the shape of the resonance and to a shift in the peak location. In view of the high accuracy with which the mass and width will probably be measured (± 20 MeV [2]) we are forced to go beyond the $O(\alpha)$ contributions in these observables aiming an accuracy of 10 MeV. The effect of $O(\geq \alpha^2)$ initial state radiation on the Z shape has been studied in [3]. It was found that the 2-loop QED corrections reduce the shift of the Z peak by about 90 MeV. The combination of the weak corrections in the non-radiative amplitude (with the s-dependence of the width and 2-loop contributions to the imaginary part of the Z self energy) and the initial state bremsstrahlung has also been performed [4] to give the most complete result for the prediction of the line shape.

The higher order corrections to the Z width are of twofold importance:

- They influence the shape of the resonance and have consequently to be considered for precision measurements of the Z mass. Both the reduction of the peak height and the shift of the peak maximum depend on the width:

$$\sigma \approx \sigma_0 \cdot \left(\frac{\Gamma_Z}{M_Z}\right)^\beta,$$

(1)

$$\sqrt{\sigma_{max}} \approx M_Z + \frac{\pi}{8}\beta\Gamma_Z - \frac{\Gamma_Z^2}{2M_Z} \tag{2}$$

with

$$\beta = \frac{2\alpha}{\pi}\left(\log\frac{M_Z^2}{m_e^2} - 1\right).$$

- Being a prediction of the Standard Model after the Z mass is known, the width serves as a first test of theory. The partial widths for $Z \to f\bar{f}$ will allow the investigation of the weak coupling constants of the various fermions at the level of quantum corrections.

In this talk we discuss in detail the radiative corrections to the fermionic partial widths $\Gamma(Z \to f\bar{f})$, $f = \nu$, l, $q(\neq t)$, and the total Z width which enters the Z line shape [5]. Other calculations have been performed for the leptonic widths [6] and also for $Z \to q\bar{q}$ [7,8]. In [8] the influence of the top quark on the $Z \to b\bar{b}$ decay width has been considered in a unitary gauge calculation.

The basis for our calculation is the on-shell scheme in the version specified in [9,10]. In contrast to [8] we perform our calculation in the renormalizable t'Hooft-Feynman gauge. Since we have to include virtual top quarks and unphysical Higgs bosons in the $Z \to b\bar{b}$ decay vertex corrections finite mass effects of the type m_t^2/M_W^2 have to be kept.

Electroweak corrections to open top final states in case of $m_t < M_Z/2$, a possibility which is experimentally not completely ruled out, have been considered in [11]. They are less important in view of the uncertainties from the top mass in the phase space factors and from large QCD corrections near threshold [12]. Therefore we restrict ourselves to the case $m_t > M_Z/2$.

The presentation is organized as follows: Section 2 contains the lowest order dicussion. A brief description of the underlying strategy for the next order calculation is given in section 3. In section 4 the one-loop electroweak corrections to the fermionic width are summarized, together with the QCD corrections and other decay channels of higher than $O(\alpha)$ in the coupling constants which enter the total width as well. This is followed by a discussion of the Z line shape (section 5) and of some none-standard effects in models with charged Higgs bosons.

2. The Z width in lowest order

In lowest order the Z propagator has the Breit-Wigner form

$$D_Z^0(s) = \frac{1}{s - M_Z^2 + iM_Z\Gamma_Z^0}. \tag{3}$$

The lowest order total width Γ_Z^0 is related to the one-loop self energy $\Sigma^Z(s)$ of the Z boson by

$$M_Z\Gamma_Z^0 = \operatorname{Im}\Sigma^Z(s = M_Z^2). \tag{4}$$

It can be written as the sum of the partial fermionic decay widths $\Gamma_Z^0(f\bar{f})$ with $m_f < M_Z/2$:

$$\Gamma_Z^0 = \sum_f \Gamma_Z^0(f\bar{f}). \tag{5}$$

These partial widths can be expressed in terms of the vector and axialvector coupling constants of the fermion f to the Z

$$v_f = \frac{I_f^3 - 2Q_f s_W^2}{2 s_W c_W} \quad a_f = \frac{I_f^3}{2 s_W c_W} \tag{6}$$

with

$$s_W = \sin\theta_W, \quad c_W = \cos\theta_W$$

as follows:

$$\Gamma_Z^0(f\bar{f}) = N_C^f \frac{\alpha}{3} M_Z \sqrt{1 - 4\mu_f} \left[v_f^2(1 + 2\mu_f) + a_f^2(1 - 4\mu_f) \right] \tag{7}$$

with $N_C^f = 3$ for quarks, $N_C^f = 1$ for leptons, and

$$\mu_f = \frac{m_f^2}{M_Z^2}.$$

The mixing angle is used in the standard on-shell definition in terms of the boson masses:

$$s_W^2 = 1 - \frac{M_W^2}{M_Z^2}. \tag{8}$$

Making use of the tree level relation between s_W^2 and M_Z by means of the moun decay constant

$$M_Z^2 = \frac{\pi\alpha}{\sqrt{2}G_\mu} \cdot \frac{1}{s_W^2 c_W^2} \tag{9}$$

we obtain another possible tree level representation of the partial decay width

$$\bar{\Gamma}_Z^0(f\bar{f}) = N_C^f \frac{G_\mu M_Z^3}{24\pi\sqrt{2}} \sqrt{1 - 4\mu_f} \left[1 - 4\mu_f + (2I_3^f - 4Q_f s_W^2)^2 (1 + 2\mu_f) \right] \tag{10}$$

leading to the Born total width in the G_μ representation:

$$\bar{\Gamma}_Z^0 = \sum_f \bar{\Gamma}_Z^0(f\bar{f}) . \tag{11}$$

In both parametrizations no dependence on the unknown standard model parameters, Higgs and top mass (M_H, m_t), is present if the mixing angle is derived from the experimental boson mass ratio (8).

For actual calculations the dependence on M_W is usually eliminated in favor of the precisely measured Fermi constant G_μ by means of the radiatively corrected form [13] of relation (9)

$$M_W^2(1 - M_W^2/M_Z^2) = \frac{A}{1 - \Delta r(\alpha, M_W, M_Z, M_H, m_t)} \tag{12}$$

with

$$A = \frac{\pi\alpha}{\sqrt{2}G_\mu} = (37.281 GeV)^2.$$

In this way a top and Higgs mass dependence is induced.

In table 1 we give the values for the total width obtained by this method for the two parametrizations specified above (for $M_Z = 92$ GeV, $M_H = 100$ GeV) as functions of the top mass. This shows the differences in the values as well as the different behaviour with m_t which are a clear indication for the need of including next order contributions.

Table 1: Lowest order total width

m_t (GeV)	Γ_Z^0 (GeV)	$\bar{\Gamma}_Z^0$ (GeV)
50	2.307	2.487
100	2.358	2.501
200	2.506	2.540

3. On-shell renormalization

The Standard Model has a certain number of free parameters which are not fixed by the theory. The definition of these parameters and their relation to measurable quantities is the task of a renormalization scheme, which completes the definition of the quantized theory.

The favoured renormalization scheme in QED is the on-shell scheme with the fermion masses m_f and the fine structure constant (the on-shell $ee\gamma$ coupling) as input parameters. The most direct and natural extension to the electroweak theory leads to the on-shell (OS) scheme of $SU(2) \times U(1)$, which has been widely used for practical applications (see e.g ref's [10,14]). Differences in the treatment of field renormalization and in the unphysical sector disappear in the final relations between physical quantities. Here we follow the OS scheme as specified in [9,10].

Starting point is the classical Lagrangian

$$L_{cl} = L_G(\vec{W}, B, g_2, g_1) + L_H(\phi, \mu^2, \lambda) + L_{FG}(\psi_L, \psi_R, \vec{W}, B) + L_{FH}(\psi_L, \psi_R, \phi, g_f). \tag{13}$$

L_G is the gauge part with the $SU(2)$ and $U(1)$ fields $\vec{W}$ and B and the corresponding gauge couplings g_2, g_1; L_H is the Higgs part with the scalar doublet ϕ and the potential parameters μ^2, λ; L_{FG} describes the fermion–gauge field interaction with left and right handed fermion fields $\psi_{L,R}$, and L_{FH} is the Higgs-fermion Yukawa term that induces the fermion masses.

In the fields and parameters of (13) the $SU(2) \times U(1)$ symmetry of L_{cl} is manifestly apparent. The physical content, however, becomes more transparent after switching to the "physical" fields and parameters

$$W^{\pm}, \ Z, \ \gamma, \ e, \ M_W, \ M_Z, \ m_f. \tag{14}$$

There is no room for $\sin^2\theta_W$ as an additional independent quantity. The simplest choice in terms of (14) is to maintain relation (8) which will be used throughout the forthcoming discussion.

Since it is convenient to work in a renormalizable gauge (t'Hooft-Feynman gauge) the gauge fixing term L_{fix} and the corresponding Fadeev-Popov ghost term L_{gh} have to be added to L_{cl} in order to obtain the Lagrangian for the quantized theory. Multiplicative field and parameter renormalization introduces renormalization constants $\sqrt{Z_2^i}$ for each field multiplet and Z_1^i for each free parameter in the original manifest symmetric version. These renormalization constants are then determined by the renormalization conditions.

The renormalization conditions give the parameters in (14) their correct physical meaning. The first subset consists of the OS conditions for the 2-point functions which make the particle content of the theory evident:

(The bubbles mean the one-loop contributions together with the counter terms.)

$$\left. \widetilde{Z} \!-\!\!\bigcirc\!\!-\! \widetilde{Z} \right|_{k^2 = M_{\widetilde{Z}}^2} = 0$$

$$\left. \text{W} \!-\!\!\bigcirc\!\!-\! \text{W} \right|_{k^2 = M_W^2} = 0$$

$$\left. f \!\longrightarrow\!\!\bigcirc\!\!\longrightarrow\! f \right|_{k^2 = m_f^2} = 0$$

The second subset defines the electric charge in the Thomson limit and allows to recover the ordinary QED as a simple substructure:

$$\left. \gamma \!\sim\!\!\bullet\!\!<^{\,e}_{\,e} \right|_{\substack{k^2=0 \\ k^0 \to 0}} = e\,\gamma_\mu \,, \qquad \left. \gamma \!-\!\!\bigcirc\!\!-\! Z \right|_{k^2=0} = 0$$

$$\left. \frac{\partial}{\partial k^2}\left(\gamma \!-\!\!\bigcirc\!\!-\! \gamma \right) \right|_{k^2=0} = 0$$

$$\left. \frac{1}{\not{k}-m_f}\left(f \!\longrightarrow\!\!\bullet\!\!\longrightarrow\! f \right) \right|_{\not{k}=m_f} = 0$$

The results can be summarized in terms of renormalized self energies $\hat{\Sigma}^j$ (with $j = \gamma, Z, W, \gamma Z$) dressing the propagators as described in the next section, and the vector and axialvector formfactors $F_{V,A}^{jf}(q^2)$ for the Zff, γff and Wff' vertices. A complete list can be found in [10].

Field renormalization ensures that we obtain finite Green functions. For physical S matrix elements the results are equivalent to those derived without field renormalization, as done in [13,18] . Our field renormalization is performed according to the gauge symmetry by introducing the minimal number of field renormalization constants. The price for this, however, is that not all residues of the propagators can be normalized to one. As a consequence, any calculation with the renormalized Lagrangian will have to include finite multiplicative wave function renormalization factors for some of the external lines in S matrix elements.

It is of course possible to perform the renormalization in such a way that these finite corrections do not appear [15,16,17]. But then the Lagrangian will contain many constants which have to be calculated in terms of the few fundamental parameters.

The advantages of the OS scheme are obvious:

- The input parameters have a clear physical meaning and can be measured directly.

- Except the Higgs and top mass M_H, m_t all parameters are known.

- It has a natural separation into "QED corrections" (virtual and real bremsstrahlung) and infrared finite "weak corrections for NC processes. This is of practical importance since the QED corrections in a realistic experiment are in general detector dependent.

The W mass M_W is not known as precisely as to make the uncertainty in the radiative corrections negligible ($\Delta M_W = 100$ MeV with LEP200). This drawback can easily be overcome by including the OS radiative correction to the μ lifetime: The non-QED correction Δr in (12) can be written in terms of the renormalized W self energy $\hat{\Sigma}^W$ (which depends on all particle masses of the model) and the sum of vertex, box, and wave function renormalization contributions:

$$\Delta r = \frac{\hat{\Sigma}^W(0)}{M_W^2} + \frac{\alpha}{4\pi \sin^2 \theta_W} \left[6 + \frac{7 - 4\sin^2 \theta_W}{2\sin^2 \theta_W} \log(\cos^2 \theta_W) \right] \tag{15}$$

where $\hat{\Sigma}^W(0)$ can be expressed in terms of the unrenormalized self energies as follows:

$$\begin{aligned} \frac{\hat{\Sigma}^W(0)}{M_W^2} &= \frac{\Sigma^W(0) - \Sigma^W(M_W^2)}{M_W^2} \\ &+ \Pi^\gamma(0) + 2\frac{\cos\theta_W}{\sin\theta_W}\frac{\Sigma^{\gamma Z}(0)}{M_Z^2} \\ &- \frac{\cos^2\theta_W}{\sin^2\theta_W}\left(\frac{\delta M_Z^2}{M_Z^2} - \frac{\delta M_W^2}{M_W^2}\right) \end{aligned} \tag{16}$$

with

$$\delta M_Z^2 = \mathrm{Re}\,\Sigma^Z(M_Z^2), \quad \delta M_W^2 = \mathrm{Re}\,\Sigma^W(M_W^2).$$

4. Higher order contributions to the Z width

4.1 The dressed Z propagator

In lowest order, after diagonalization of the neutral boson mass matrix, the propagator matrix is diagonal. But mixing due to quantum corrections prohibits the photon and Z boson from propagating independently of each other in higher orders. Consequently, the propagator of the γZ system has to be considered as a 2×2 matrix. The radiative corrections to the propagator system can be obtained by inversion of the matrix (transverse parts only)

$$(\mathbf{D}_{\mu\nu})^{-1} = i\, g_{\mu\nu} \begin{pmatrix} k^2 + \hat{\Sigma}^\gamma(k^2) & \hat{\Sigma}^{\gamma Z}(k^2) \\ \hat{\Sigma}^{\gamma Z}(k^2) & k^2 - M_Z^2 + \hat{\Sigma}^Z(k^2) \end{pmatrix} \tag{17}$$

with the 1-particle irreducible (1PI) renormalized self energies specified in section 3 to one-loop order, yielding:

$$\mathbf{D}_{\mu\nu} = -i\, g_{\mu\nu} \begin{pmatrix} D_\gamma & D_{\gamma Z} \\ D_{\gamma Z} & D_Z \end{pmatrix} \tag{18}$$

where ($s = k^2$)

$$D_\gamma(s) = \frac{1}{s + \hat{\Sigma}^\gamma(s) - \frac{\hat{\Sigma}^{\gamma Z}(s)^2}{s - M_Z^2 + \hat{\Sigma}^Z(s)}} \tag{19}$$

$$D_Z(s) = \frac{1}{s - M_Z^2 + \hat{\Sigma}^Z(s) - \frac{\hat{\Sigma}^{\gamma Z}(s)^2}{s + \hat{\Sigma}^\gamma(s)}} \tag{20}$$

$$D_{\gamma Z}(s) = -\frac{\hat{\Sigma}^{\gamma Z}(s)}{[s + \hat{\Sigma}^\gamma(s)][s - M_Z^2 + \hat{\Sigma}^Z(s)] - \hat{\Sigma}^{\gamma Z}(s)^2}. \tag{21}$$

Obviously the matrix (17) can be diagonalized only for one specific value of k^2. This has been done by fixing the mixing counter term in such a way that (17) is diagonal for $k^2 = 0$. In $O(\alpha)$, with the leading log terms resummed to all orders, the propagators are simplified to

$$D_\gamma \;=\; \frac{1}{s + \hat\Sigma^\gamma(s)} \tag{22}$$

$$D_Z \;=\; \frac{1}{s - M_Z^2 + \hat\Sigma^Z(s)}$$

$$D_{\gamma Z} \;=\; -\frac{1}{s}\hat\Sigma^{\gamma Z}(s)\,\frac{1}{s - M_Z^2 + i\mathrm{Im}\hat\Sigma^Z(s)}$$

The further approximation of the Z propagator in (20)

$$\mathrm{Re}\ \hat\Sigma^Z \approx 0, \quad \mathrm{Im}\ \hat\Sigma^Z(s) \approx \mathrm{Im}\ \hat\Sigma^Z(M_Z^2)$$

leads to the Breit-Wigner form

$$D_Z^0(s) \;=\; \frac{1}{s - M_Z^2 + i\,M_Z\Gamma_Z^0}$$

which corresponds to our Born formula (3).

Off resonance, in the continuum region, the approximation (22) is adequate. Around the Z peak, however, (22) becomes insufficient:

1. The on-resonance value of the amplitude for $e^+e^- \rightarrow f\bar{f}$ in lowest order is of $O(1)$ and not of $O(\alpha)$ as in the continuum: The tree level width which is given by the imaginary part of the one-loop Z self energy $\mathrm{Im}\ \Sigma^Z(M_Z^2)$ cancels the coupling constants in the numerator of the matrix element. For the next order corrections to the cross section around the Z peak the $O(\alpha^2)$ contributions to the Z width have to be included. One part of them is given by the imaginary part of the $\left(\hat\Sigma^{\gamma Z}\right)^2$ term in (20).

2. The real part of $\left(\hat\Sigma^{\gamma Z}\right)^2$ in (20) gives a $O(\alpha^2)$ correction to the resonance amplitude. In a systematic expansion up to $O(\alpha)$ it would therefore not appear. A numerical study shows that it is indeed negligible if the top quark is not too heavy (< 150 GeV). For a large mass splitting in the (t,b) doublet, however, the $O(\alpha^2)$ term matches the experimental accuracy aimed in LEP experiments.

4.2 The corrected Z width

For an appropriate discussion of the Z width we proceed as follows:

1st step:
The resummed form (20) is still insufficient for the imaginary part of the Z propagator: besides the reducible $O(\alpha^2)$ term $\mathrm{Im}\left(\hat\Sigma^{\gamma Z}\right)^2$ we need also the 2-loop irreducible part $\mathrm{Im}\hat\Sigma^Z_{(2)}$ of the diagonal Z self energy contributing to the $O(\alpha^2)$ width as well. Note that the s-dependence of the imaginary part is also significant: the replacement

$$M_Z\,\Gamma_Z(M_Z^2) \;\rightarrow\; \sqrt{s}\,\Gamma_Z(s) = \mathrm{Im}\ \hat\Sigma^Z(s)$$

causes a shift of the resonance peak on the energy scale of about 35 MeV [4,19] to lower values.

Altogether, we have to replace the imaginary part of the denominator in (20) by the proper expression

$$\text{Im}\left[\hat{\Sigma}^Z(s) - \frac{\left(\hat{\Sigma}^{\gamma Z}(s)\right)^2}{s + \hat{\Sigma}^{\gamma}(s)}\right] + \frac{s}{M_Z^2} \cdot \text{Im}\,\hat{\Sigma}^Z_{(2)}(M_Z^2) \tag{23}$$

where $\hat{\Sigma}^Z$ still denotes the 1PI one-loop part (a similar discussion has been given by Wetzel in [7]). For $s = M_Z^2$ we obtain the usual on-shell Z width.

2nd step:
The term $\text{Im}\,\hat{\Sigma}^Z_{(2)}(M_Z^2)$ is related by unitarity to the corrections to the Z width which are not of the reducible type:

$$\text{Im}\,\hat{\Sigma}^Z_{(2)}(M_Z^2) = M_Z\,\Delta\Gamma_Z \tag{24}$$

The term $\Delta\Gamma_Z$ summarizes all $O(\alpha^2)$ contributions which are missing in (23):
– weak vertex corrections to the decays $Z \to f\bar{f}$;
– QED corrections to the decays $Z \to f\bar{f}$, $f \neq \nu$;
– QCD corrections to the decays $Z \to q\bar{q}$;
– other decay channels of higher order in the coupling constants. In practice only the decay $Z \to \sum_f H f\bar{f}$ is of some importance for a light Higgs (≈ 5 MeV for $M_H = 10$ GeV); other decay channels can be neglected (see [22] and the references given there).

With the weak form factors $F_{V,A}^{Zf}$ we can write for $\Delta\Gamma_Z$

$$\begin{aligned}
\Delta\Gamma_Z = & \sum_f N_C^f \frac{2}{3}\alpha M_Z \left[v_f\,\text{Re}F_V^{Zf}(M_Z^2) + a_f\,\text{Re}F_A^{Zf}(M_Z^2)\right] \\
& + \sum_f N_C^f \frac{\alpha}{3} M_Z \left(v_f^2 + a_f^2\right) \cdot \delta_{QED} \\
& + \sum_{f=q} N_C^f \frac{\alpha}{3} M_Z \left(v_f^2 + a_f^2\right) \cdot \delta_{QCD} \\
& + \sum_f \Gamma(Z \to H f\bar{f}),
\end{aligned} \tag{25}$$

The weak form factors (where the photon exchange has bee removed) are listed in [10]. They correspond to the diagrams in Figure 1 (neutral Higgs boson exchange can be neglected because of small Yukawa couplings). For the light fermions ($\neq b, t$) also the charged "Higgs" diagrams are negligible. For $Z \to b\bar{b}$, however, due to the virtual top presence, they become important for large top masses exhibiting also a quadratic rise with m_t. The corresponding form factors have been calculated in [5] and [8].

The second and third terms in (25) are the QED and QCD corrections from virtual and real photon resp. gluon contributions, in the massless limit given by

$$\delta_{QED} = \frac{3\alpha}{4\pi} Q_f^2 \tag{26}$$

$$\delta_{QCD} = \frac{\alpha_s(M_Z^2)}{\pi} + 1.405\left(\frac{\alpha_s(M_Z^2)}{\pi}\right)^2. \tag{27}$$

Taking into account the mass dependence of the QCD corrections in the partial width for $Z \to b\bar{b}$ increases Γ_Z by 2 MeV if $\alpha_s(M_Z^2) = 0.12 \pm 0.02$ [21] is used. The uncertainty in α_s induces an uncertainty in the total width of about 12 MeV.

The fourth term in (25), the decay width into Higgs and fermion pairs, is taken from [20] (for earlier work see [21]). Other 3- and 4-body decay channels, which contribute in principle also to the Z width in higher order, remain below 1 MeV. The ratios of the corresponding decay channels to the partial muonic width $\Gamma(Z \to \mu^+\mu^-)$ are [22]:

$$Z \rightarrow H\gamma \qquad\qquad < 10^{-4}$$

$$Z \rightarrow ggg,\ gg\gamma \qquad \sim 10^{-4}$$

$$Z \rightarrow \textstyle\sum_{f_1 f_2} W f_1 f_2 \quad \sim 10^{-6}$$

$$Z \rightarrow \textstyle\sum_{l,q} llqq \qquad \sim 10^{-3}$$

3rd step:
Neglecting for the moment the real part of the $(\hat{\Sigma}^{\gamma Z})^2$ term in the Z propagator (20) and performing a Taylor expansion in $\mathrm{Re}\,\hat{\Sigma}^Z$ around the Z mass

$$\mathrm{Re}\,\hat{\Sigma}^Z(s) = (s - M_Z^2)\,\Pi_Z \tag{28}$$

with

$$\Pi_Z = \mathrm{Re}\,\frac{\partial\hat{\Sigma}^Z}{\partial s}(M_Z^2) \tag{29}$$

we can write for the propagator after step 1 and 2:

$$D_Z(s) = \frac{1}{1 + \Pi_Z} \cdot \frac{1}{s - M_Z^2 + i\,\frac{s}{M_Z}\cdot\frac{\Gamma_Z^0 + \Delta\Gamma_Z}{1 + \Pi_Z}} \cdot \tag{30}$$

Therein the expression

$$\frac{\Gamma_Z^0 + \Delta\Gamma_Z}{1 + \Pi_Z} \equiv \Gamma_Z \tag{31}$$

can now be identified with the physical Z width. The factor in front gives a correction to the overall normalization of the Z exchange amplitude; it has to be combined with the other corrections at the external vertices for the complete one-loop matrix element for $e^+e^- \rightarrow f\bar{f}$. The appearance of this factor in the imaginary part has to be understood as the correction to the Z width coming from the wave function renormalization of the Z line in the decay matrix elements. Writing it in the denominator

$$\frac{1}{1 + \Pi_Z} = 1 - \Pi_Z + \Pi_Z^2 \cdots$$

takes into account the leading log summation for the light fermions in Π_Z which appear in terms of the photon vacuum polarization $\Pi_\gamma(M_Z^2)$.

The combination of the wave function renormalization with the parametrization (5) - (7) leads to the following common factor in the fermionic decay channels and hence in the total fermionic width:

$$\frac{\alpha}{3}M_Z \cdot \frac{1}{4s_W^2 c_W^2} \cdot \frac{1}{1 + \Pi_Z} = \frac{\sqrt{2}G_\mu M_Z^3}{12\pi} \cdot \frac{1 - \Delta r}{1 + \Pi_Z} \tag{32}$$

The structure of Δr and Π_Z in the leading terms from the light fermions and a potentially heavy top quark

$$\Pi_Z = \Pi_\gamma(M_Z^2) + \frac{c_W^2 - s_W^2}{s_W^2}\Delta\rho + \cdots \tag{33}$$

$$\Delta r = -\Pi_\gamma(M_Z^2) - \frac{c_W^2}{s_W^2}\Delta\rho + \cdots$$

with

$$\Delta\rho = \frac{\alpha}{4\pi} \cdot \frac{3}{4s_W^2 c_W^2}\left(\frac{m_t}{M_Z}\right)^2 \tag{34}$$

shows that the correction factor in (32) is very close to 1 unless the top is very heavy. In that case

$$\frac{1 - \Delta r}{1 + \Pi_Z} \approx 1 + \Delta\rho \tag{35}$$

yields an approximation of better than 0.5% (see Figure 2).

The parametrization of the Z width in terms of G_μ, as given in (10) and (11) in lowest order, approximates the total fermionic width with an accuracy of better than 10 MeV if the top is not too heavy, as shown explicitly in table 2 for the fermionic deacy width (10) and with weak corrections. For top masses above 200 GeV, however, the corrections are bigger than 20 MeV.

Table 2: Total width. Born, eq. (11), and with inclusion of weak corrections ($M_Z = 92$ GeV, $M_H = 100$ GeV)

m_t (GeV)	Born (with G_μ)	with weak corrections
50	2.4869	2.4876
100	2.5010	2.4978
200	2.5401	2.5240

The variation with the Higgs mass is much weaker: keeping M_Z and m_t fixed, the variation of the total width is 6 MeV for M_H between 10 GeV and 1 TeV. This is too small to be experimentally visible.

4th step: Finally we have to discuss the effect of the $(\hat{\Sigma}^{\gamma Z})^2$ term in the Z propagator (20). The inclusion of this real part takes care of the fact that the Z mass gets a contribution in higher order originating from mixing with the photon. In our iterative approach, where $\hat{\Sigma}^{\gamma Z}$ is the renormalized mixing including the corresponding counter term with a piece proportional to $\Delta\rho$ in (34), the presence of $(\hat{\Sigma}^{\gamma Z})^2$ means a contribution of the type $\sim \alpha^2(m_t/M_Z)^4$ and can become of some influence in case of a very heavy top. The effect on the physical Z mass is absorbed in an additional contribution to the Z mass counter term δM_Z^2. Since δM_Z^2 enters the quantity Δr in (15),(16) we have a modification of Δr according to:

$$\Delta r \rightarrow \Delta r + \frac{c_W^2}{s_W^2} \mathrm{Re}\left(\frac{\hat{\Sigma}^{\gamma Z}(M_Z^2)}{M_Z^2[M_Z^2 + \hat{\Sigma}^{\gamma}(M_Z^2)]}\right). \tag{36}$$

The same additive contribution appears in Π_Z as well (with the opposite sign) which means that the correction factor in the Z width $(1 - \Delta r)/(1 + \Pi_Z)$ is insensitive to this term. The tiny effect observed in the total width comes from the slight change in the mixing angle when it is calculated from (12) with help of the modified Δr in (36). For more details we refer to [10,24]. Table 3 shows quantitatively the influence of the new term in the total width. The upper number is always derived without this second order term. The differences remain below 2.7 MeV.

Table 3: Total Z width ($M_Z = 92$ GeV, $M_H = 100$ GeV)

m_t (GeV)	50	100	150	200	230
Γ_Z (GeV)	2.5589	2.5699	2.5820	2.5982	2.6101
	2.5593	2.5694	2.5807	2.5961	2.6074

4.3 Partial widths

The total width shows an increase with increasing top mass which is due to the quadratic top mass term in (32) via $\Delta\rho$, eq. (34), and to the decrease of the mixing angle in the vector coupling constants. This behaviour is encountered also in each fermionic partial width with exception of the $Z \to b\bar{b}$ decay where the top mass dependence is much weaker. The reason for this is the additional top dependence of the vertex corrections in $Z \to b\bar{b}$ which cancels the top contributions from the gauge boson 2-point functions. This is shown in more detail in table 4 where the partial deacy widths for $Z \to d\bar{d}$ and $Z \to b\bar{b}$ are compared. For the b quark channel, the partial width remains practically constant over the whole range of the top mass, whereas the d partial width increases by about 10 MeV. A global difference is present due to the finite b mass in the phase space factors.

Table 4: Partial widths, no QED and QCD corrections ($M_Z = 92$ GeV, $M_H = 100$ GeV)

m_t (GeV)	$\Gamma(Z \to d\bar{d})$ (MeV)	$\Gamma(Z \to b\bar{b})$ (MeV)
50	378.6	374.6
100	380.4	375.3
150	382.7	374.7
200	385.9	373.6
230	388.2	372.8

5. The Z line shape

Besides α and G_μ we need the Z mass M_Z as an experimental quantity for completion of our input to fix the theory. M_Z will be measured from the shape of the resonance, in particular from the location of the maximum. The relation between $\sqrt{s_{max}}$ and M_Z is sizably influenced by the initial state QED corrections and is to a good approximation described by eq. (2). For a final answer the QED corrections have to be combined with the non-QED weak corrections. This can be done by a convolution of the total non-radiative cross section σ_W (which contains the weak corrections) with the spectrum $\rho(k)$ of the energy carried away by the radiated photons:

$$\sigma(s) = \int_0^{k_{max}} dk\, \rho(k)\, \sigma_W(s(1-k)). \tag{37}$$

The energy spectrum has been calculated up to $O(\alpha^2)$ in the hard photon part and to all orders in the leading soft photon contributions (see F.A. Berends, these procedings). σ_W contains the dressed propagators and vertices. The form factor contributions to the coupling constants are practically energy independent over the resonance range; therefore they do not influence the location of the peak maximum. On the other hand, the width in the Z propagator has a direct effect on the displacement of the resonance peak. Hence, in principle, the extraction of M_Z from the line shape depends on the values given to the unknown parameters.

The main results following from (37) are [4,25]:

- The position of the peak maximum is shifted to lower values by about 35 MeV. This shift comes from the s-dependence of the width and has been confirmed by Bardin et al. [19].

- The dependence of the maximum position on the unknown standard model parameters M_H, m_t is insignificant. This can also be understood in terms of eq. (2), where the variation of Γ_Z is too small to be of experimental importance. This is demonstrated explicitly in table 5 for the total cross section in $e^+e^- \to \mu^+\mu^-$.

Table 5: Peak maximum and position in $e^+e^- \to \mu^+\mu^-$ ($M_Z = 92$ GeV, $\alpha_s = 0.12$)

m_t (GeV)	M_H (GeV)	σ_{max} (nb)	$\sqrt{s_{max}}$ (GeV)
40	100	1.369	92.094
60	100	1.452	92.094
90	10	1.446	92.094
90	100	1.453	92.094
90	1000	1.454	92.094
230	10	1.454	92.096
230	100	1.461	92.096
230	1000	1.461	92.095

The large difference in the peak cross section between the first two lines in table 5 is due to the open top production for $m_t = 40$ GeV which contributes to Γ_Z already at the tree level. The location of the maximum, however, is not influenced. It remains stable within 2 MeV.

6. Non-standard effects from a second Higgs doublet

As an example for new physics effects which can manifest themselves in Z decays we consider the minimal extension of the Standard Model which has two Higgs doublets in $SU(2) \times U(1)$ leaving the relation

$$\rho = \frac{M_W^2}{M_Z^2 \cos^2 \theta_W} = 1$$

unchanged. The strongest motivation for extending the Higgs sector may come from supersymmetry. But also non-supersymmetric arguments advocate two Higgs doublets, such as the Peccei-Quinn mechanism to solve the strong CP problem [26], and the discussion of CP violation [27]

The vacuum expectation values v_1, v_2 of the complex doublets ($j = 1, 2$)

$$\Phi_j = \begin{pmatrix} \phi_j^+(x) \\ (v_j + \eta_j(x) + i\,\chi_j(x))/\sqrt{2} \end{pmatrix}$$

induce the masses of the vectorbosons in the following way:

$$M_W = \frac{1}{2}g_2\sqrt{v_1^2 + v_2^2}, \quad M_Z = \frac{1}{2}\sqrt{g_1^2 + g_2^2}\,\sqrt{v_1^2 + v_2^2}$$

3 of the eight degrees of freedoms of the doublet fields are absorbed in forming the longitudinal polarization states of the $W^\pm$, Z, and 5 remain as physical particles: a pair of charged Higgs bosons $H^\pm$, two neutral scalars H_0, H_1, and a single neutral pseudoscalar H_2. These physical states are obtained by diagonalizing the mass matrix coming from the Higgs potential:

$$\begin{aligned} H^+ &= -\phi_1^+ \sin\beta + \phi_2^+ \cos\beta \\ H_2 &= -\chi_1 \sin\beta + \chi_2 \cos\beta. \end{aligned} \tag{38}$$

for the charged Higgs and the neutral pseudoscalar, and

$$H_0 = \eta_1 \cos\alpha + \eta_2 \sin\alpha$$
$$H_1 = -\eta_1 \sin\alpha + \eta_2 \cos\alpha$$

for the 2 neutral scalars. The mixing angle β is determined by the ratio

$$\tan\beta = v_2/v_1 \tag{39}$$

whereas α depends on all parameters of the Higgs potential.

The structure of the Yukawa couplings that would arise from a supersymmetric model implies that Φ_1 gives masses to the down type quarks and Φ_2 to the up type quarks. A non-SUSY argument suggesting such a pattern is the absence of flavor changing neutral currents at the tree level.

The appearence of charged physical scalar states and the looser constraints on the Yukawa couplings may give rise to phenomenologically appealing consequences in the decay modes of the neutral vector boson. Two different scenarios yielding a set of enhanced Yukawa couplings are possible: The situation $v_1 > v_2$ which enhances the d couplings to $H^{\pm}$ by v_2/v_1, and $v_2 > v_1$ enhancing the u couplings by v_1/v_2, according to the Yukawa Lagrangian

$$\mathcal{L}_{Yu} = \frac{g_2}{\sqrt{2}} \left(\frac{m_d}{M_W} \tan\beta \cdot \bar{d}\frac{1+\gamma_5}{2}u + \frac{m_u}{M_W} \cot\beta \cdot \bar{u}\frac{1-\gamma_5}{2}d \right) H^+ + h.c. \tag{40}$$

A situation of particular interest, also for the non-enhanced case $\tan\beta = 1$, is encountered in the $Z - b\bar{b}$ vertex where the charged Higgs coupling to the (t,b) family involves the term

$$\frac{g_2}{\sqrt{2}} \cdot \frac{m_t}{M_W} \cdot \frac{v_1}{v_2} \cdot \frac{1 \pm \gamma_5}{2}$$

which becomes large if the top is heavy. The non-standard $H^{\pm}$ bosons enter the $Z - b\bar{b}$ vertex in connection with virtual t quarks as follows:

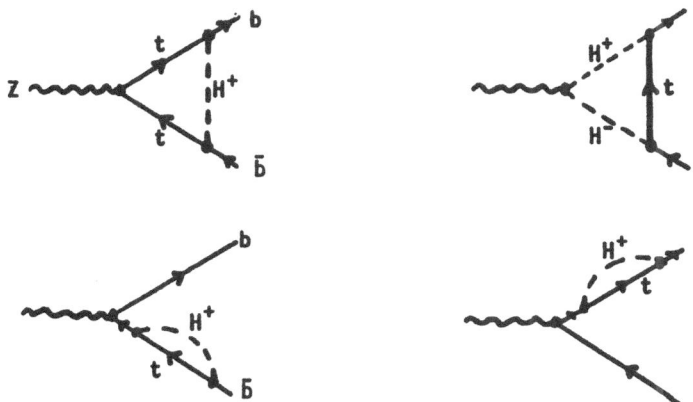

They give a negative contribution to the partial decay width $\Gamma(Z \to b\bar{b})$ which is displayed in

156

Figure 3 for the minimal model and for the presence of charged Higgs bosons (with $M_{H^+} =$ 100 GeV). The standard model result is practically top independent, as discussed already in section 4.3, whereas the charged Higgs diminuish the width if the top becomes heavy. The dashed line shows the $d\bar{d}$ partial width for illustration. Such an increase would also be expected from extra fermion generations with large doublet mass splitting, which in a conventional extension of the minimal model couple only to the gauge bosons and increase $\Delta\rho$ in (34) similarly as the top quark does.

Flavor changing Z decays

In the minimal model the decay rates for flavor changing decays of the Z boson like $Z \to b\bar{s}$ are to small to be experimentally detectable [28]. The reason can be found in a twofold suppression: the higher order in the coupling constants and the small non-diagonal Kobayashi-Maskawa matrix elements. The (virtual) presence of charged Higgs bosons and the possibility $v_1 > v_2$ may enhance these decays rates by several orders of magnitude. The following diagrams involving the top quark yield the dominant contribution:

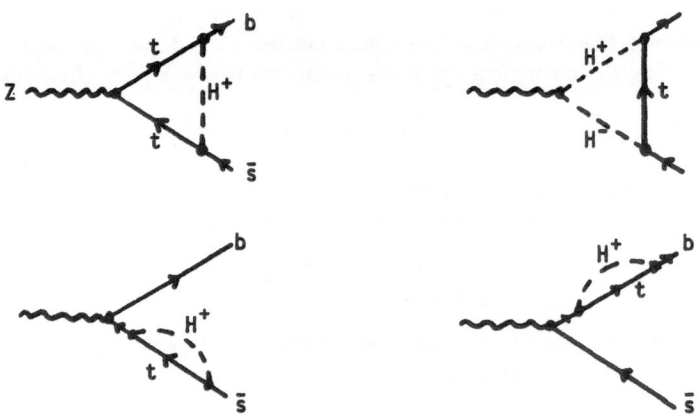

The branching ratio

$$\frac{\Gamma(Z \to b\bar{s}) + \Gamma(Z \to \bar{b}s)}{\Gamma_Z} \sim |\, U_{ts}U_{tb}^* \,|^2 \left(\frac{m_t}{M_W} \cdot \frac{v_1}{v_2}\right)^4 \qquad (41)$$

with the KM matrix elements [32]

$$U_{tb} \cong 1, \quad U_{ts} \le 0.05$$

contains the enhancement factor in the fourth power. Obeying the constraint from the experimental mass splitting between the neutral B mesons [29] the branching ratio becomes larger than 10^{-6} for top masses above 100 GeV [30].

7. Summary

In the Standard Model, the Z width is a prediction after the Z mass has been measured and the values of m_t, M_H have been specified. Together with the QED initial state bremsstrahlung corrections, the width is the essential ingredient in forming the Z resonance line shape.

The high precision in the experimental determination of the line shape at LEP and SLC requires a theoretical treatment of Γ_Z aiming an uncertainty of ± 10 MeV. To this end we have to include, together with non-fermionic decay channels, the next order corrections in the fermionic partial widths which in lowest order are the only contributions to the total width Γ_Z. These consist of the electroweak one-loop corrections (including the QED part) and the QCD corrections for the hadronic partial widths. The unknown parameters of the Standard Model, M_H and m_t, influence the prediction in a calculable way. Whereas the variation of Γ_Z with the top mass is sizable (more than 20 MeV) the dependence on the Higgs mass is not very significant (≤ 6 MeV for M_H between 10 GeV and 1 TeV). The position of the peak maximum, from which M_Z will be determined experimentally, is stable within 2 MeV for all values of M_H and m_t in the considered range. This is important for M_Z being an independently measurable input parameter.

Among the other decay channels contributing to Γ_Z in higher order the decay into Higgs and fermion pairs is the only significant one yielding several MeV if the Higgs is light (< 10 GeV); others are below 1 MeV. The largest uncertainty in Γ_Z is induced by α_s in the hadronic decay modes yielding $\Delta\Gamma_Z = 12$ MeV. The recent re-evaluation [31] $\alpha_s = 0.11 \pm 0.01$ would reduce this error to $\Delta\Gamma_Z = 6$ MeV.

Among the partial widths the $Z \rightarrow \bar{b}b$ decay channel is of specific interest since the top quark in the vertex corrections cancels the increase for large m_t observed in the other partial widths. As a result, $\Gamma(Z \rightarrow \bar{b}b)$ is constant within 2 MeV over the whole top mass range up to 250 GeV. For this reason it is an ideal probe for several kinds of new physics objects: a sequential fourth generation with mass splitting in the the doublet would increase the b partial width, whereas charged Higgs bosons predicted by the extended Standard model with two scalar doublets diminuish it, in particular in the case of enhanced Yukawa couplings. Such non-standard charged Higgs bosons with enhanced Yukawa couplings can also give rise to flavor changing Z decays with branching ratios $> 10^{-6}$ if the top mass is above 100 GeV.

References:

1. M. Greco, G. Pancheri, Y. Srivastava, Nucl. Phys. B 171 (1980) 118; E: Nucl. Phys. B 197 (1982) 543;
   F.A. Berends, R. Kleiss, S. Jadach, Nucl. Phys. B 202 (1982) 63;
   M. Böhm, W. Hollik, Nucl. Phys. B 204 (1982) 45

2. Physics with LEP, CERN 86-02, eds. J. Ellis and R. Peccei

3. F.A. Berends, G. Burgers, W.L. van Neerven, Phys. Lett. 185 B (1987) 395;
   G. Altarelli, G. Martinelli, in [2];
   O. Nicrosini, L. Trentadue, Phys. Lett. 196 B (1987) 551; Z. Phys. C 39 (1988) 479;
   E.A. Kuraev, V.S. Fadin, Sov. J. Nucl. Phys. 41 (1985);
   V.S. Fadin, V.A. Khoze, Yad. Fyz. 47 (1988) 1693

4. F.A. Berends, G. Burgers, W. Hollik, W.L. van Neerven, Phys. Lett. 203 B (1988) 177

5. W. Beenakker, W. Hollik, Z. Phys. C 40 (1988) 141

6. P. Antonelli, M. Consoli, C. Corbo, Phys. Lett. 99 B (1981) 475;
   M. Consoli, S. Lo Presti, L. Maiani, Nucl. Phys. B 223 (1983) 474;
   F. Jegerlehner, Z. Phys. C 32 (1986) 425

7. W. Wetzel, in [2] and Nucl. Phys. B 227 (1983) 1

8. A.A. Akhundov, D.Yu. Bardin, T. Riemann, Nucl. Phys. B 276 (1986) 1

9. M. Böhm, W. Hollik, H. Spiesberger, Fortschr. Phys. 34 (1986) 687

10. W. Hollik, DESY 88-188 (1988), to appear in Fortschr. Phys.

11. W. Beenakker, W. Hollik, in: CERN 87-08, ECFA 87-08 (1987), eds. A. Böhm, W. Hoogland

12. J Jersak, E. Laerman, P.M. Zerwas, Phys. Rev. D 25 (1980) 1218

13. A. Sirlin, Phys. Rev. D 22 (1980) 971

14. A. Barroso et al., in CERN/ECFA 87-08 (1987), eds. A. Böhm and W. Hoogland

15. D. Yu. Bardin, P. Ch. Christova, O.M. Fedorenko, Nucl. Phys. B 175 (1980) 435; Nucl. Phys. B 197 (1982) 1

16. F. Fleischer, F. Jegerlehner, Phys. Rev. D 23 (1982) 2001

17. K.I. Aoki, Z. Hioki, R. Kawabe, M. Konuma, T. Muta, Suppl. Progr. Theor. Phys. 73 (1982) 1

18. D.C. Kennedy, B.W. Lynn, SLAC-PUB 4039 (1986, revised 1988)

19. D.Yu. Bardin, A. Leike, T. Riemann, M. Sachwitz, Phys. Lett. 206 (1988) 539;
    D.Yu. Bardin, M.S. Bilenky, G.V. Mithselmakher, T.Riemann, M. Sachwitz, Zeuthen preprint PHE 89-05

20. F.A. Berends, R. Kleiss, Nucl. Phys. B 260 (1985) 32

21. J.D. Bjorken, SLAC-198 (1977);
    J Finjord, Phys. Scripta 21 (1980) 143

22. E. Franco, in [2]

23. W. de Boer, SLAC-PUB-4428 (1988)

24. M. Consoli, W. Hollik, F. Jegerlehner, CERN-TH.5395/89 (1989)

25. G. Burgers, in "Polarization at LEP", CERN 88-06, eds. G. Alexander et al.

26. R.D. Peccei, H.R. Quinn, Phys. Rev. Lett. 38 (1977) 1440; Phys. Rev. D 16 (1977) 1719

27. A.J. Buras, Proc. of the EPS Conference, Bari 1985

28. G. Mann, T. Riemann, Annalen der Physik 40 (1983) 334;
    M. Clemens, C. Footman, A. Kronfeld, S. Navasimhan, D. Photiadis, Phys. Rev. D 27 (1983) 570;
    V. Ganapathi, T. Weiler, E. Laerman, I. Schmitt, P.M. Zerwas, Phys. Rev. D 27 (1983) 579

29. H. Albrecht et al. (ARGUS Coll.), DESY 87-029 (1987)

30. J.L. Hewett, S. Nandi, T. Rizzo, Ames preprint OSU 201, IS-J-2983 (1988);
    M.J. Savage, CALTECH preprint CALT-68-1496 (1988);
    C. Busch, DESY 88-148 (1988)

31. G. Altarelli, CERN-TH.5290/89 (1989)

32. Particle Data Group, Phys. Lett. 204 B (1988) 1

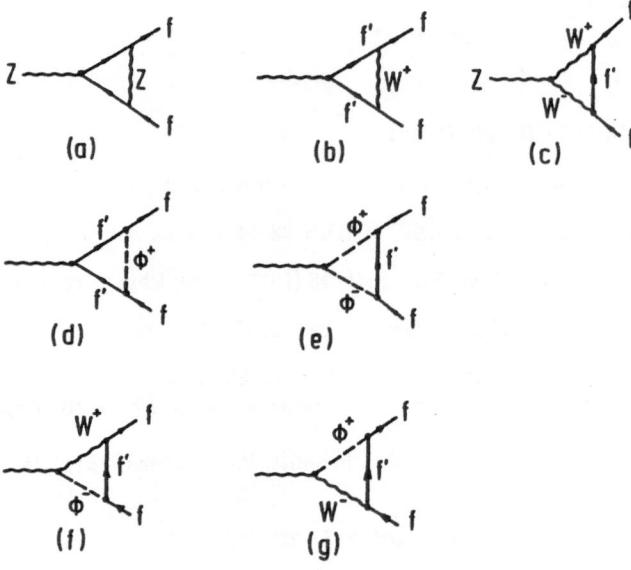

Figure 1: Vertex corrections (neutral Higgs contributions neglected)

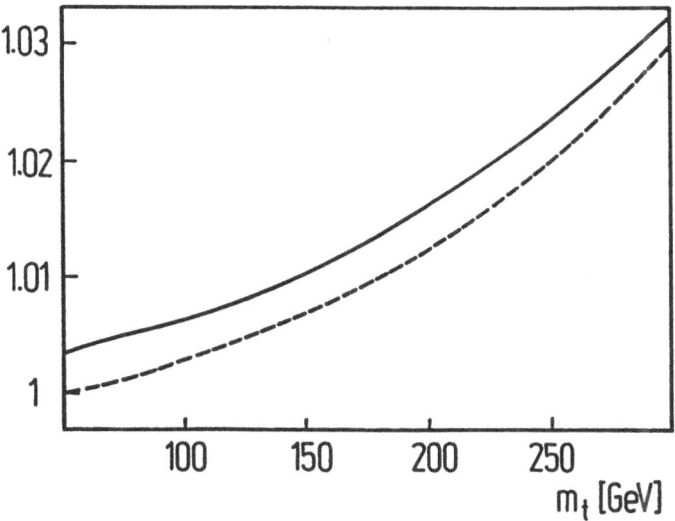

Figure 2: $(1 - \Delta r)/(1 + \Pi_Z)$ (solid line) and approximate form (35) (dashed line)

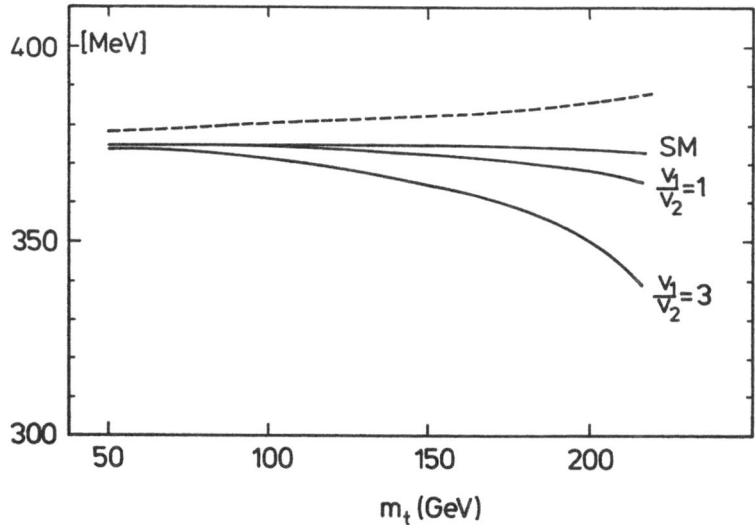

Figure 3: $\Gamma(Z \to \bar{d}d)$ in the Standard Model (dashed) and $\Gamma(Z \to \bar{b}b)$ in the Standard Model (SM, solid) and in the presence of charged Higgs bosons

Radiative Corrections for e⁺e⁻ Collisions   Editor: J.H. Kühn
© Springer-Verlag Berlin, Heidelberg 1989

On the derivation of Standard Model Parameters from the Z Peak

T. Riemann[1], D. Bardin[2], M. Bilenky[2], M. Sachwitz[1]

[1]Institut für Hochenergiephysik der AdW der DDR, Berlin-Zeuthen, GDR
[2]Joint Institute for Nuclear Research, Dubna

CONTENTS

1. Introduction
2. Form factor approach to electroweak radiative corrections
3. QED corrections on resonance and its interplay with EWRC
4. Determining standard model parameters
5. Conclusions

1. INTRODUCTION

The first complete calculation of electroweak radiative corrections (EWRC) in the standard theory /1/ for the reaction

$$e^+ e^- \longrightarrow (\gamma, Z) \longrightarrow \mu^+ \mu^- , \qquad (1.1)$$

has been published ten years ago /2/. Later on, there have been several recalculations covering also the energy region of resonance production of the weak neutral Z-boson /3-6/. From reaction (1.1) one hopes to measure mass M_z and width Γ_z of the Z-boson with absolute errors as small as 20 - 50 MeV. Consequently, the combined error of cross section and asymmetry calculations due to theoretical uncertainties and due to the software used (Monte Carlo programs) should not exceed 0.1 %. It takes much effort of different kind in order to guaranty such an accuracy; e.g.:

1. Numerical results on EWRC and QED corrections based on equal assumptions and approximations but realised in different approaches have to be shown to agree for a large variety of input parameters and kinematic variables;

2. An adequate choice of necessary approximations, input parameters

etc. has to be found out;

3. Certain higher order corrections have to be taken into account;

4. The software has to be organised very efficiently in order to reach the necessary precision without wasting computer time - this raises the problem of sufficiently precise approximate analytic formulae.

In Chapter 1, we present the form factor approach to EWRC /6/ in e^+e^--annihilation into fermion pairs. Some observations on QED corrections to reaction (1.1) at the Z resonance and on the interplay between EWRC and QED are contained in Chapter 2. In Chapter 3, some aspects of the determination of standard theory parameters from the Z peak are discussed.

2. FORM FACTOR APPROACH TO ELECTROWEAK RADIATIVE CORRECTIONS

There have been proposed several possibilities to compose cross section formulae out of the large amount of contributing Feynman diagrams. In search of an approach whose language is sufficiently close to Born physics and which depends only on observable, gauge-invariant building blocks we generalised a notation which had been first introduced in neutrino electron scattering /7/.

The one-loop corrected matrix element for the e^+e^--annihilation into two light fermions depends on only five dynamic variables F_A, ρ_Z, v_e, v_f, v_{ef} :

$$M = M_\gamma + M_z , \tag{2.1}$$

$$M_\gamma \sim \alpha(0) \, F_A \cdot \gamma_\alpha \otimes \gamma_\alpha , \tag{2.2}$$

$$M_z \sim G_\mu \, \rho_z \left[\gamma_\alpha\gamma_5 \otimes \gamma_\alpha\gamma_5 + v_e\gamma_\alpha \otimes \gamma_\alpha\gamma_5 + v_f\gamma_\alpha\gamma_5 \otimes \gamma_\alpha + v_{ef}\gamma_\alpha \otimes \gamma_\alpha \right]. \tag{2.3}$$

The one-loop corrected vector couplings are:

$$v_a = 1 - 4 \sin^2\theta_W |Q_a| \, \kappa_a(s), \qquad a = e,f , \tag{2.4}$$

$$v_{ef} = v_e + v_f - 1 + 16 \sin^4\theta_W |Q_e Q_f| \, \kappa_{ef}(s) . \tag{2.5}$$

The function $F_A(s)$ consists of the vacuum polarisation insertions of the photon propagator and renormalizes the finestructure constant, while $\rho_Z(s)$ and $\kappa_a(s)$ contain the genuine weak loop corrections /6/. They may be interpreted as complex-valued finite, process dependent renormalizations of the muon decay constant and of the weak mixing angle. From (2.4 - 5) it is evident that it is not very useful to leave the on mass shell scheme definition of the weak mixing angle,

$$\sin^2 \theta_W = 1 - M_W^2 / M_Z^2 , \qquad (2.6)$$

in favor of an effective parameter $\sin^2 \theta_{eff}$ - one would not really know which one out of the three candidates should be chosen.

A calculation of EWRC for a given process consists of several steps. In brackets we indicate the corresponding Fortran codes developed by the Dubna-Zeuthen EWRC group:

1. Determination of the necessary input parameters (α, G_μ, M_Z ; M_W, $\sin^2 \theta_W$; m_t, M_H, α_s) taking into account that not all of them are independent (SETCON);

2. Calculation of the electroweak form factors (ROKAP, ROKAPP);

3. Calculation of the Z width (ZWRATE);

4. Calculation of cross sections or spin amplitudes from the electroweak form factors (DZEWCS);

5. Calculation of the apparatus dependent $0(\alpha)$ QED corrections (no cut: BREMU);

6. Inclusion of higher order QED corrections.

As will be discussed in more detail in the next chapter, any of the calculation steps depends numerically on the exact implementation of all foregoing steps. Evidently, the introduction of electroweak form factors on the matrix element level has the advantage of great flexibility not only because they are the natural degrees of freedom of the problem, but also for technical reasons. They allow an effective organisation of the theoretical material even beyond the minimal theory. It is easy to include additional

contributions or to change e.g. from cross sections to spin
amplitudes. This makes the form factor approach also well suited for
applications in Monte Carlo programs. The differential cross section
for (1.1) reads as follows ($c=\cos\theta$):

$$\frac{d\sigma}{dc} = \frac{\pi\alpha^2}{2s}\left\{\sum_{B=\gamma,I,Z}\ \sum_{b=0,e,i,f} c_b \cdot \text{Re}\left[\ K(B)\cdot\left[V(B)\cdot R_b^V(B)+A(B)\cdot R_b^A(B)\right]\right]\right\} +$$

$$+ \frac{d\sigma_{box}}{dc}\ , \qquad\qquad\qquad (2.7)$$

$$c_b = \left\{\ 1,\ Q_e^2,\ Q_e Q_f,\ Q_f^2\ \right\}\ , \qquad\qquad b = o,e,i,f, \qquad (2.8)$$

$$K(B) = \left\{\ |Q_e Q_f F_A|^2,\ 2|Q_e Q_f|\chi\cdot\rho_Z\cdot F_A^*,\ |\chi\cdot\rho_Z|^2\ \right\}\ , \qquad (2.9)$$

$$V(B) = \left\{\ 1,\ v_{ef},\ 1+|v_e|^2+|v_f|^2+|v_{ef}|^2\ \right\}\ , \qquad\qquad (2.10)$$

$$A(B) = \left\{\ 0,\ 1,\ 2\text{Re}\left(\ v_e v_f^*+ v_{ef}\right)\ \right\}\ , \qquad\qquad B = \gamma,I,Z, \qquad (2.11)$$

$$\chi = \frac{G_\mu}{\sqrt{2}}\frac{M_Z^2}{8\pi\alpha}\frac{s}{s-M^2}\ , \qquad\qquad (2.12)$$

$$M^2 = M_Z^2 - i\Gamma_Z(s)\cdot M_Z\ , \qquad\qquad (2.13)$$

$$\Gamma_Z(s) = \frac{s}{M_Z^2}\Gamma_Z\ . \qquad\qquad (2.14)$$

The case b = 0 corresponds to the Born cross section plus weak loop
corrections while b = e, i, f represent QED contributions with
initial (e) or final (f) state radiation or their interference (i).
The angular dependence is contained in the dynamical variables R:

$$R_o^{V,A}(B) = \frac{1}{2}\left[\ (1+c)^2 \pm\ (1-c)^2\ \right]\ , \qquad\qquad B = \gamma,I,Z. \qquad (2.15)$$

The EWRC presented here have been undertaken intensive numerical comparisons /8,9/ and agree within one or several permille with those calculated in /5,10/.

Such an accuracy makes necessary to take into account tiny effects. One of them is the difference between d-quark and b-quark production due to weak loop effects of the finite t-quark mass which influences only the b-quark channel (small quark mixing matrix elements). As an example, we scetch the calculation of this effect. Photon exchange and box diagrams give completely negligible contributions so that there remain the finite t-mass corrections to an off shell Z-boson /11/ (Fig. 1). They are contained in an additional term $V(s, m_t)$ in (2.3):

$$M_Z^{(t)} \sim G_\mu \left[\gamma_\alpha (v_e + \gamma_5) \otimes \gamma_\alpha ((v_f + \gamma_5) + \Delta V(s, m_t)(1 + \gamma_5)) \right]. \qquad (2.16)$$

The ΔV vanishes for vanishing m_t. From (2.16) one immediately derives

$$\Delta \rho = \Delta V,$$
$$\Delta \kappa_e = 0,$$
$$\Delta \kappa_f = \Delta \kappa_{ef} = -\Delta V. \qquad (2.17)$$

The ΔV depends only logarithmically on s/M_Z^2 so that at LEP energies it is sufficient to use its value at $s = M_Z^2$. This allows to use instead the corresponding correction ΔV^Γ for the Z width /12/ which has exactly the same value. In Table 1 we quote the resulting corrections for $M_Z = 93 GeV$, $M_H = 100 GeV$, $\alpha_s = .12$ (no QED included; for comparison we show also the corresponding values for small m_t).

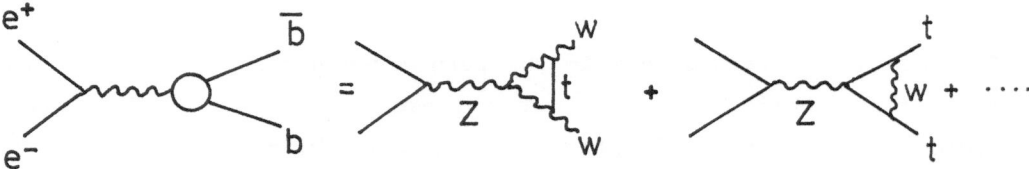

Fig. 1. Additional vertex corrections to the b-quark channel due to finite t-quark mass

Table 1.   Numerical influence of vertex corrections due to finite t-quark mass on b-quark production in comparison to the unaffected d,s-quark channels (σ_T^{peak} - peaking cross section, $\sqrt{s}_{max}$ - its position, $\sqrt{s}_{\pm 1/2}$ - beam energies with half the peak cross section)

channel		d,s	b	d,s	b
m_t	(GeV)	230	230	60	60
σ_T^{peak}	(nbarn)	9.010	8.707	8.926	8.876
$\sqrt{s}_{max}$	(GeV)	91.983	91.984	91.984	91.984
$\sqrt{s}_{-1/2}$	(GeV)	90.691	90.691	90.713	90.714
$\sqrt{s}_{1/2}$	(GeV)	93.298	93.299	93.274	93.275

3. QED CONTRIBUTIONS ON RESONANCE AND THEIR INTERPLAY WITH THE WEAK LOOP CORRECTIONS

The differential cross section (2.7) has been written such that QED corrections are included in a natural way. In the most general case - polarisation of both initial and final state fermions - there are 18 different QED contributions. Due to symmetry properties /12,13/,

$$R_i^{V,A} (I) = \frac{1}{2} \left[R_i^{V,A} (\gamma) + R_i^{V,A} (Z) \right] , \qquad (3.1)$$

$$R_f^A(B) = 0 \qquad \text{(if no cuts applied)}, \qquad (3.2)$$

$$R_b^{V,A} (B) = R_b (B,c) + s_b^{V,A} \cdot R_b (B,-c) , \qquad B=\gamma,I,Z,$$
$$\text{(for cuts symmetric in } \cos \theta) \qquad (3.3)$$

$$s_b^V = - s_b^A = - s_i^V = s_i^A = 1 , \qquad b=0,e,f, \qquad (3.4)$$

only six bremsstrahlung functions remain as independent observables:

$$R^A_e(B,c), \ B=\gamma,I,Z, \qquad R^A_i(B,c), \ B=\gamma,Z, \qquad R^A_f(\gamma,c). \qquad (3.5)$$

These bremsstrahlung functions depend on kinematics (here s and $\cos\theta$) and on the complex parameter M^2, (2.13). For initial state radiation, there are also logarithmic dependencies on the electron mass $(B=\gamma,I,Z)$ and on the final fermion mass $(B=\gamma)$. For the total cross section σ_T and the integrated forward-backward asymmetry A_{FB}, the integrated bremsstrahlung functions (3.5) have been analytically calculated in order $O(\alpha)$ /13/. The angular dependent functions (3.5) and those containing an additional cut on the photon energy are also analytically known to order $O(\alpha)$ /14/.

When including bremsstrahlung terms into (2.7) we made use of the fact that the electroweak form factors do not depend on the scattering angle. The very small angular dependent contributions due to ZZ- and WW-box diagrams which in principle could also be included into the form factors /15/ have been separated for this purpose. Further, the form factors have a very smooth dependence on the energy variable s (of order $\ln(s/M^2_z)$). Both properties together allow to factorize the form factors out of the steeply verying bremsstrahlung integrands such simplifying the corresponding integrations, e.g. over the effective beam energy. Though this is of minor importance for Monte Carlo calculations it becomes substantial in the analytic approach.

Another problem for analytic calculations is the energy dependence of the Z width. In Fig. 2 a Z boson propagator in an arbitrary Feynman diagram is shown.

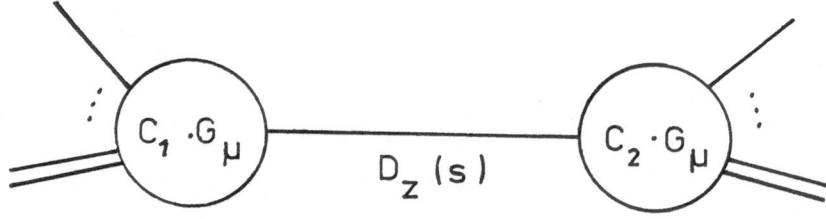

Fig. 2. Z-boson propagator in an arbitrary Feynman diagram

Using the quite accurate approximation (2.14), one can apply the following Z-boson tranformation /16/:

$$M_z' = \left[1 + \left(\Gamma_z / M_z \right)^2 \right]^{-1/2} \cdot M_z = M_z + \delta M_z , \qquad (3.6)$$

$$\Gamma_z' = \left[1 + \left(\Gamma_z / M_z \right)^2 \right]^{-1/2} \cdot \Gamma_z \sim \Gamma_z , \qquad (3.7)$$

$$G_\mu' = \left(1 + i\Gamma_z / M_z \right)^{-1} \cdot G_\mu , \qquad (3.8)$$

$$\delta M_z = - \frac{1}{2} \Gamma_z^2 / M_z = -35 \text{ MeV} . \qquad (3.9)$$

Then, the following identity is fulfilled:

$$G_\mu D_z(s) = G_\mu \left[s - M_z^2 + i M_z \Gamma(s) \right]^{-1} = G_\mu' \left[s - M_z'^2 + i M_z' \Gamma'(s) \right]^{-1} \qquad (3.10)$$

In fact, the Z-boson transformation replaces $\Gamma(s)$ by a constant width parameter. Consequently, one can use analytic results obtained for the constant width case for the more accurate approach with $\Gamma(s)$ by the replacements (3.6-8) in final expressions. This way one obtains the dashed curves in Figs.3 and 4 which correspond to analytic integrations of bremsstrahlung to order $0(\alpha)$ /13,16/ (Fig.3) and to an approximate integration of the convoluted Born cross section (Fig.4) /17/. The additional peak shift due to the width function is in both cases exactly equal to the parameter δM_z (3.9) which effectively describes a shift of the center of gravity of the resonance from $s^{1/2} = M_z$ to $s^{1/2} = M_z'$. The numerical value of the shift is of the order of the anticipated experimental error for M_z and has also been observed in /18/ where higher order QED effects have been studied.

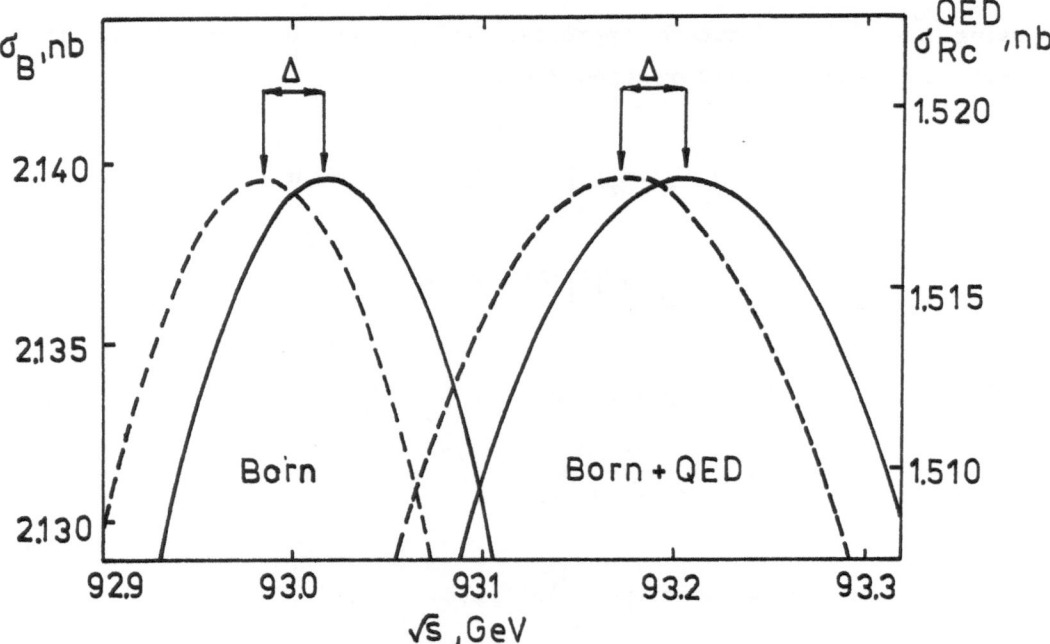

Fig 3. Muon pair production cross section in the Z-peak region both in born approximation and including $O(\alpha)$ QED corrections (no cut; M_Z=93GeV, Γ_Z=2.5GeV, $\sin^2\theta_W$=0.23).Solid line- Γ_Z, dashed line- $\Gamma_Z(s)$

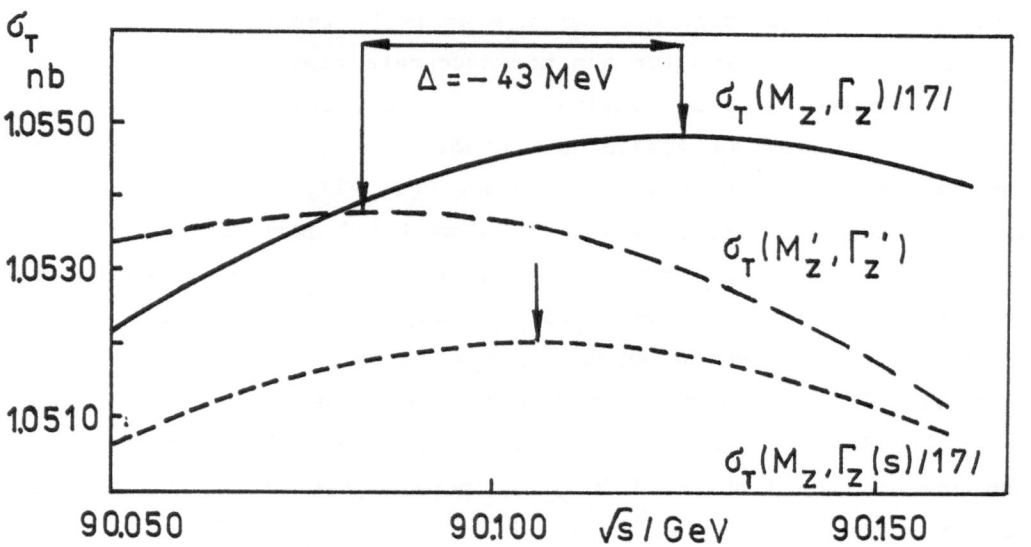

Fig 4. Muon pair production cross section in the Z-peak region with higher order QED corrections, integrations approximated (no cut; M_Z=93GeV, Γ_Z=2.8GeV, $\sin^2\theta_W$=0.221).Solid line- Γ_Z, dashed line- $\Gamma_Z(s)$

Table 2. Counting rates expected at LEP for $L=10^{31}/(\text{cm}^2\text{sec})$ and $d=10$ days running time per energy. Bremsstrahlung is treated as explained in the text. Further, statistical errors $N^{1/2}$, and the changes if neglecting weak loops (ΔN_{1L}) or energy dependence of the width (ΔN_Γ) on the counting rates are shown; $M_z=93$, $m_t=90$, $M_H=100$ (in GeV), $\alpha_s=0.12$

energy point	1	2	3	4	5	6	A	7
$\sqrt{s}$,GeV	81	83	85	87	89	91	92.1	93.1
N(I)	298	374	517	830	1681	5110	11354	16339
N(II)	350	402	504	723	1301	3503	7465	11752
N(III)	356	412	522	762	1409	3952	8592	13101
$N^{1/2}$	19	20	23	28	38	63	93	114
ΔN_{1L}	-20	-18	-16	-13	-6	19	67	117
ΔN_Γ	-1	-1	-3	-6	-19	-93	-224	-38

energy point	B	8	9	10	11	12	13
$\sqrt{s}$,GeV	94.1	95	97	99	101	103	105
N(I)	9824	5151	1784	905	565	399	301
N(II)	9105	5957	2883	1792	1275	981	793
N(III)	9291	5802	2884	1825	1315	1005	814
$N^{1/2}$	96	76	48	43	36	32	29
ΔN_{1L}	81	45	13	3	2	2	5
ΔN_Γ	166	102	33	15	8	5	3

As may be seen from Table 2, the influence of the width (ΔN_{Γ}) is of the same order as the net effect of all the other electroweak one loop corrections (ΔN_{1L}). Both are small compared to the QED contributions. The counting rates in the tables are:

  N(I)   - Born;

  N(II)   - Born plus weak loops plus $0(\alpha)$ QED;

  N(III) - additionally soft photon exponentiation.

Nevertheless, they cannot be neglected if accuracies of the order of 0.1 % are to be ensured. This scale even makes necessary a careful study of the interplay of QED and weak one-loop corrections, though both are at least of order $0(\alpha)$. The cross section formula (2.7) which rests on the matrix elements (2.2-3) suggests the treatment of bremsstrahlung as being radiated from charged fermion propagators which are dressed by weak interactions. This is evidently necessary in the region of the radiative tail where bremsstrahlung substantially dominates over the Born cross section and a naive handling of weak loops would by far underestimate their influence. A closer inspection shows that already at smaller energies the common treatment of QED and weak loops improves numerics by at least several permille both for σ_T and A_{FB} /6/, see Fig. 5. This has also been observed in /19/.

Another tiny but not negligible higher order QED effect is due to the careful treatment of all the imaginary parts which are introduced by $M^2 = M_Z^2 - iM_Z\Gamma_Z$ (Γ_Z is of order $0(\alpha)$). An instructive example is the $0(\alpha)$ QED contribution to A_{FB} as calculated in /13/ where in the first reference a resonating γZ-box approximation is used while the second one is exact in this respect. At the peak, the difference is of order $0(\alpha^2)$. The exact contribution from the interference of initial and final state radiation is negligibly small: 0.007 %, -0.000 %, 0.027 % for B=γ,I,Z corr., compared to a net asymmetry of -0.393 % (M_Z = 93GeV, Γ_Z = 2.5GeV, $\sin^2\theta_W$ = 0.23). The approximate γZ-box formulae lead to the following contributions:0.007 %, 0.000 %, 0.133 %; net asymmetry: -0.237 %. A closer inspection of the corresponding analytic bremsstrahlung and γZ-box formulae (eqs.

(4.12,13) of the second ref. in /13/) shows sensible relative cancellations at $s = M_z^2$. Consequently, one has to expect a more or less pronounced influence of soft photon cuts diminishing these cancellations /14/. The discussed differences have to be taken into account and are a typical measure of the general influence of the various small imaginary parts.

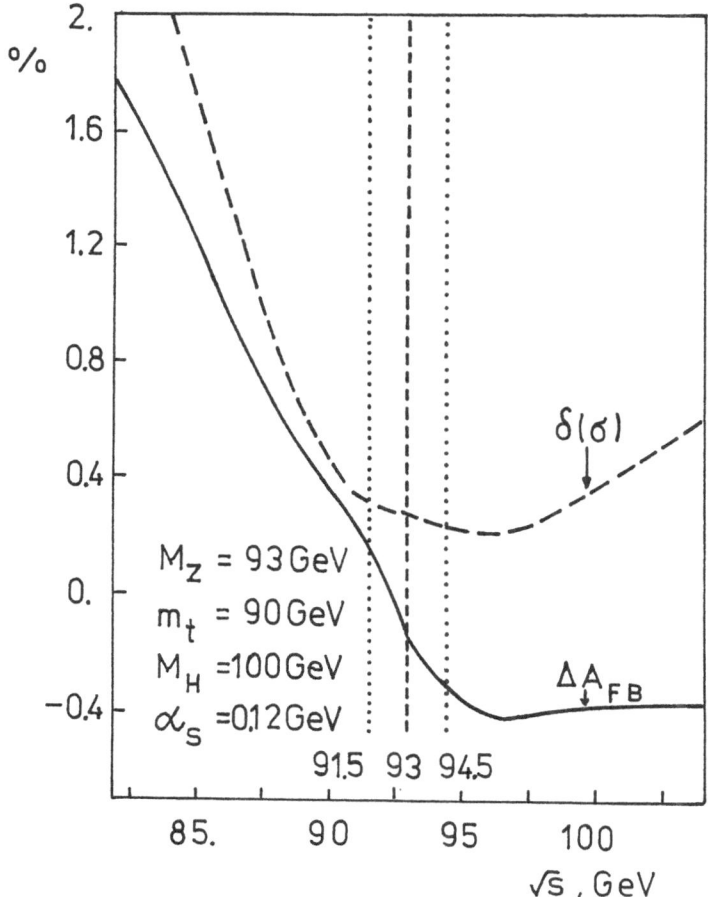

Fig 5. Changes $\delta(\sigma_T)$ of $\sigma_T(s)$ and ΔA_{FB} of A_{FB} (both in %) due to use of Born couplings in the QED contributions instead of the dressed, complex couplings introduced in eqs. (2.4-5)

At the end of this chapter on some features of QED corrections (which have to be taken into account for precision measurements) we would like to give a general comment on analytic bremsstrahlung calculations. They are often very cumbersome but, at the same time, to many physicists they seem to be of small value because real experiments with no doubt will be analysed with Monte Carlo programs. Of course, the value of analytic results does not rest on flexibility. Instead, they allow the study of qualitative aspects. They very clearly exhibit general structures. But they are also of great value for certain numerical purposes. A central problem for precision experiments is to ensure that the software used in the analysis is sufficiently accurate /20/. One powerful check is their comparison with corresponding analytic results as far as those exist. Further, analytic exact $0(\alpha)$ results may serve both as testing ground for leading log approximations and as input for higher order calculations.

4. DETERMINING STANDARD MODEL PARAMETERS

The analysis of the Z line shape is a multi parameter problem because the cross section (1.1) depends at least on α, G_μ, M_z ; m_t , M_z, α_s . This raises the question of the possibility of an unbiased determination of M_z. An accurate knowledge of the Z-boson mass is very important as input for forthcoming tests of the standard theory. Modelling a line shape scan based on data as given in Table 2 /6/ yielded Table 3 based on statistical errors only. The additional inclusion of systematic errors makes yet more evident - the Z mass determination is not influenced by the present lack of knowledge on m_t and M_H.

This has also been suspected in /18/ in view of the weak dependence of the total cross sections, especially of the Z peak position, on these parameters. But one must be careful in drawing definite conclusions - the accuracy to be reached for M_z is of the order of 30 MeV/90 GeV = 0.03%. Often the peak shift due to some

Table 3: Results of line shape scans each based on thirteen measurements of σ_T (see Table 2). Statistical errors are taken into account

fixed	-	α_s	M_H	M_H,α_s	m_t,M_H	m_t,M_H,α_s
M_z	93.000± ±.017	93.000± ±.017	93.000± ±.016	93.000± ±.017	93.000± ±.016	93.000± ±.016
m_t	119±80	94±78	93±76	90±70	-	-
M_H	846±2192	142±926	-	-	-	-
α_s	.124±.027	-	.120±.024	-	.120±.020	-

phenomenon is taken as a measure of the error induced on the Z mass measurement phenomenon is not properly taken into account. Intuitively it is clear that indeed such a peak shift is correlated with the corresponding peak deformation as a whole. Nevertheless, an adequate estimation can only be obtained from a realistic line shape analysis /21,6/.

As an instructive example, it has been demonstrated in /16/ that a fit to data using a wrong analytical ansatz for the Z-width (constant width in the resonance function (2.12)) leads to a wrong estimation of M_z while Γ_z remains unaffected. The corresponding fit results (in GeV) to the assumed data are:

$$\Gamma_z(s): \quad M_z = 93.000 \pm 0.016 \quad \Gamma_z = 2.498\pm0.011 \qquad (4.1)$$

$$\Gamma_z : \quad M_z = 92.966 \pm 0.016 \quad \Gamma_z = 2.498\pm0.011. \qquad (4.2)$$

In (4.2), the net error of M_z became larger compared to (4.1) by the Z mass parameter shift δM_z of (3.9) - but this is not reflected in an enlarged fit error estimate. The reason is that one uses in case of (4.2) a wrong theoretical ansatz, measures in fact the M_z' of (3.7) but interpretes the M_z' as M_z. This simple exercise shows the potential defects of any simplified analytical formula.

Another example may be taken from /17/ where has been stated that the difference of using $\Gamma_z(s)$ or Γ_z is minor. Though, the corresponding peak shift, $\Delta = -43 MeV$, is not too small (see Fig. 4). This fact remained hidden in that work due to the use of an unappropriate Born ansatz using a constant width.

At the other hand, one can overestimate errors if one uses an ansatz too close to the pure theoretical formalism. In /6,23/ it has been performed an estimate of the expected error of the Z width from (1.1) with floating M_z and Γ_z . As a result, it has been obtained:

$$\Delta\Gamma_z = 14 \text{ MeV}_{stat} \oplus (20 + 35 + 70) \text{ MeV}_{syst} . \qquad (4.3)$$

$$\Delta L_a = \qquad\qquad (1 + 2 + 3) \% , \qquad\qquad (4.4)$$

with ΔL_a the absolute luminosity error which has been considered to be the dominant systematic error. Why such an unexpected strong correlation of absolute luminosity and width ? It is artificially introduced because the cross section normalisation has not been left open as a free parameter. Introducing a normalisation factor ρ immediately diminishes the width error:

$$\Delta\Gamma_z = 17 \text{ MeV}_{stat} \oplus (2 + 4) \text{ MeV}_{syst} , \qquad (4.5)$$

for $\Delta L_a = 1 \div 3$ %.

An interesting possibility to leave open the cross section normalisation without introducing an additional "technical" degree of freedom has been proposed in /21/: Instead of the input parameters (α, G_μ, M_z) one can also use $(\alpha, M_z, \sin^2\theta_w)$. This leads to a modified resonance function (2.12):

$$\kappa = \frac{1}{16\sin^2\theta_w \cos^2\theta_w} \frac{s}{s - M^2} . \qquad (4.6)$$

If one now allows for a floating weak mixing angle, the overall normalisation remains open in a theoretically clean electroweak one loop approach. As a by-product of the improved width measurement one realises an independent measurement of $\sin^2\theta_W$ from the line shape. Strictly speaking this determination of $\sin^2\theta_W$ is a measurement of the Fermi constant G_Z at an energy scale $s \sim M_Z^2$ while G_μ has been determined at low energy from muon decay. It could yield information on new phenomena which show in the cross section normalisation ρ at higher energy. The possible accuracy for the weak mixing angle depends on the accuracy of the absolute luminocity calibration and could reach $\Delta\sin^2\theta_W \sim 0.003$ ($\Delta L_a = 1\%$).

At present, there is great interest in well-adapted approximate analytic formulae as proposed in /17/ (see also /22-24/) because the simulation and reconstruction programs for data analysis should not be too complicated. Much work in this direction is now in progress.

5. SUMMARY

In the present contribution, we propose the use of electroweak form factors for the analysis of e^+e^--annihilation. On several examples we demonstrate potential problems for the analysis of precision experiments. In principle, the theoretical description of standard physics around the Z peak is not a source of surprising discoveries. Though, work in this field is very important because any search for new interactions and/or particles seems only possible if all standard phenomena are very precisely understood. The unexpected complexity of the problems to be solved in this connection stimulates rapid progress in the handling of standard physics, including old-fashioned quantum electrodynamics. This makes life in standard physics interesting enough though we all hope for phantastic discoveries at the Z peak.

References

1. S.L. Glashow: Nucl. Phys. 22 579 (1961)
   S. Weinberg: Phys. Rev. Letters 19 1264 (1967)
   A. Salam: in Elementary Particle Theory, ed. N. Svartholm
   (Stockholm 1968) p.361
2. G Passarino, M. Veltman: Nucl. Phys. B160 151 (1979)
3. W. Wetzel: Nucl. Phys. B227 1 (1982)
4. B.W. Lynn, R.G. Stuart: Nucl. Phys.B253 216 (1985)
5. W. Hollik: Phys. Letters B152 121 (1985)
6. D. Bardin,M. Bilenky,G. Mitselmakher,T. Riemann,M. Sachwitz:
   Berlin-Zeuthen prepr. PHE 89-05(1989),
   subm. to Z.Physik C .
7. S. Sarantakos,A. Sirlin,W. Marciano: Nucl. Phys. B217 84 (1983)
8. D. Bardin, G. Burgers, B.W. Lynn: private communication
   D. Bardin, M. Bilenky, G. Mitselmakher: CERN-DELPHI note 88-43
   Phys 27 (1988)
9. F.A.Berends, W.Hollik, M.Sachwitz: private communication
   F.A.Berends: talk presented at this conference
10. G. Burgers: in: Polarization at LEP, eds. G. Alexander et al:
    CERN report CERN 88-06 (1988),p. 121 and CERN-TH 5119/88(1988)
11. G. Mann, T. Riemann: Annalen d. Phys. 40 334 (1983)
12. T. Riemann, D. Bardin, M. Sachwitz: in New Theories in
    Physics (World Scientific, 1988, Ed. Z. Ajduk et al) and
    Berlin-Zeuthen prepr. PHE 88-11 (1988)
13. D. Bardin, M. Bilenky, O. Fedorenko, T. Riemann: JINR Dubna
    prepr. E2-87-663 (1987), E2-88-324 (1988)
14. D. Bardin et. al: in preparation
15. D. Bardin, C. Burdik, P. Christova, T. Riemann: Berlin-Zeuthen
    prepr. PHE 88-15 (1988), to appear in Z.Physik C (1989)
16. D. Bardin, A. Leike, T. Riemann, M. Sachwitz: Phys. Lett.
    206B 539 (1988)
17. R. N. Cahn: Phys. Rev. D36 2666 (1987)
18. F.A. Berends, G. Burgers, W. Hollik, W.L. van Neerven: Phys.
    Lett. 203B 177 (1988)

19. F. Boillot, Z. Was: prepr. MPI-PAE/Exp. El. 196 (1988), to appear
    in Z.Physik C (1989)
20. R. Kleiss: talk presented at this conference
21. T. Riemann, M. Sachwitz: Phys. Lett. 212B 488 (1988)
22. G. Altarelli, S. Ganguli, A. Gurtu, K. Mazumuar:Tata Inst. prepr.
    TIFR/ IHEP-89/1 (1989)
23. R. Leiste et al: in Proc. Seminar "Physics of e^+e^- Interactions",
    Nov. 1988, Dubna, E2-89-205 (1989)
24. W. de Boer: prepr. SLAC-PUP-4682 (1988), to appear in Nucl.
    Instr. Meth. (1989)

Radiative Corrections for e⁺e⁻ Collisions   Editor: J.H. Kühn
© Springer-Verlag Berlin, Heidelberg 1989

Radiative Corrections in the Standard Model
And the Rho Parameter

Giampiero Passarino

Dipartimento di Fisica Teorica, Università di Torino,Torino, Italy
INFN, Sezione di Torino, Italy

Abstract: The problem of electroweak radiative corrections in the standard model is considered. Renormalization is discussed by fitting the bare parameters of the Lagrangian in terms of three data points. The $\rho-$ parameter is considered and its importance for confrontation with experimental data is reviewed, expecially for the case where vector boson masses are input parameters.

Recent years have shown a general consensus on the fact that electroweak interactions are entering in a new era of precision experiments. In a systematic investigation of the validity of the standard model we must therefore consider the radiative corrections as an indispensable part of the verification of the theory.

The central point in this issue is the comparison between theory and experiments, to be performed with the e^+e^- colliders going into operation, which requires at least three measurements. The model indeed contains three free parameters, not counting fermion masses and mixing angles in the quark sector. The Higgs mass is also a free parameter but the measurements relevant to SLC/LEP are weakly dependent on m_H, the so called Veltman's screening theorem.

Most radiative corrections are less than 1% but as soon as the data will be better than 1% we must follow a well defined and unique procedure [1]. Given the data, subtract radiative corrections and determine the parameters g, θ_W and M. Once the parameters are fixed we are left with the problem of finding a gauge invariant quantity which represents the standard reference point. Indeed in gauge theories we are facing a constraint of internal consistency rather than a problem of parametrizing the data. If three measurements are obtained the rest must follow because of gauge invariance. A convenient way to realize this unambiguous procedure is to study the $\rho-$ parameter [2].

A well known definition of ρ requires the knowledge of vector boson self-energies at low momentum. If $S_{ij}(p^2)$ represents the $i-j$ transition ($i = W, Z^0, \gamma$), we have

$$\rho = 1 - \frac{S_{WW}(0) - c_\theta^2 S_{ZZ}(0)}{M_W^2}$$

$c_\theta = \cos \theta_W$ being the cosine of the weak mixing angle. The deviation of ρ from one signals the degree of isospin breaking as observed in the form of mass differences between the members of fermion doublets as well as a drastic departure from the minimal Higgs sector. In short ρ is sensitive to the interaction of vector bosons with fermions and possibly other objects and the corresponding effects are visible even if the masses of these objects are larger than 100 GeV. Disregarding radiative corrections the result $\rho \geq 1/(2\,I)$ was proven in [2], where I is the highest weak isospin found among Higgs particles. In the minimal version of the standard model $\rho = 1$ at lowest order. This expression for the $\rho-$ parameter follows from using low energy data as input. Since in this case masses are not measured directly we have to consider low energy cross sections as $\mu-$ decay and $\nu_\mu e^-$ scattering. However the original definition goes well beyond the low q^2 region. If M, M_0 and c_θ are the bare W and Z^0 masses together with the bare $\cos\theta_W$ then with

$$\rho = \frac{M^2}{c_\theta^2 M_0^2}$$

the Higgs $\Delta I = 1/2$ rule implies $\rho = 1$. The bare parameters can be fixed by the knowledge of three experimental data, each of which is in principle defined at a different scale. It is evidently important to extend the $q^2 \sim 0$ considerations to $q^2 \sim M_W^2, M_Z^2$. From this point of view the precise values of M_W and M_Z will be a crucial test of the whole structure of the standard model, while until now the measurement of ρ from low energy data has represented our best window on unexplored energy scales.

M_Z will be measured to within a fraction of a percent and we may expect that in the future precise measurements will be available from angular distributions in W decay or total cross section in W^+W^- production. For the three data points we may use α, G_F and M_W. The measure of M_Z represents a constraint on the internal consistency of the theory reflected by the value of ρ. With the values of the masses at our disposal $\rho - 1$, computed from the previous definition, signals new physics and its importance must be reevaluate well beyond considering ρ as a phenomenological parameter to enter νe or ν hadron interactions.

For these reasons we advocate the reinstatement of the original interpretation as given in [1-3] instead of considering ρ as an additional parameter to fit the data.

The $\rho-$ parameter is given by the bare parameters of the theory which are fixed in terms of experimental quantities and in a near future we will be in a position where to test the region $q^2 \sim M_Z^2, M_W^2$. This is not the same as defining a $\rho(q^2)$ with $\rho(0)$ fixed by low energy data. First consider $q^2 = 0$. If we only include fermions into the radiative corrections then $S_{ZZ}(0)$ is actually equivalent to $S_{33}(0)$, namely we may put s_θ^2 to zero in the corresponding self energy. This remains true when we include vector bosons as long as the renormalization is such that for the $Z^0 - \gamma$ transition we have $S_{Z\gamma}(0) = 0$. Now self energies appear in ρ because by considering low energy cross sections the measured M_Z^2 is actually $M_Z^2 - S_{ZZ}(0)$ and M_W^2 is $M_W^2 - S_{WW}(0)$. Therefore

$$\rho = \frac{M_W^2 - S_{WW}(0)}{c_\theta^2 [M_Z^2 - S_{ZZ}(0)]} = 1 - \frac{S_{WW}(0) - c_\theta^2 S_{ZZ}(0)}{M_W^2}$$

$$= 1 - \frac{S_{WW}(0) - S_{33}(0)}{M_W^2}$$

All of this corresponds to a physically well defined quantity. As soon as q^2 increases $c_\theta^2 S_{ZZ}(q^2) \neq S_{33}(q^2)$ and the full expressions must be used. The correct definition requires for instance knowledge of $S_{ZZ}(-M_Z^2)$.

Having outlined our strategy we now proceed in the implementation of a specific renormalization procedure, namely we construct a scheme for fitting the parameters. A major ingredient is given by the one-loop corrected propagators of the theory. Basically what we have to do amounts to the following [5]. Consider for a certain particle all self-energy diagrams that do not contain any further self-energy part in the internal lines. The sum of the contributions will be denoted by $S(\Delta_F, p^2)$, Δ_F being bare propagators (for simplicity we consider scalar particles). Next we define functions $\Delta_F^{(n)}$

$$i \Delta_F^{(n)}(p^2) = i \Delta_F(p^2) \left\{ 1 - i \Delta_F(p^2) S \left[\Delta_F^{(n-1)}, p^2 \right] \right\}^{-1}$$

Here $\Delta_F^{(0)} = \Delta_F$. Assuming that the limit exists we have for the complete propagator

$$\overline{\Delta}_F(p^2) = \lim_{n \to \infty} \Delta_F^{(n)}(p^2)$$

In actual calculations the propagators must be calculated with a certain accuracy in the coupling constant. For our considerations the recipe is simple. We consider diagrams with one closed loop (no self-energy loops) and bare propagators, and tree diagrams with dressed propagators where S is computed by considering one loop self-energy diagrams. Internal consistency requires that we neglect the two loop reducible diagrams, because two loop renormalization should be done according to the previous definition.

Before considering dressed propagators however, we discuss some technical details connected with renormalization. The way we perform renormalization avoids the explicit introduction of renormalization constants. Three quantities are computed, electric charge, Fermi coupling constant and position of the Z^0 peak. By comparison g, s_θ and M are fixed.

$$g = f_g\left(\alpha, G_F, M_Z\right)$$
$$s_\theta = f_s\left(\alpha, G_F, M_Z\right)$$
$$M = f_M\left(\alpha, G_F, M_Z\right)$$

These quantities are by themselves infinite and scale dependent, but infinities and the arbitrary mass scale cancel out of all physical results, including ρ as it should be. Consider the electric charge defined to be the residue of the pole at zero momentum transfer in $e\mu$ scattering. In a one loop renormalization what we get is a replacement of the type [3]

$$4\pi\alpha = g^2\, s_\theta^2\,(1 + \delta\alpha)$$

$\delta\alpha$ denotes the contributions of wave-function factors, two vertex diagrams, γ propagator insertions and $Z^0 - \gamma$ transition. It is a well known fact that purely e.m. contributions, exclusive of photon propagator diagrams, cancel. Vector bosons however modify this situation. First we compute infinities (and finite parts as well) containing m_μ and m_e. They cancel as expected and therefore we only consider the massless limit. Consider

$$\delta\alpha = \delta^W + \delta^V + \delta^\gamma + \delta^{Z\gamma}$$

δ^W is given in ref. [3] in terms of B form factors [6]. The two vertex diagrams give δ^V which can be cast into the form

$$\delta^V = \frac{g^2}{32\,\pi^2} \sum_{f=e,\mu} \left[C_f(0) - \frac{v_\theta + 1}{4\,c_\theta^2}\, C_{ff}(0) \right]$$

where $v_\theta = 4\, s_\theta^2 - 1$ and using the C form factors [6]

$$C_{ff}(s) = -2\,C_{24} + (C_{11} + C_{23})\,s + 1$$
$$C_f(s) = 6\,C_{24} - (C_0 + C_{11} + C_{23})\,s - 1$$

The non e.m. contributions add up to a non zero result

$$\delta^W + \delta^V = \frac{g^2}{16\,\pi^2} \left(\Delta - \ln M^2\right)$$

where $\Delta = -2/(n-4) + \gamma - \ln \pi$. Also vector bosons give rise to a novel feature

$$\delta^{Z\gamma} = -v_\theta\, \frac{c_\theta}{s_\theta}\, \frac{S_{Z\gamma}(0)}{M^2} \neq 0$$

Indeed the $Z^0 - \gamma$ transition at zero momentum is

$$S_{Z\gamma}(0) = -\frac{g^2}{8\pi^2}\frac{s_\theta}{c_\theta}M^2\left(\Delta - \ln M^2\right)$$

Being $\delta^W + \delta^V + \delta^{Z\gamma}$ different from zero we seem to loose a very appealing aspect of QED, namely the possibility of expressing electric charge renormalization via the summation of photon self- energy diagrams alone.

$$4\pi\alpha = \frac{g^2 s_\theta^2}{1 - \frac{g^2 s_\theta^2}{16\pi^2}\Pi_{\gamma\gamma}(0)}, \qquad S_{\gamma\gamma} = \frac{g^2 s_\theta^2}{16\pi^2}p^2\,\Pi_{\gamma\gamma}$$

The same problem has been considered and solved in a different approach to radiative corrections by the authors of ref. [4]. Here we supply our solution. See however also the approach of ref [7]. Start with a triplet of vector bosons B_μ^i and a singlet B_μ^0. The relevant terms in the interaction Lagrangian are

$$\mathcal{L} = \frac{i}{4}\,(\bar\nu,\bar e)\,\gamma^\mu\begin{pmatrix} g\,B_\mu^3 + g''\,B_\mu^0 & \sqrt{2}\,g\,W_\mu^+ \\ \sqrt{2}\,g\,W_\mu^- & -g\,B_\mu^3 + g''\,B_\mu^0 \end{pmatrix}\left(1+\gamma^5\right)\begin{pmatrix}\nu\\e\end{pmatrix}$$
$$+ \frac{i}{4}g'''\,B_\mu^0\,\bar e\gamma^\mu\left(1-\gamma^5\right)e - \frac{1}{2}g^2F^2\,W_\mu^+W_\mu^- - \frac{1}{4}F^2\left(g\,B_\mu^3 + g'\,B_\mu^0\right)^2$$

where for simplicity we considered only one fermion doublet. Moreover F is the Higgs vacuum expectation value. Before diagonalizing the mass matrix we introduce a new coupling constant $\bar g$, related to g by

$$\bar g = g\left(1 + g^2\,\Gamma\right)$$

Where Γ is for the moment a free coefficient. Next we define fields A_μ and Z_μ^0 by

$$Z_\mu^0 = \frac{\bar g\,B_\mu^3 + g'\,B_\mu^0}{\left(\bar g^2 + g'^2\right)^{1/2}}, \qquad A_\mu = \frac{-g'\,B_\mu^3 + \bar g\,B_\mu^0}{\left(\bar g^2 + g'^2\right)^{1/2}}$$

The final result follows by using

$$g' = g'' = -\frac{s_\theta}{c_\theta}\,\bar g, \qquad g''' = g' + g'', \qquad F = \sqrt{2}\frac{M}{\bar g}$$

For the quadratic terms we get

$$\mathcal{L}_2 = -M^2\,W_\mu^+W_\mu^- - \frac{1}{2}\frac{M^2}{c_\theta^2}\,Z_\mu^0 Z_\mu^0 + \bar g^2 M^2\Gamma\left(Z_\mu^0 Z_\mu^0 - \frac{s_\theta}{c_\theta}Z_\mu^0 A_\mu + 2\,W_\mu^+W_\mu^-\right) + O\left(\bar g^4\right)$$

Charged and neutral currents receive extra contributions

$$\mathcal{L}_{CC}^{extra} = -\frac{i\bar g^3}{2\sqrt{2}}\Gamma\left[W_\mu^+\,\bar\nu\gamma^\mu\left(1+\gamma^5\right)e + W_\mu^-\,\bar e\gamma^\mu\left(1+\gamma^5\right)\nu\right]$$
$$\mathcal{L}_{NC}^{extra} = \frac{i}{4}\bar g^3\Gamma\left(s_\theta A_\mu + c_\theta Z_\mu^0\right)\left[\bar e\gamma^\mu\left(1+\gamma^5\right)e - \bar\nu\gamma^\mu\left(1+\gamma^5\right)\nu\right]$$

At this point we fix Γ by requiring

$$\Gamma = \frac{1}{8\pi^2}\,B_0(0,M,M)$$

where B_0 is the scalar two-point function [6]. With this choice the vector boson self-energies are modified. Dropping from now on the bar we get

$$S_{\gamma\gamma} = \frac{g^2\,s_\theta^2}{16\,\pi^2}\,p^2\,\Pi_{\gamma\gamma}$$

$$S_{ZZ} = \frac{g^2}{16\,\pi^2\,c_\theta^2}\left[\left(S_{ZZ}^0 - 32\,\pi^2\,M^2\,\Gamma\right) - 2\,s_\theta^2\left(S_{Z\gamma}^0 + 16\,\pi^2\,M^2\,\Gamma\right) + s_\theta^4\,p^2\,\Pi_{\gamma\gamma}\right]$$

$$S_{Z\gamma} = \frac{g^2}{16\,\pi^2}\,\frac{s_\theta}{c_\theta}\left[\left(S_{Z\gamma}^0 + 16\,\pi^2\,M^2\,\Gamma\right) - s_\theta^2\,p^2\,\Pi_{\gamma\gamma}\right]$$

$$S_{WW} = \frac{g^2}{16\,\pi^2}\left[\left(\Pi_{WW}^0 + s_\theta^2\,\Pi_{WW}^1\right)p^2 + \left(\Sigma_{WW}^0 - 32\,\pi^2\,M^2\,\Gamma\right) + s_\theta^2\Sigma_{WW}^1\right]$$

The explicit expressions will be given in appendix. The important fact to be noted here is the following

$$S_{Z\gamma}(0) = 0$$

which makes $\delta^{Z\gamma} = 0$ in the corrections to the electric charge. Moreover the extra $ee\gamma\,(\mu\mu\gamma)$ vertex gives rise to a contribution

$$\delta^{extra} = -\frac{g^2}{16\,\pi^2}\left(\Delta - \ln M^2\right)$$

Thus

$$\delta^{Z\gamma} = 0, \qquad \delta^W + \delta^V + \delta^{extra} = 0$$

and photon propagator diagrams alone contribute to the shift in e^2. At the same time we consider the eeZ^0 coupling. The infinite parts coming from wave-function factors and vertices add up to

$$\mathcal{L}_{eeZ} = -(2\,\pi)^4\,i\,\frac{i\,g^3}{32\,\pi^2}\,c_\theta\,\Delta\,\bar{v}\gamma^\mu\left(1 + \gamma^5\right)u$$

which is exactly cancelled by the extra eeZ^0 vertex proportional to Γ. By means of the same mechanism all the required cancellations take place. There is no parity violation in low energy for the e.m. current, no ultraviolet divergence contained in the $\bar{\nu}\nu\gamma$ coupling and moreover the neutrino charge remains zero.

The second experimental quantity we need is given by G_F or rather by the muon lifetime

$$\frac{1}{\tau_\mu} = \frac{m_\mu^5}{192\,\pi^3}\,\frac{g^4}{32\,M^4}\left(1 + \delta_\mu^{e.m.} + \delta_\mu^{weak}\right)$$

where

$$\delta_\mu^{e.m.} = \frac{\alpha}{2\,\pi}\left(\frac{25}{4} - \pi^2\right)$$

If we compute δ_μ^{weak} from wave-function factors, four vertex diagrams and five box diagrams [3], we obtain

$$\delta^W + \delta^V + \delta^B = \frac{g^2}{\pi^2}\left[\frac{3}{4} + \frac{1}{2}\left(\Delta - \ln M^2\right) + \frac{1}{4}\,\frac{7 - 4\,s_\theta^2}{4\,s_\theta^2}\,\ln c_\theta^2\right]$$

The extra contribution from $\mathcal{L}_{CC}^{extra}$ gives

$$\delta_\mu^{extra} = -4\,g^2\Gamma = -\frac{g^2}{2\,\pi^2}\left(\Delta - \ln M^2\right)$$

Therefore we have an infrared and ultraviolet finite correction beside W self-energy diagrams

$$\delta_\mu = \frac{g^2}{4\pi^2}\left[3 + \frac{7 - 4s_\theta^2}{4\,s_\theta^2}\ln c_\theta^2\right] + \delta_\mu^{e.m.}$$

We are now in a position to write down the one loop dressed propagators.

$$\Delta_{\mu\nu}^{\gamma\gamma} = \frac{1}{(2\pi)^4\,i}\frac{\delta_{\mu\nu}}{p^2}\left[1 - \frac{g^2 s_\theta^2}{16\pi^2}\Pi_{\gamma\gamma}\right]^{-1}$$

$$\Delta_{\mu\nu}^{ZZ} = \frac{1}{(2\pi)^4\,i}\,\delta_{\mu\nu}\left\{p^2 + \frac{M^2}{c_\theta^2} - \frac{g^2}{16\pi^2 c_\theta^2}\left[\overline{S}_{ZZ}^0 - 2s_\theta^2\,\overline{S}_{Z\gamma}^0 + s_\theta^4\,p^2\,\Pi_{\gamma\gamma}\right]\right\}^{-1}$$

$$\Delta_{\mu\nu}^{Z\gamma} = \frac{1}{(2\pi)^4\,i}\,\delta_{\mu\nu}\frac{g^2}{16\pi^2}\frac{s_\theta}{c_\theta}\left(\frac{\overline{S}_{Z\gamma}^0}{p^2} - s_\theta^2\,\Pi_{\gamma\gamma}\right)$$

$$\times\left\{p^2 + \frac{M^2}{c_\theta^2} - \frac{g^2}{16\pi^2 c_\theta^2}\left[\overline{S}_{ZZ}^0 - 2s_\theta^2\,\overline{S}_{Z\gamma}^0 + s_\theta^4\,p^2\,\Pi_{\gamma\gamma}\right]\right\}^{-1}$$

where with $\overline{S}$ we denote the corresponding S opportunely subtracted. Before constructing ρ we discuss in some details how renormalization actually works for $e^+e^- \to \mu^+\mu^-$. With the outlined procedure we only need to worry about self-energy diagrams. In fitting the parameters the combination $gs_\theta = e$ is easily fixed in terms of α from $e\mu$ scattering

$$\frac{1}{g^2 s_\theta^2} = \frac{1}{4\pi\alpha} + \frac{1}{16\pi^2}\Pi_{\gamma\gamma}(0)$$

Thus photon exchange in $e^+e^- \to \mu^+\mu^-$ gives an amplitude proportional to

$$-(2\pi)^4 i\,\frac{4\pi\alpha}{p^2}\left\{1 - \frac{\alpha}{4\pi}\left[\Pi_{\gamma\gamma}(p^2) - \Pi_{\gamma\gamma}(0)\right]\right\}^{-1}$$

where the combination of self-energies is ultraviolet finite. For the Z^0 propagator the situation is slightly more complicated. Given

$$\Delta^{-1} = p^2 + \frac{M^2}{c_\theta^2} - S_{ZZ}(p^2)$$

we first obtain

$$\Delta = \frac{c_\theta^2}{g^2}\left[\frac{c_\theta^2}{g^2}p^2 + \frac{M^2}{g^2} - \frac{c_\theta^2}{g^2}S_{ZZ}(p^2)\right]^{-1}$$

Using the second fitting equation, obtained from $\mu-$ decay, the combination M^2/g^2 is fixed

$$8\frac{M^2}{g^2} = \frac{1 + \delta_G}{G_F} + \frac{1}{2\pi^2}\overline{\Sigma}_{WW}(0)$$

where $\delta_G = 1/2\,\delta_\mu$. Therefore we get

$$\Delta = 8\,G_F \frac{c_\theta^2}{g^2} \left\{ 8\,G_F \frac{c_\theta^2}{g^2} p^2 + 1 + \delta_G + \frac{G_F}{2\pi^2} \overline{\Sigma}_{WW}(0) \right.$$

$$\left. - \frac{G_F}{2\pi^2} \left[\overline{S}_{ZZ}^0(p^2) - 2 s_\theta^2 \overline{S}_{Z\gamma}^0(p^2) + s_\theta^4 p^2 \Pi_{\gamma\gamma}(p^2) \right] \right\}^{-1}$$

The third fitting equation, namely the position of the Z^0 pole, gives

$$\frac{M^2}{c_\theta^2} = M_Z^2 + \frac{g^2}{16\,\pi^2 c_\theta^2} \, Re \left[\overline{S}_{ZZ}^0 \left(-M_Z^2\right) - 2 s_\theta^2 \overline{S}_{Z\gamma}^0 \left(-M_Z^2\right) - s_\theta^4 M_Z^2 \Pi_{\gamma\gamma} \left(-M_Z^2\right) \right]$$

$$\equiv M_Z^2 + \frac{g^2}{16\,\pi^2 c_\theta^2} \, Re\, f \left(-M_Z^2\right)$$

The previous equation defines $f(p^2)$. By combining the two equations we easily obtain a solution for c_θ^2/g^2

$$8\,G_F M_Z^2 \frac{c_\theta^2}{g^2} = 1 + \delta_G + \frac{G_F}{2\,\pi^2} \, Re \left[\overline{\Sigma}_{WW}(0) - f\left(-M_Z^2\right) \right]$$

Finally Δ may be cast into the following form

$$\Delta = 8\,G_F M_Z^2 \frac{c_\theta^2}{g^2} \left\{ \left[1 + \delta_G + \frac{G_F}{2\,\pi^2} \overline{\Sigma}_{WW}(0) \right] \left(p^2 + M_Z^2 \right) \right.$$

$$\left. - \frac{G_F}{2\,\pi^2} Re \left[M_Z^2 f(p^2) + p^2 f\left(-M_Z^2\right) \right] - i \frac{G_F M_Z^2}{2\,\pi^2} Im\, f(p^2) \right\}^{-1}$$

$$= 8\,G_F M_Z^2 \frac{c_\theta^2}{g^2} \, P_Z^{-1}$$

The quantity in braket is ultraviolet finite. Inside the function f we used the zero'th order value for s_θ^2, which follows from the fitting equations

$$\overline{s}_\theta^2 = \tfrac{1}{2} \left[1 - \sqrt{1 - \frac{2\,\pi\alpha}{G_F M_Z^2}} \right]$$

Up to first order the solution for s_θ^2 is found to be

$$s_\theta^2 = \overline{s}_\theta^2 \left(1 + \frac{\overline{c}_\theta^2}{\overline{c}_\theta^2 - \overline{s}_\theta^2} \kappa \right)$$

$$\kappa = \delta_G - \frac{\alpha}{4\,\pi} \Pi_{\gamma\gamma}(0) + \frac{G_F}{2\,\pi^2} \left[\overline{\Sigma}_{WW}(0) - Re\, f\left(-M_Z^2\right) \right]$$

Once more the corrections to s_θ^2 are not finite. However the $Z^0 - Z^0$ propagator diagram together with the two $Z^0 - \gamma$ transition diagrams give a finite answer

$$- \left(2\,\pi^2\right) i \frac{G_F M_Z^2}{2\,P_Z} \left\{ \gamma^\mu \left(\overline{v}_\theta - \gamma^5\right) \otimes \gamma^\mu \left(\overline{v}_\theta - \gamma^5\right) - \left[4 \frac{\overline{s}_\theta^2 \overline{c}_\theta^2}{\overline{c}_\theta^2 - \overline{s}_\theta^2} \kappa - \frac{\alpha}{\pi} \left(\frac{\overline{S}_{Z\gamma}^0}{p^2} - \overline{s}_\theta^2 \Pi_{\gamma\gamma} \right) \right] \right.$$

$$\left. \times \left[\gamma^\mu \gamma^5 \otimes \gamma^\mu + \gamma^\mu \otimes \gamma^\mu \gamma^5 - 2\,\overline{v}_\theta \gamma^\mu \otimes \gamma^\mu \right] \right\}$$

In conclusion we turn to the $\rho-$parameter. As discussed at length in the first part we assume precise knowledge of four quantities, α, G_F, M_W and M_Z. Suppose to fix the parameters with

α, G_F and M_W and to consider the measure of M_Z as a constraint on the internal consistency of the theory. Solving for g^2, s_θ^2 and M^2 we get

$$g^2 = \frac{8\,G_F\,M_W^2}{1 + \frac{\alpha}{4\,\pi}\,Re\,\delta_g}, \qquad s_\theta^2 = \frac{\bar{s}_\theta^2}{1 + \frac{\alpha}{4\,\pi}\,Re\,\delta_s}$$

where

$$\bar{s}_\theta^2 = \frac{\pi\alpha}{2\,G_F M_W^2}$$

$$S_{WW}^i\left(-M_W^2\right) = -M_W^2\,\Pi_{WW}^i\left(-M_W^2\right) + \Sigma_{WW}^i\left(-M_W^2\right)$$

and

$$\delta_g = \frac{4\,\pi}{\alpha}\,\delta_G - \frac{1}{M_W^2}\left\{\frac{1}{\bar{s}_\theta^2}\left[\bar{S}_{WW}^0\left(-M_W^2\right) - \bar{\Sigma}_{WW}^0(0)\right] + \bar{S}_{WW}^1\left(-M_W^2\right) - \bar{\Sigma}_{WW}^1(0)\right\}$$

$$\delta_s = \Pi_{\gamma\gamma}(0) - \delta_g$$

Also

$$M^2 = \frac{M_W^2}{1 + \frac{\alpha}{4\,\pi}\,Re\,\delta_M}$$

$$\delta_M = -\frac{1}{M_W^2}\left[\frac{1}{\bar{s}_\theta^2}\,\bar{S}_{WW}^0\left(-M_W^2\right) + \bar{S}_{WW}^1\left(-M_W^2\right)\right]$$

Now let M_0 be the Z^0 bare mass, we get

$$c_\theta^2\,M_0^2 = \frac{\bar{c}_\theta^2\,M_Z^2}{1 + \frac{\alpha}{4\,\pi}\,Re\,\delta_M'}$$

$$\delta_M' = \frac{1}{\bar{s}_\theta^2\,\bar{c}_\theta^2}\left\{-\frac{1}{M_Z^2}\left[\bar{S}_{ZZ}^0\left(-M_Z^2\right) - 2\,\bar{s}_\theta^2\,\bar{S}_{Z\gamma}^0\left(-M_Z^2\right)\right] - \frac{\bar{s}_\theta^2}{M_W^2}\left[\bar{S}_{WW}^0\left(-M_W^2\right) - \bar{\Sigma}_{WW}^0(0)\right]\right.$$

$$\left. + \bar{s}_\theta^4\left[\frac{4\,\pi}{\alpha}\,\delta_G + \Pi_{\gamma\gamma}\left(-M_Z^2\right) - \Pi_{\gamma\gamma}(0) - \frac{\bar{S}_{WW}^1\left(-M_W^2\right) - \bar{\Sigma}_{WW}^1(0)}{M_W^2}\right]\right\}$$

From the previous results it follows

$$\rho = \frac{M_W^2}{M_Z^2\,\bar{c}_\theta^2}\left[1 + \frac{\alpha}{4\,\pi}Re\left(\delta_M' - \delta_M\right)\right]$$

$$= \frac{M_W^2}{M_Z^2\,\bar{c}_\theta^2}\left[1 + \delta\rho\right]$$

In $\delta_M' - \delta_M$ we use $M_0 = M_W/\bar{c}_\theta$. Although the various quantities appearing in δ_M' and δ_M are by themselves infinite, the combination occurring for the $\rho-$ parameter is finite, as it should be. In conclusion we consider the dependence of the $\rho-$ parameter on a new as yet undiscovered quark doublet (or top-bottom as well). From low energy data we obtain the well known result

$$\rho_{ud} = 1 + \tfrac{3}{8}\frac{G_F}{\pi^2}\left[m_u^2 + m_d^2 + 2\frac{m_u^2 m_d^2}{m_u^2 - m_d^2}\ln\frac{m_u^2}{m_d^2}\right]$$

If masses are input parameters then

$$\frac{\pi^2\, \bar{c}_\theta^2}{G_F}\,\delta\rho_{ud} = \left[\tfrac{3}{4}\left(m_u^2 + m_d^2\right)\bar{c}_\theta^2 - M_W^2\left(\tfrac{1}{2} - \bar{s}_\theta^2 + \tfrac{10}{9}\bar{s}_\theta^4\right)\right]\ln\bar{c}_\theta^2 + \tfrac{3}{4}\left(m_u^2 + m_d^2\right)\bar{s}_\theta^2\ln\frac{m_d^2}{M_W^2}$$

$$+ \left[M_W^2\left(\tfrac{3}{4} - 2\bar{s}_\theta^2 + \tfrac{8}{3}\bar{s}_\theta^4\right) - \tfrac{3}{4}m_u^2\bar{c}_\theta^2\right]\sum_i Re\, G_1\left(x_{z_u}^i\right)$$

$$+ \left[M_W^2\left(\tfrac{3}{4} - \bar{s}_\theta^2 + \tfrac{2}{3}\bar{s}_\theta^4\right) - \tfrac{3}{4}m_d^2\bar{c}_\theta^2\right]\sum_i Re\, G_1\left(x_{z_d}^i\right)$$

$$+ 2\left[\tfrac{3}{4}\left(m_d^2 - m_u^2\right) - \tfrac{3}{2}M_W^2\right]\left(\bar{c}_\theta^2 - \bar{s}_\theta^2\right)\sum_i Re\, G_2\left(x_w^i\right)$$

$$- 3\,M_W^2\left(\tfrac{1}{2} - \tfrac{4}{3}\bar{s}_\theta^2 + \tfrac{16}{9}\bar{s}_\theta^4\right)\sum_i Re\, G_3\left(x_{z_u}^i\right)$$

$$- 3\,M_W^2\left(\tfrac{1}{2} - \tfrac{2}{3}\bar{s}_\theta^2 + \tfrac{4}{9}\bar{s}_\theta^4\right)\sum_i Re\, G_3\left(x_{z_d}^i\right)$$

$$+ 3\,M_W^2\left(\bar{c}_\theta^2 - \bar{s}_\theta^2\right)\sum_i Re\, G_3\left(x_w^i\right) + 3\,m_u^2\left(\tfrac{1}{2} - \bar{s}_\theta^2\right)\sum_i Re\, G_1\left(x_w^i\right)$$

$$+ \tfrac{2}{9}M_W^2\,\bar{s}_\theta^4\left(\ln\frac{m_u^2}{m_d^2} - 5\ln\frac{m_u^2}{M_W^2}\right) + \tfrac{3}{2}m_u^2\bar{s}_\theta^2\,F_1(x) + \tfrac{3}{4}\left(m_d^2 - m_u^2\right)\bar{s}_\theta^2\,F_2(x)$$

Where we have defined the following quantities

$$x_W^{1,2} \quad \text{are the roots of} \quad M_W^2 x^2 + \left(m_u^2 - m_d^2 - M_W^2\right)x + m_u^2 - i\epsilon = 0$$
$$x_{Z u(d)}^{1,2} \quad \text{are the roots of} \quad M_0^2 x^2 - M_0^2 x + m_{u(d)}^2 - i\epsilon = 0$$

and

$$x = \frac{m_u^2}{m_u^2 - m_d^2}$$

The previous result follows when we express the B form factors in terms of F and G functions [6]

$$B_0\left(p^2, m_1, m_2\right) = i\,\pi^2\left[\Delta - \ln(-p^2 - i\epsilon) - \sum_i G_1(x_i)\right]$$

$$B_1\left(p^2, m_1, m_2\right) = -\tfrac{1}{2}i\,\pi^2\left[\Delta - \ln(-p^2 - i\epsilon) - 2\sum_i G_2(x_i)\right]$$

$$B_{21}\left(p^2, m_1, m_2\right) = \tfrac{1}{3}i\,\pi^2\left[\Delta - \ln(-p^2 - i\epsilon) - 3\sum_i G_1(x_i)\right]$$

Δ is the ultraviolet factor, namely $\Delta = -2/(n-4) + \gamma - \ln \pi$, x_i are the roots of $-p^2 x^2 + (p^2 + m_2^2 - m_1^2)x + m_1^2 - i\epsilon = 0$ and

$$G_n(z) = \int_0^1 dx \, x^{n-1} \ln(x-z), \qquad F_n(z) = -\int_0^1 dx \frac{x^n}{x-z}$$

For $m_u \gg M_W, m_d$ we use the asymptotic behavior of the $F, G-$functions [2]

$$G_n\left(x_W^1\right) \sim \frac{1}{n} \ln\left(-\frac{m_u^2}{M_W^2}\right), \quad G_1\left(x_W^2\right) \sim -1 + i\pi, \quad G_2\left(x_W^2\right) \sim -\tfrac{3}{4} + \tfrac{i}{2}\pi$$

$$F_1(x) \sim \ln\left(\frac{m_u^2}{m_d^2}\right) - 1, \qquad F_2(x) \sim \ln\left(\frac{m_u^2}{m_d^2}\right) - \tfrac{3}{2}$$

$$G_1\left(x_{Zu}^{1,2}\right) \sim \tfrac{1}{2}\left(\ln\frac{m_u^2}{M_W^2} + \ln \bar{c}_\theta^2\right) \mp \tfrac{i}{2}\pi$$

As a consequence we find a cancellation of the logarithms in the asymptotic behavior of $\delta\rho_{ud}$ and the expected result emerges

$$\delta\rho_u \sim -\tfrac{3}{8}\frac{G_F m_u^2}{\pi^2}$$

For a very heavy quark the same result is obtained, independently from the input parameters, i.e. low energy data or vector boson masses. However for $m_{top} \approx 100 - 200\,\mathrm{GeV}$ we have to use the full expression for ρ_{ud} and not its asymptotic form. For degenerate fermions ($m_u = m_d = m$) we easily find that terms proportional to $G_F m^2 \ln(m^2/M_W^2)$ and to $G_F m^2$ cancel out in the final answer when $m \gg M_W$, as expected. Actually we get $\delta\rho_{ud} = O(|\,m_u^2 - m_d^2\,|)$ for $m_u \approx m_d \gg M_W$. For light quarks, $m_u, m_d \ll M_W$ we only need the behavior of $G_n(x)$ near 0 and 1

$$x \to 1, \quad G_1(x) \sim -1 + i\pi, \quad G_2(x) \sim -\tfrac{3}{4} + \tfrac{i}{2}\pi$$

$$x \to 0, \quad G_n(x) \sim -\frac{1}{n^2}$$

This gives

$$\delta\rho_{ud} = \frac{G_F M_W^2}{\pi^2 \bar{c}_\theta^2}\left[\left(-\tfrac{8}{9}\ln\frac{m_u^2}{M_W^2} - \tfrac{2}{9}\ln\frac{m_d^2}{M_W^2} - \tfrac{50}{27}\right)\bar{s}_\theta^4 + \left(-\tfrac{1}{2} + \bar{s}_\theta^2 - \tfrac{10}{9}\bar{s}_\theta^4\right)\ln \bar{c}_\theta^2\right]$$

For completeness we give the explicit expressions for the functions $\Pi_{\gamma\gamma}, ..., \Sigma_{WW}^1$ in terms of $B-$form factors.

$$\Pi_{\gamma\gamma} = \tfrac{2}{3} - 12\,B_{21}(M,M) + 7\,B_0(M,M)$$
$$+ 4\sum_f \left[B_f(m_e,m_e) + \tfrac{4}{3}\,B_f(m_u,m_u) + \tfrac{1}{3}\,B_f(m_d,m_d) \right]$$

$$S_{Z\gamma}^0 = p^2 \left\{ \tfrac{2}{3} - 10\,B_{21}(M,M) + \tfrac{13}{2}\,B_0(M,M) \right.$$
$$\left. + \sum_f \left[B_f(m_e,m_2) + 2\,B_f(m_u,m_u) + B_f(m_d,m_d) \right] \right\} - 2\,M^2 B_0(M,M)$$

$$S_{ZZ}^0 = p^2\,\Pi_{ZZ}^0 + \Sigma_{ZZ}^0$$

$$\Pi_{ZZ}^0 = \tfrac{2}{3} - 9\,B_{21}(M,M) + \tfrac{25}{4}\,B_0(M,M) - B_{21}(M_0,m_H)$$
$$- B_1(M_0,m_H) - \tfrac{1}{4}\,B_0(M_0,m_H)$$
$$+ \tfrac{1}{2}\sum_f \left[B_f(m_e,m_e) + B_f(m_\nu,m_\nu) + 3\,B_f(m_u,m_u) + 3\,B_f(m_d,m_d) \right]$$

$$\Sigma_{ZZ}^0 = -2\,M^2 B_0(M,M) + \tfrac{1}{2}\,M_0^2 B_1(M_0,m_H) + \tfrac{5}{4}\,M_0^2 B_0(M_0,m_H)$$
$$- \tfrac{1}{2}\,m_H^2 B_1(M_0,m_H) - \tfrac{1}{4}\,m_H^2 B_0(M_0,m_H)$$
$$- \tfrac{1}{2}\sum_f \left[m_\nu^2 B_0(m_\nu,m_\nu) + m_e^2 B_0(m_e,m_e) \right.$$
$$\left. + 3\,m_u^2 B_0(m_u,m_u) + 3\,m_d^2 B_0(m_d,m_d) \right]$$

$$\Sigma_{WW}^0 = \tfrac{9}{2}(M_0^2 - M^2)B_1(M_0,M) + \tfrac{1}{4}(13M_0^2 - 21M^2)B_0(M_0,M)$$
$$+ \tfrac{1}{2}(M^2 - m_H^2)B_1(M,m_H) + \tfrac{1}{4}(5M^2 - m_H^2)B_0(M,m_H)$$
$$+ \sum_f \left[(m_e^2 - m_\nu^2)B_1(m_\nu,m_e) - m_\nu^2 B_0(m_\nu,m_e) + 3\,(m_d^2 - m_u^2)B_1(m_u,m_d) \right.$$
$$\left. - 3\,m_u^2 B_0(m_u,m_d) \right]$$

$$\Sigma_{WW}^1 = 2(M^2 - M_0^2)\,[2B_1(M_0,M) + B_0(M_0,M)] - 2\,M^2\,[2B_1(\lambda,M) + B_0(\lambda,M)]$$

$$\Pi_{WW}^0 = \tfrac{2}{3} - 9B_{21}(M_0,M) - 9B_1(M_0,M) + \tfrac{7}{4}\,B_0(M_0,M)$$
$$- B_{21}(M,m_H) - B_1(M,m_H) - \tfrac{1}{4}\,B_0(M,m_H)$$
$$+ 2\sum_f \left[B_{21}(m_\nu,m_e) + B_1(m_\nu,m_e) + 3\,B_{21}(m_u,m_d) + 3\,B_1(m_u,m_d) \right]$$

$$\Pi_{WW}^1 = 8\,B_{21}(M_0,M) - 2\,B_0(M_0,M) + 8\,B_1(M_0,M) - 8\,B_{21}(\lambda,M)$$
$$- 8\,B_1(\lambda,M) + 2\,B_0(\lambda,M)$$

where $B_f = 2\,B_{21} - B_0$ and λ is the photon mass. Ultraviolet finiteness for physical quantities can be verified very easily by using the following infinite parts

190

$$P.P. \left(\overline{S}_{ZZ}^0 \right) = \left(r_z p^2 + \mu^2 \right) \Delta \qquad P.P. \left(\overline{S}_{Z\gamma}^0 \right) = r_z p^2 \, \Delta$$

$$P.P. \left(\overline{S}_{WW}^0 \right) = \left(r_z p^2 + \mu^2 \right) \Delta \qquad P.P. \left(\Pi_{\gamma\gamma} \right) = r_\gamma \, \Delta$$

where $P.P.$ stands for pole part and

$$r_\gamma = 3 - \tfrac{32}{9} N_f, \qquad r_z = \tfrac{16}{9} - \tfrac{4}{3} N_f$$

$$\mu^2 = M_0^2 - 6 M^2 - \tfrac{1}{2} \sum_f \left(m_\nu^2 + m_e^2 + 3 m_u^2 + 3 m_d^2 \right)$$

References

[1] M. Veltman, Vector Boson and Higgs System, TASI Summer School 1984;
    F. Antonelli, M. Consoli and G. Corbò, Phys. Lett. 91B(1980) 90;
    M. Veltman, Phys. Lett. 91B(1980)95.
[2] D. Ross and M. Veltman, Nucl. Phys. B95(1975)135;
    M. Veltman, Nucl. Phys. B123(1977)89.
[3] M. Green and M. Veltman, Nucl. Phys. B169(1980)137.
[4] D.C. Kennedy and B.W. Lynn, SLAC-PUB-4039(1988)..
[5] M. Veltman, Physica 29(1963)186.
[6] G. Passarino and M. Veltman, Nucl. Phys. B160(1979)151.
[7] W. Hollik, Desy preprint, Desy 88-188, December 1988.

Radiative Corrections for e^+e^- Collisions   Editor: J.H. Kühn
© Springer-Verlag Berlin, Heidelberg 1989

Radiative Corrections to Boson Pair Production in e^+e^- Annihilation

M. Böhm

Physikalisches Institut der Universität Würzburg
Am Hubland, D-8700 Würzburg, Fed. Rep. of Germany

1. Introduction

The expected accuracy of the experiments of *SLC* and especially *LEP I* is a great challenge for theorists to produce adequate predictions for e^+e^- annihilation into fermion pairs at the Z resonance. Indeed their response for the physics of the Z peak is remarkable: Complete 1-loop calculations and summations of leading higher order terms are available [16]. This opens the possibility for a precise determination of the Z mass - one of the fundamental parameters of the standard model (SM) - and for first accurate tests by measuring the partial and total widths of the Z. Compared to this, physics which is not supported by such a fantastic resonance bonus looks not so exciting, but in any case has its merits. Z physics is directly sensitive only to the more convenient subsector of the SM. Parts like the non-Abelian couplings, the mass generation mechanism can be tested more directly in other reactions. Therefore, W pair production at *LEP II* will be one of the other fields where parameters of the SM — the W mass, the non-Abelian couplings — can directly be investigated. For the desired accuracy 1-loop corrections are required. Therefore, it is important to have complete $O(\alpha)$ corrections to $e^+e^- \rightarrow W^+W^-$ but also for the annihilation into neutral boson pairs: $e^+e^- \rightarrow \gamma\gamma$, γZ, ZZ. The cross sections for these processes (see fig. 1.1) are comparable to the μ-pair QED cross section and therefore can be measured with good accuracy. Especially the $\gamma\gamma$ process is interesting since it is in lowest order still a pure QED reaction. Moreover the radiative corrections to it do not depend on the unknown parameters of the SM like the Higgs mass or the masses of the quarks or the hadronic contribution to the vacuum polarization. Therefore, this process is very clean. Any deviation from the QED (plus its tiny weak correction) results is the sign for new physics.

In this talk we present the results for the lowest order and radiatively corrected cross sections for $e^+e^- \rightarrow$ *pairs of gauge bosons* together with the technical tools necessary for the calculation of the corresponding Feynman diagrams. The results for W-pair production will be discussed in the following talk by F. Jegerlehner [16].

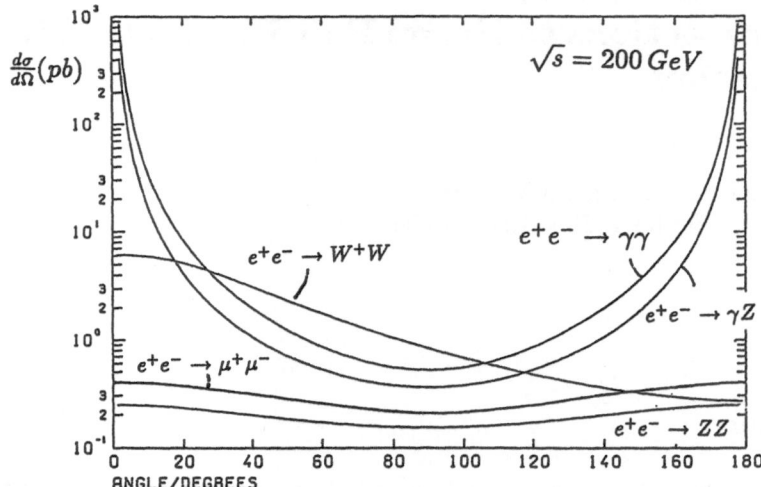

Figure 1.1: The lowest order differential cross section for e^+e^--annihilation into $\mu^+\mu^-$(QED part only), $\gamma\gamma$, γZ, ZZ, W^+W^- at 200 GeV.

2. The Born cross section

We choose our normalization such that the invariant matrix element for the production of gauge boson pairs $(k_1\lambda_1, k_2\lambda_2)$ in e^+e^- annihilation $(p_1\sigma_1, p_2\sigma_2)$

$$\mathcal{M}(p_1\sigma_1, p_2\sigma_2; k_1\lambda_1, k_2\lambda_2) = \langle k_1\lambda_1, k_2\lambda_2 | T | p_1\sigma_1, p_2\sigma_2 \rangle = \bar{v}(p_1)T_{\mu\nu}u(p_2)\varepsilon^\mu(k_1\lambda_1)\varepsilon^\nu(k_2\lambda_2) \tag{2.1}$$

is in the following way related to the differential cross section:

$$\frac{d\sigma}{d\Omega} = \frac{\alpha^2}{4s}\sqrt{\left(1 - \frac{M_1^2 + M_2^2}{s}\right)^2 - 4\frac{M_1^2 M_2^2}{s^2}} \, |\mathcal{M}|^2 \,. \tag{2.2}$$

For $s \gg m_e^2$ chiral symmetry is obeyed leading to the restrictions: $\sigma_1 = -\sigma_2 = \sigma = \pm 1/2$ and allowing us to write for (2.1) $\mathcal{M}(\sigma, \lambda_1\lambda_2; s, t)$. Because of Lorentz invariance the general matrix element can be decomposed into the following (overcomplete) set of standard matrix elements and formfactors $F_i^\sigma(s, t)$:

$$\mathcal{M}(\sigma, \lambda_1, \lambda_2; s, t) = \sum_{i=0}^{9} \mathcal{M}_i^\sigma F_i^\sigma(s, t)$$

with $(\omega_\sigma = (1 + 2\sigma\gamma_5)/2)$

$$\mathcal{M}_0^\sigma = \bar{v}(p_1) \not{\varepsilon}_1 (\not{k}_1 - p_1) \not{\varepsilon}_2 \, \omega_\sigma \, u(p_2)$$

$$\mathcal{M}_1^\sigma = \bar{v}(p_1) \not{k}_1 (\varepsilon_1\varepsilon_2) \, \omega_\sigma \, u(p_2)$$

$$\mathcal{M}_2^\sigma = \bar{v}(p_1) [\not{\varepsilon}_1 (\varepsilon_2 k_1)] \, \omega_\sigma \, u(p_2)$$

$$\mathcal{M}_3^\sigma = \bar{v}(p_1) \, [\,\not{\varepsilon}_2 \, (\varepsilon_1 k_2)\,] \, \omega_\sigma \, u(p_2)$$

$$\mathcal{M}_4^\sigma = \bar{v}(p_1) \, [\,\not{\varepsilon}_1 \, (\varepsilon_2 p_2)] \, \omega_\sigma \, u(p_2)$$

$$\mathcal{M}_5^\sigma = \bar{v}(p_1) \, [\,\not{\varepsilon}_2 \, (\varepsilon_1 p_1)\,] \, \omega_\sigma \, u(p_2)$$

$$\mathcal{M}_6^\sigma = \bar{v}(p_1) \, \not{k}_1 \, \omega_\sigma \, u(p_2) \, (\varepsilon_1 k_2) \, (\varepsilon_2 k_1)$$

$$\mathcal{M}_7^\sigma = \bar{v}(p_1) \, \not{k}_1 \, \omega_\sigma \, u(p_2) \, (\varepsilon_1 p_1) \, (\varepsilon_2 p_2)$$

$$\mathcal{M}_8^\sigma = \bar{v}(p_1) \, \not{k}_1 \, \omega_\sigma \, u(p_2) \, [\, (\varepsilon_1 k_2) \, (\varepsilon_2 p_2)]$$

$$\mathcal{M}_9^\sigma = \bar{v}(p_1) \, \not{k}_1 \, \omega_\sigma \, u(p_2) \, [\, (\varepsilon_1 p_1) \, (\varepsilon_2 k_1)\,] \, . \tag{2.3}$$

where ε_1^μ, ε_2^μ denote the polarization four vectors of the bosons ($k_\mu \varepsilon^\mu(k, \lambda) = 0$; $\varepsilon^\mu(k, \lambda)\varepsilon_\mu(k, \lambda') = -\delta_{\lambda\lambda'}$) and $u(p_1), \bar{v}(p_2)$ are the spinors of the incoming e^+, e^- respectively.

The Born matrix elements contain the couplings of the left- and right-handed electrons to the gauge bosons which we write as:

$$\begin{aligned}
g_\gamma^- &= -1/2, & g_\gamma^+ &= -1/2 \\
g_Z^- &= (-\tfrac{1}{2} + s_w^2)/s_w c_w, & g_Z^+ &= s_w/c_w \\
g_W^- &= 1/\sqrt{2}s_w, & g_W^+ &= 0.
\end{aligned}$$

In the case of $e^+ e^- \rightarrow \gamma\gamma, \gamma Z, ZZ$ we have contributions of t- and u-channel exchange

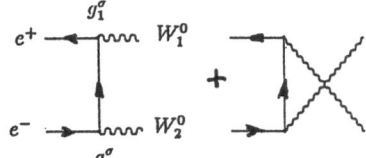

giving:

$$\mathcal{M}_B^\sigma = g_1^\sigma g_2^\sigma \left(\frac{1}{t} \mathcal{M}_0^\sigma + \frac{1}{u} \mathcal{M}_{0,u}^\sigma \right)$$

with

$$\mathcal{M}_{0,u}^\sigma = \mathcal{M}_0^\sigma + 2\mathcal{M}_1^\sigma - 2\mathcal{M}_2^\sigma + 2\mathcal{M}_3^\sigma.$$

W-pairs are produced from t-channel ν_e-exchange and s-channel γ- and Z-exchange

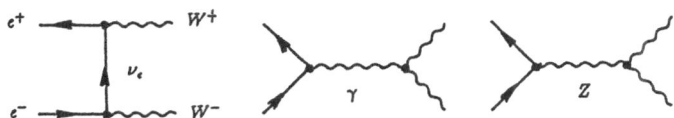

containing the non-Abelian 3-boson couplings and yielding:

$$\mathcal{M}_B(-,\lambda_+,\lambda_-,s,t) = \mathcal{M}_0^- \frac{1}{2s_w^2} \frac{1}{t} - \left[\mathcal{M}_1^- - \mathcal{M}_2^- - \mathcal{M}_3^-\right] \left[\frac{2}{s} - \frac{c_w}{s_w} g_-^- \frac{2}{s - M_Z^2}\right]$$

$$\mathcal{M}_0(+,\lambda_+,\lambda_-,s,t) = -\left[\mathcal{M}_1^+ - \mathcal{M}_2^+ - \mathcal{M}_3^+\right] \left[\frac{2}{s} - \frac{2}{s - M_Z^2}\right].$$

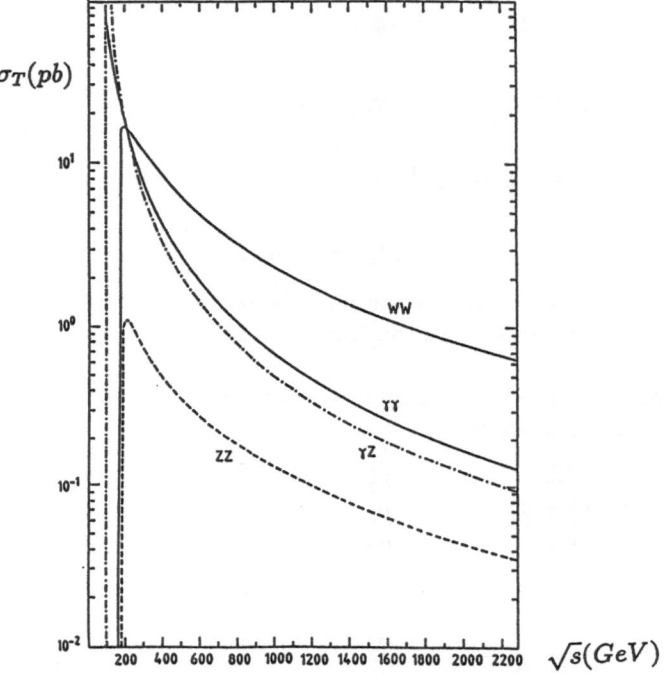

Figure 2.1: The lowest order total cross section for $e^+e^- \rightarrow WW$, $\gamma\gamma$, γZ, ZZ up to 2000 GeV.

For $s \gg M_Z^2$, i.e. in the case where unbroken $SU(2)_w$-symmetry is realized, $\mathcal{M}^+ \rightarrow 0$ since then only pure W^0 exchange contributes.

In fig. 1.1 we have shown already the differential cross section at *LEP II* following from these expressions and compared them to the μ-pair cross section.

Fig. 2.1 shows the total cross section up to 2 TeV. Above 500 GeV we see unification of electromagnetic and weak interactions in the sense that all these cross sections show the same energy dependence. Their numerical values differ just by the corresponding couplings.

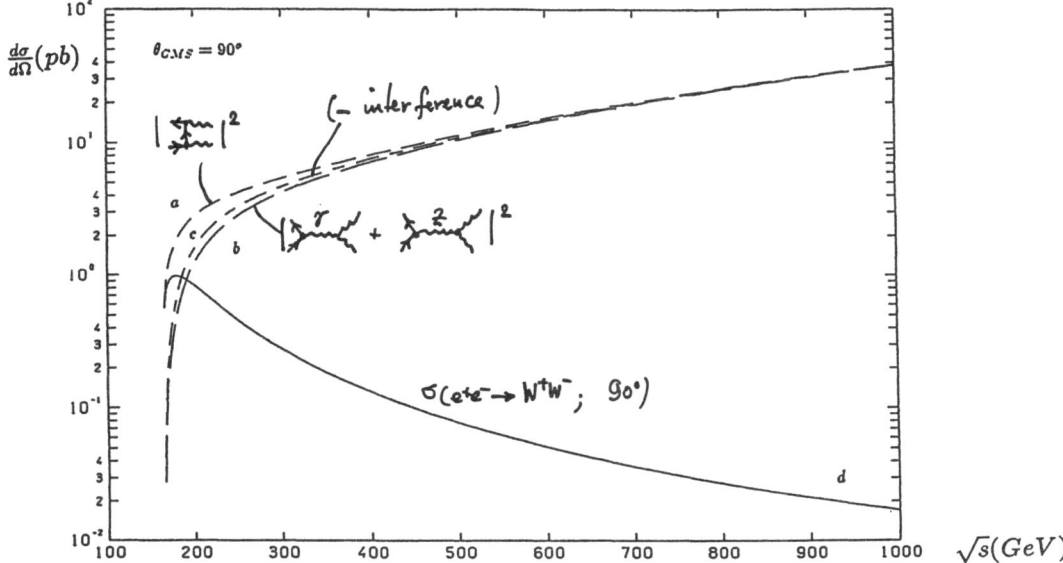

Figure 2.2: The lowest order cross section for $e^+e^- \to W^+W^-$ at 90° scattering angle. Shown are the t-channel (a), the s- channel (b) cross section and their interference (c) yielding a result (d) which respects unitarity.

Finally, we present in fig. 2.2 the separate contributions to W-pair production at 90° scattering angle as a function of energy. Whereas t-channel exchange alone, as well as s-channel by itself rise fast with energy, their interference is strongly destructive yielding a cross section which behaves like $1/s$ thereby respecting unitarity. This demonstrates how the non-Abelian coupling leads to big compensations at high energies which is necessary in order to have a consistent scattering theory.

3. Organization and calculation of RC's

For all the reactions $e^+e^- \to W_1W_2$ the complete virtual 1-loop electroweak radiative corrections together with the real 1-photon bremsstrahlung correction in the soft photon approximation have been reliably calculated. For $e^+e^- \to \gamma\gamma$ also a Monte Carlo program for hard bremsstrahlung exists.

The virtual corrections are calulated in an on-shell renormalization scheme [1] using the fine structure constant α and the masses of the gauge bosons M_W, M_Z, the fermions m_f and the Higgs M_H as physical parameters. Real photons are completely decoupled from the Z, such that QED is present as a simple substructure. As numerical input data the fine structure constant α, the Fermi constant G, the lepton masses and the hadronic vacuum polarization are used. Since the top mass m_t and the Higgs mass M_H are unknown the calculations are performed with "typical values" of these parameters like $m_t = 60\,GeV, M_H = 100\,GeV$, and the dependence of physical observables on m_t, M_H is discussed. The separation into electromagnetic and weak virtual corrections which is possible and convenient in $e^+e^- \to f\bar{f}$ reactions makes sense only for the productions of neutral boson pairs $e^+e^- \to \gamma\gamma,\ \gamma Z,\ ZZ$. This is not so in W^+W^- production, since there the electric charge can flow through the whole diagram. W-pair production is also a process where many diagrams contribute. Whereas in Born approximation we have 3 of them, the 1-loop corrections are given by more than 250 diagrams. This requires a systematic treatment and in order to have reliable results several independent calculations, which meanwhile exist and agree with each other [8,9]. The use of the machinery of computer algebra ($SCHOONSHIP,\ MACSYMA,\ \dots$) is of great help since the basic rules are simple and easy to implement. The complications arise from the great number of diagrams and the huge number of terms which are generated by every diagram especially in the case of the box diagrams.

4. Calculation of a typical box diagram

The techniques for the systematic evaluation of 1-loop diagrams were developed by [17,18]. We use them in a slightly modified form and present them with help of an example: the (Z, W^-, W^+, ν_e) box for $e^-e^+ \to W^-W^+$:

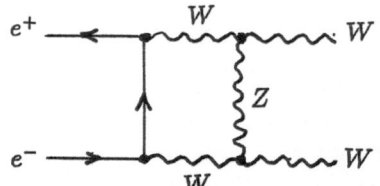

The corresponding matrix element contains besides the $We\nu$-couplings $(e\gamma^\mu(1-\gamma_5)/2\sqrt{2}s_w)$ also the non-Abelian W^+W^-Z coupling ($e\,c_w/s_w\,\Gamma_{\lambda\mu\rho}$):

$$\delta\mathcal{M}^- = \frac{\alpha c_w^2}{(4\pi)2s_w^4} \int \frac{d^4q}{i\pi^2} \frac{\bar{v}(p_1)\gamma^\mu(\slashed{q} + \slashed{k}_1 - \slashed{p}_1)\gamma^\nu\omega_-u(p_2)\,\Gamma_{\lambda\mu\rho}\,\Gamma_{\sigma\nu}^\lambda\,\varepsilon_1^\rho\,\varepsilon_2^\sigma}{[q^2 - M_Z^2]\,[(q+k_1)^2 - M_W^2]\,(q+k_1-p_1)^2\,[(q-k_2)^2 - M_W^2]}$$

(4.1)

with

$$\Gamma_{\lambda\mu\rho} = g_{\lambda\mu}(2q+k_1)_\rho + g_{\mu\rho}(-q-2k_1)_\lambda + g_{\rho\lambda}(k_1-q)_\mu$$
$$\Gamma_{\lambda\sigma\nu} = g_{\lambda\sigma}(-k_2-q)_\nu + g_{\sigma\nu}(2k_2-q)_\lambda + g_{\nu\lambda}(2q-k_2)_\sigma$$

We insert the Γ-tensors in the numerator, work out the tensor and Dirac algebra thereby making use of the mass shell conditions (e.g. $\slashed{p}_1 u(p_1) = m_e u(p_1) \approx 0,\ k_\mu\varepsilon^\mu = 0$) and organize the result with respect to tensors of the integration momentum: $q_\mu, q_\mu q_\nu, q_\mu q_\nu q_\rho$. q^2-terms may be written as $q^2 - M_Z^2 + M_Z^2$ leading to simpler integrals. In this way $\mathcal{M}$ can be expressed with help of the standard scalar and tensor integrals

$$A = \frac{\mu^{D-4}}{i\,\pi^2} \int \frac{d^Dq}{(2\pi)^{D-4}} \frac{1}{[q^2 - m^2]}$$

$$\{B_0; B_\mu; B_{\mu\nu}\} = \frac{\mu^{D-4}}{i\,\pi^2} \int \frac{d^D q}{(2\pi)^{D-4}} \frac{\{1;\, q_\mu;\, q_\mu q_\nu\}}{[q^2 - m_1^2]\,[(q - p_1)^2 - m_2^2]}$$

$$\{C_0; C_\mu; C_{\mu\nu}; C_{\mu\nu\rho}\} = \frac{\mu^{D-4}}{i\,\pi^2} \int \frac{d^D q}{(2\pi)^{D-4}} \frac{\{1;\, q_\mu;\, q_\mu q_\nu;\, q_\mu q_\nu q_\rho\}}{[q^2 - m_1^2]\,\prod_{j=1}^{2}[(q - p_j)^2 - m_{j+1}^2]}$$

$$\{D_0; D_\mu; D_{\mu\nu}; D_{\mu\nu\rho}; D_{\mu\nu\rho\sigma}\} = \frac{1}{i\,\pi^2} \int d^4 q \frac{\{1;\, q_\mu;\, q_\mu q_\nu;\, q_\mu q_\nu q_\rho;\, q_\mu q_\nu q_\rho q_\sigma\}}{[q^2 - m_1^2]\,\prod_{j=1}^{3}[(q - p_j)^2 - m_{j+1}^2]} \qquad (4.2)$$

in the following way:

$$t^- = \frac{\alpha c_w^2}{(4\pi)2 s_w^4}\,\bar{v}(p_+)\Big\{-8\gamma^\mu\,\varepsilon_+^\nu\,\varepsilon_-^\rho\,D_{\mu\nu\rho}$$

$$+ D_{\mu\nu}\,\big[\,\not{k}_+ \gamma^\mu\,\not{k}_- - 2(k_+ - k_-)^\nu$$

$$+ \not{k}_+(-2\varepsilon_-^\mu\,k_-^\nu - 8 p_+^\mu\,\varepsilon_-^\nu + 2 p_-^\mu\,\varepsilon_-^\nu) + \not{k}_+ \gamma^\mu\,\not{k}_-\,8\varepsilon_-^\nu$$

$$- \not{k}_-(-2\varepsilon_+^\mu\,k_+^\nu - 8 p_-^\mu\,\varepsilon_+^\nu + 2 p_+^\mu\,\varepsilon_+^\nu) - \not{k}_+ \gamma^\mu\,\not{k}_-\,8\varepsilon_+^\nu$$

$$+ \not{k}_+\,8\varepsilon_+^\mu\,\varepsilon_-^\nu + \gamma^\mu(-16\varepsilon_+^\nu\,p_-\varepsilon_- + 16\varepsilon_-^\nu\,p_+\varepsilon_+ + 2(k_+ - p_+)^\nu\,\varepsilon_+\varepsilon_-)\big]$$

$$+ D_\mu\,\big[\,\not{k}_+(\not{k}_+ - \not{p}_+)\,\not{k}_-\,2(k_+ - k_-)^\mu + \not{k}_+ \gamma^\mu\,\not{k}_-(M_Z^2 - 4 k_+ k_-)$$

$$+ \not{k}_+(-\varepsilon_-^\mu(3 M_Z^2 + 3 M_W^2 + 2 p_+ k_+) - 4 k_-^\mu\,\varepsilon_- k_+ + 4 p_-^\mu\,\varepsilon_- k_+)$$

$$- \not{k}_-(\varepsilon_+^\mu(3 M_Z^2 + 3 M_W^2 + 2 p_- k_-) + 4 k_+^\mu\,\varepsilon_+ k_- - 4 p_+^\mu\,\varepsilon_+ k_-)$$

$$+ \not{k}_+(2(k_+ - p_+)^\mu\,\varepsilon_+\varepsilon_- - 8\varepsilon_+^\mu\,\varepsilon_- p_- + 8\varepsilon_-^\mu\,\varepsilon_+ p_+)$$

$$+ \gamma^\mu \varepsilon_+\varepsilon_-(M_Z^2 - 3 M_W^2 + 6 p_+ k_+) + \not{k}_+ \gamma^\mu\,\not{k}_-\,4\varepsilon_+ k_- + 4\,\not{k}_+ \gamma^\mu\,\not{k}_-\,4\varepsilon_- k_+\big]$$

$$+ D_0\,\big[\,\not{k}_+(\not{k}_+ - \not{p}_+)\,\not{k}_-(M_Z^2 - 4 k_+ k_-) + \not{k}_+\,\varepsilon_+\varepsilon_-(M_W^2 - 2 p_+ k_+ - 3 M_Z^2)$$

$$+ \not{k}_+\,\varepsilon_- k_+(-2 M_W^2 + 4 p_- k_- + 2 M_Z^2) - \not{k}_-\,\varepsilon_+ k_-(-2 M_W^2 + 4 p_+ k_+ + 2 M_Z^2)\big]$$

$$+ C_\mu\,\big[\,\not{k}_+ \gamma^\mu\,\not{k}_- - \not{k}_-\,3\varepsilon_+^\mu - \not{k}_+\,3\varepsilon_-^\mu + \gamma^\mu \varepsilon_+\varepsilon_-\big]$$

$$+ C_0\,\big[\,\not{k}_+\,2\varepsilon_- k_+ - \not{k}_-\,2\varepsilon_+ k_- + \not{k}_+\,3\varepsilon_- p_- - \not{k}_-\,3\varepsilon_+ p_+ - \not{k}_+\,4\varepsilon_+\varepsilon_-\big]\Big\}\omega_- u(p_-).$$

In the next step the tensor integrals $B_\mu \cdots D_{\mu\nu\rho\sigma}$ are built up from the metric tensor $g_{\mu\nu}$, tensors of the external momenta p^i_μ and invariant functions using Lorentz covariance:

$$B_\mu = p_{1_\mu} B_1 \tag{4.3}$$

$$B_{\mu\nu} = g_{\mu\nu} B_{00} + p_{1_\mu} p_{1_\nu} B_{11} \tag{4.4}$$

$$C_\mu = p_{1_\mu} C_1 + p_{2_\mu} C_2 = \sum_{i=1}^{2} p_{i_\mu} C_i \tag{4.5}$$

$$C_{\mu\nu} = g_{\mu\nu} C_{00} + \sum_{i\leq j=1}^{2} p_{[i_\mu} p_{j_\nu]} C_{ij} \tag{4.6}$$

$$C_{\mu\nu\rho} = \sum_{i=1}^{2} g_{[\mu\nu} p_{i_\rho]} C_{00i} + \sum_{i\leq j\leq k=1}^{2} p_{[i_\mu} p_{j_\nu} p_{k_\rho]} C_{ijk} \tag{4.7}$$

$$D_\mu = p_{1_\mu} D_1 + p_{2_\mu} D_2 + p_{3_\mu} D_3 = \sum_{i=1}^{3} p_{i_\mu} D_i \tag{4.8}$$

$$D_{\mu\nu} = g_{\mu\nu} D_{00} + \sum_{i\leq j=1}^{3} p_{[i_\mu} p_{j_\nu]} D_{ij} \tag{4.9}$$

$$D_{\mu\nu\rho} = \sum_{i=1}^{3} g_{[\mu\nu} p_{i_\rho]} D_{00i} + \sum_{i\leq j\leq k=1}^{3} p_{[i_\mu} p_{j_\nu} p_{k_\rho]} D_{ijk} \tag{4.10}$$

$$D_{\mu\nu\rho\sigma} = g_{[\mu\nu} g_{\rho\sigma]} D_{0000} + \sum_{i\leq j=1}^{3} g_{[\mu\nu} p_{i_\rho} p_{j_\sigma]} D_{00ij} + \sum_{i\leq j\leq k\leq l=1}^{3} p_{[i_\mu} p_{j_\nu} p_{k_\rho} p_{l_\sigma]} D_{ijkl} \tag{4.11}$$

(The $[\cdots]$ brackets mean summation over those permutations which yield different terms).

Using this decomposition and contracting the tensors we get:

$$
\begin{aligned}
\delta \mathcal{M}^- = \frac{\alpha\, c_w^2}{8\,\pi\, s_w^4} \Big\{ & \mathcal{M}_0^- && [20\, D_{00} + 2\,(2\, D_{23} + D_{22} + D_2)\, t \\
& && -2\,(D_{33} + D_{23} + D_{13} + 3\, D_3 + D_2 + D_0)\, s + (2\, D_3 + D_2 + D_0) \\
& && +2\,(4\, D_{33} + 6\, D_{23} + D_{22} + 4\, D_{13} + 8\, D_3 + 3\, D_2 + 2\, D_0)\, M_W^2] \\[6pt]
+ & \mathcal{M}_1^- && [+ (2\, D_{33} + 6\, D_{23} + 2\, D_{22} + 2\, D_{13} - 4\, D_3 - D_2 + D_0)\, t \\
& && + (2\, D_3 + D_2 - 3\, D_0)\, M_Z^2 + 2\,(D_{33} + D_{23} + D_{13} + D_3)\, M_W^2 \\
& && +10\, D_{00} - 16\, D_{003} - 8\, D_{002} - 4\, C_0] \\[6pt]
+ & (\mathcal{M}_2^- - \mathcal{M}_3^-) && [-(3\, D_{33} + 2\, D_{23} - 5\, D_{13} - D_3 + 2\, D_0)\, t + 4\, D_{13}\, s \\
& && -(3\, D_3 - 2\, D_0)\, M_Z^2 - (3\, D_{33} + 4\, D_{23} + 11\, D_{13} + 4\, D_3)\, M_W^2 \\
& && -8\, D_{003} - 8\, D_{00} - 3\, C_2 + 2\, C_0] \\[6pt]
+ & (\mathcal{M}_4^- - \mathcal{M}_5^-) && [-(2\, D_{23} + 2\, D_{22} - D_2)\, t + 2\,(D_{33} + 2\, D_{23} + D_{13} + 2\, D_3)\, s \\
& && -2\,(4\, D_{33} + 9\, D_{23} + 2\, D_{22} + 4\, D_{13} + 4\, D_3 + 2\, D_2)\, M_W^2 \\
& && -(2\, D_3 + 3\, D_2)\, M_Z^2 - 8\, D_{002} - 26\, D_{00} + 4\, C_2 + 3\, C_0] \\[6pt]
+ & \mathcal{M}_6^- && 8\,[2\, D_{133} + D_{123} - D_{13}]
\end{aligned}
$$

$$+ \quad \mathcal{M}_7^- \quad 8\left[2\,D_{223} + D_{222} + 4\,D_{23} + 3\,D_{22} + 2\,D_2\right]$$

$$+ \left(\mathcal{M}_8^- + \mathcal{M}_9^-\right)\, 8\left[D_{233} + D_{223} + D_{123} + D_{23} + 2\,D_{13}\right]\}$$

Here $\mathcal{M}_i^-$ denote the standard matrix elements (eq. 2.3), s, t the Mandelstam variables and $D_0 \ldots D_{123}, \ldots$ the invariant functions. The same result may be obtained using algebraic computer programs [19] allowing a check of these calculations.

Our aim is to express everything with help of the scalar integrals $A_0, \ldots, D_0$ which are completely known as functions containing besides square roots and logarithms also dilogarithms $Li_2(z)$ [18]. In order to facilitate our presentation we show this reduction for the vector triangle integral:

$$C^\mu = \int d^D q \; q^\mu \Delta_1 \Delta_2 \Delta_3 = \langle q^\mu \Delta_1 \Delta_2 \Delta_3 \rangle = \sum p_i^\mu C_1^i \qquad (4.12)$$

with $\Delta_1 = (q^2 - m_1^2)^{-1}$, $\Delta_2 = [(q + p_1)^2 - m_2^2]^{-1}$, $\Delta_3 = [(q + p_2)^2 - m_3^2]^{-1}$. Multiplying by $p_{j\mu}$ gives:

$$p_{j\mu} C^\mu = \sum (p_j p_i) C_1^i = \langle q\, p_j\; \Delta_1 \Delta_2 \Delta_3 \rangle \qquad (4.13)$$

or

$$C_1^i = (p_i p_j)^{-1} \langle q\, p_j\; \Delta_1 \Delta_2 \Delta_3 \rangle. \qquad (4.14)$$

We express $q\, p_j$ by the Δ's:

$$q\, p_j = \frac{1}{2}(\Delta_{j+1}^{-1} - \Delta_1^{-1} - f_j) \qquad\qquad f_j = p_j^2 - m_{j+1}^2 + m_1^2 \qquad (4.15)$$

and obtain:

$$C_1^i = \frac{1}{2}\,(p_i p_j)^{-1}\left[\underbrace{\langle \Delta_1 \Delta_2 \Delta_3 \Delta_{j+1}^{-1}\rangle}_{B_0} - \underbrace{\langle \Delta_2 \Delta_3 \rangle}_{B_0} - f_j \underbrace{\langle \Delta_1 \Delta_2 \Delta_3 \rangle}_{C_0}\right]. \qquad (4.16)$$

Thereby the C_1^i are expressed by the scalar 2-point integrals B_0 and the scalar triangle integral C_0. Similar but more complicated calculations allow to express all tensor integrals by $A_0, \ldots, D_0$. In this way all the virtual 1-loop diagrams of the SM can be evaluated and provide after renormalization together with the corresponding calculations for the real bremsstrahlung the formulas for the numerical evaluation of the radiative corrections.

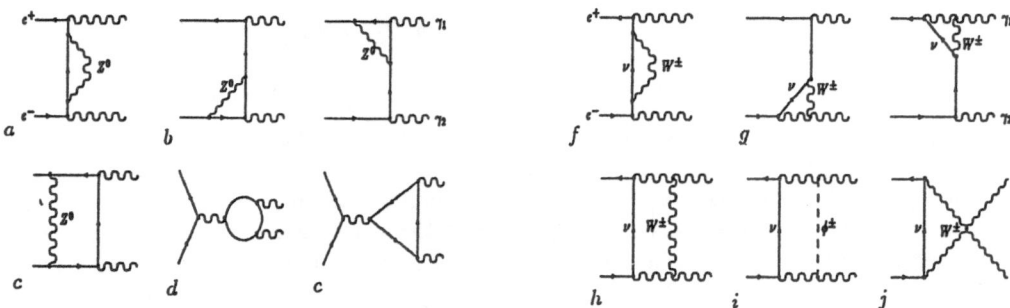

Figure 5.1: One-loop Feynman diagrams containing a virtual $Z^0/W^\pm$:, a,f) self energy; b,g) vertex correction; c,h,i) box diagrams; d) boson loops with $W^\pm, \Phi^\pm$, and ghosts; e) anomaly. The remaining diagrams are obtained by crossing $\gamma_1 \leftrightarrow \gamma_2$

5. Results and discussion

In this section we present the results of the calculations of radiative corrections for the processes $e^+e^- \to \gamma\gamma$, γZ, ZZ. Included are the complete 1-loop electroweak virtual corrections and soft bremsstrahlung. The contribution of hard bremsstrahlung is more sensitive to the experimental cuts. It should be treated with help of Monte Carlo programs.

5.1 $e^+e^- \to \gamma\gamma$

We write the differential cross section in the form

$$\frac{d\sigma}{d\Omega} = \left(\frac{d\sigma}{d\Omega}\right)_B \quad (1 + \delta_{QED} + \delta_Z + \delta_W)\,. \tag{5.1}$$

The lowest order cross section is purely electromagnetic also at LEP energies. The QED corrections were calculated by [13,14,15]. A complete treatment which is free of misprints and including hard bremsstrahlung was performed by [3]. The weak virtual corrections were treated by [10] and [4]. We follow here the work of [4] since it contains explicit analytical results and a detailed numerical evaluation of the weak correction effects. The weak corrections have the remarkable property that there is no contribution of the Higgs boson, therefore the results do not depend on M_H and consequently are free from the uncertainties of this unknown parameter.

Fermions contribute only via the anomaly diagram (e). Since the standard model is anomaly free the corresponding contributions are negligible $\left(\sim \frac{m_e}{\sqrt{s}} \cdot \frac{\alpha}{\pi}\right)$ and do not contain the Z-peak. Therefore the weak fermionic corrections may be neglected. The uncertainties coming from the hadronic vacuum polarization or the light quark masses respectively do not enter the predictions for $e^+e^- \to \gamma\gamma$. Neither do uncertainties in the top quark mass or contributions by further families of leptons and quarks. The diagrams which do contribute contain the Z-bosons (fig. 5.1 a-e), the $W^\pm$-bosons (fig. 5.1 f-j) with their 't Hooft-Feynman gauge partners $\Phi^\pm$ (fig. d,i) and the corresponding charged Fadeev-Popov ghosts $u^\pm$ (fig. d). The result is shown in (fig. 5.2) for 90° scattering, that angle where their relative magnitude is biggest.

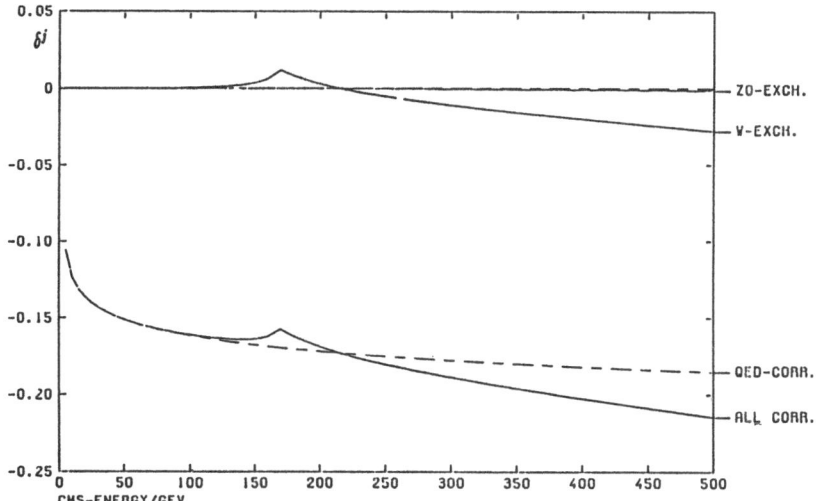

Figure 5.2: The relative e.m. and weak corrections to $e^+e^- \to \gamma\gamma$ at $90°$ scattering for energies up to 500 GeV.

The neutral weak corrections δ_Z are very small. For all energies below 500 GeV they remain smaller than $1^0/_{00}$. The charged correcions δ_W increase with energy until they reach a maximum of $+1.3$ % at the W^+W^--threshold. Then they go through zero around 220 GeV and become negative. As can be seen from fig. 5.2 the QED corrections dominate by far.

At $LEP\,I$ the total weak corrections are smaller than $1^0/_{00}$. Therefore $e^+e^- \to \gamma\gamma$ is at $LEP\,I$ practically a pure QED process, allowing clean QED tests at 100 GeV CMS energy. At $LEP\,II$ the weak corrections are in the 1% region, i. e. small but already with our present knowledge of parameters of the standard model completely predictable since they are free of M_H, m_t, unknown generations as long as they couple in the standard model way. We may resume the situation by saying that from the theoretical point of view $e^+e^- \to \gamma\gamma$ is very clean, almost pure QED and therefore an excellent hunting ground for new physics like non-standard couplings, heavy electrons, additional gauge bosons and so on [11].

5.2 $e^+e^- \to \gamma Z, \; ZZ$

In this section we will neglect the effects which are related with the instability i. e. the finite width of the Z-boson. Their influence is important around threshold only and will be discussed in 5.3.

The QED corrections to $e^+e^- \to \gamma Z$ have been calculated by [12] and [5]. The two results agree. For $e^+e^- \to ZZ$ the QED corrections where given by [6]. The weak corrections are more complicated than in the $\gamma\gamma$-case and depend now on all parameters of the standard model. One source of these additional corrections is

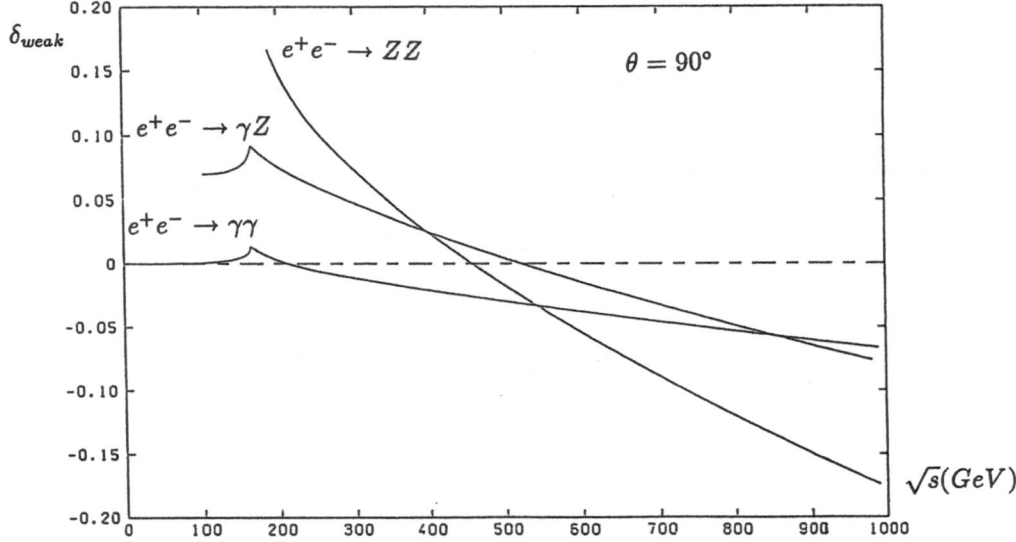

Figure 5.3: Weak radiative corrections to $e^+e^- \to \gamma\gamma, \gamma Z, ZZ$ at 90° scattering.

the Z-boson wave function renormalization together with γZ-mixing which does not vanish for on-shell Z-bosons. The corresponding contributions are in the γZ case:

$$\delta\mathcal{M}^\sigma = -\mathcal{M}_B^\sigma \left(\frac{1}{2} \frac{\partial \hat{\Sigma}^{ZZ}(k^2)}{\partial k^2} \bigg|_{k^2=M_Z^2} + \frac{1}{g_e^\sigma} \frac{\hat{\Sigma}^{\gamma Z}(M_Z^2)}{M_Z^2} \right), \qquad (5.2)$$

where $\hat{\Sigma}^{ZZ}(k^2), \hat{\Sigma}^{ZZ}(k^2)$ are the renormalized, transverse Z self energy and $Z\gamma$ mixing energy, respectively. These depend on all fermion properties and the Higgs mass, but are partly compensated by using G_F as an input quantity. For the ZZ matrix element we get twice this contribution. The γZ mixing effect is numerically small. For $M_H = 100\,GeV, m_t = 60\,GeV$ it is less than $1^0/_{00}$.

In fig. 5.3 we compare the weak radiative corrections for 90° scattering of all these neutral boson pair production processes using $M_Z = 93\,GeV$ and $M_W = 82\,GeV$ as well as $M_H = 100\,GeV, m_t = 45\,GeV$ as input. A more detailed discussion will be given after inclusion of the finite width effects.

5.3   Finite width effects

More realistic calculations of reactions where Z- or W-bosons are produced must include the effects which are caused by the instability of these particles. These are especially important around the thresholds of the corresponding reactions. In order to discuss this problem for $e^+e^- \to ZZ \to final\ states$ we consider those diagrams which contain 2 Z poles leading to cross sections which have a Breit-Wigner resonance behaviour in the invariant masses of the final states. Doing this background processes are neglected. But since $M_Z/\Gamma_Z \gg 1$ we consider this as a good approximation.

In this sense we start following the work of Denner and Sack [7] with:

$$e^+e^- \to Z(k_1)\,Z(k_2) \to \sum_i f_i(q_i) + \sum_j f_j(q_j) \qquad (5.3)$$

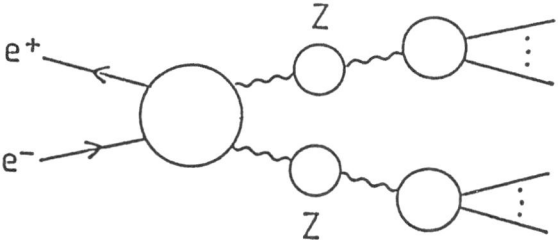

The matrix element for this process can be written:

$$\mathcal{M} = \int d^4k_1\, d^4k_2\, A_{\mu\nu}(p_1p_2, k_1k_2)\, \Delta_{\mu\rho}^Z(k_1)\, \Delta_{\nu\sigma}^Z(k_2)\, B_\rho(k_1, q_i)\, B_\sigma(k_2, q_j) \qquad (5.4)$$

where $A_{\mu\nu}$ is the amplitude of the annihilation of an e^+e^- pair into two virtual Z's of invariant masses k_1^2, k_2^2; $\Delta_{\mu\nu}^Z$ denotes the Z propagator and B_ρ the decay matrix element of the off-shell Z into the final state with momenta q_i.

The propagator consists of a transverse and a longitudinal part:

$$\Delta_{\mu\nu}(k) = \left(-g_{\mu\nu} + \frac{k_\mu k_\nu}{k^2}\right) \frac{1}{k^2 - M_Z^2 + \Sigma_Z^T(k^2)} - \frac{k_\mu k_\nu}{k^2} \frac{1}{k^2 - M_Z^2 + \Sigma_Z^L(k^2)}. \qquad (5.5)$$

The longitudinal part does not contribute for on-shell Z's since it describes an unphysical degree of freedom. Otherwise stated it does not show resonance behaviour. Moreover it gives together with the production amplitude $A_{\mu\nu}$ contributions which are smaller by factors $m_e/\sqrt{s}$ than the transverse part. Therefore we are allowed to neglect the longitudinal part, keeping the transverse which means three polarizations of the off-shell Z. This gives

$$\mathcal{M} = \sum_{\lambda_1\lambda_2} \int d^4k_1\, d^4k_2 A^{\mu\nu} \frac{\varepsilon_\mu(k_1, \lambda_1)\, \varepsilon_\rho(k_1, \lambda_1)}{k_1^2 - M_Z^2 + \Sigma_Z^T(k_1^2)} B^\rho(k_1, q_i) \frac{\varepsilon_\nu(k_2, \lambda_2)\, \varepsilon_\sigma(k_2, \lambda_2)}{k_2^2 - M_Z^2 + \Sigma_Z^T(k_2^2)} B^\sigma(k_2, q_j).$$

$$(5.6)$$

Our on-shell renormalization conditions for the Z-boson propagator are:

$$\left(\Sigma_Z^T(M^2)\right)_{Ren} = 0 \qquad \text{mass renormalization} \qquad (5.7)$$

Defining

$$\Pi(k^2) = \frac{Re\ \Sigma_Z^T(k^2)}{k^2 - M_Z^2} \qquad (5.8)$$

we can introduce the correct wave function renormalization which is given by $\Pi(M_Z^2)$. Therefore we write for the cross section

$$d\sigma = \int dk_1^2 \int dk_2^2\, d\sigma(s, k_1^2, k_2^2) \frac{1}{\pi} \frac{\sqrt{k_1^2}\,\Gamma(k_1^2)}{\left[(k_1^2 - M_Z^2)^2 + k_1^2\Gamma^2(k_1^2)\right]} \frac{1}{\pi} \frac{\sqrt{k_2^2}\,\Gamma(k_2^2)}{\left[(k_2^2 - M^2)^2 + k_2^2\Gamma^2(k_2^2)\right]}$$

$$(5.9)$$

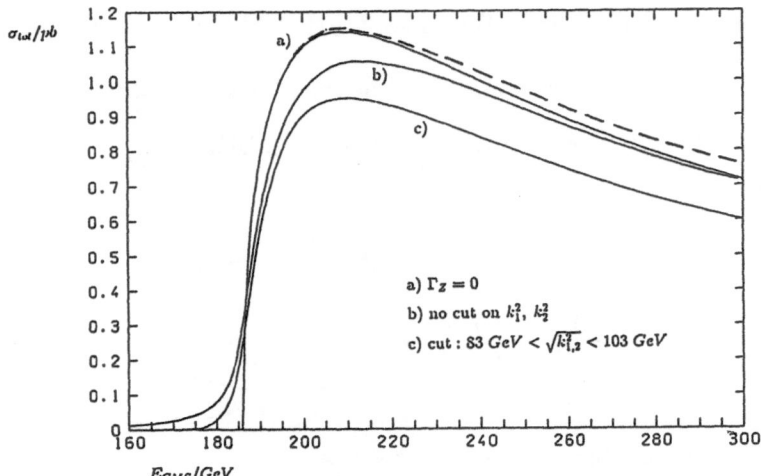

Total cross section for $e^+e^- \to Z^0Z^0$ including all 1-loop corrections: a) for stable Z^0-bosons; b) including the finite decay width; c) including the finite decay width and a cut on the invariant masses of the decay products.

Figure 5.4:

with:

$$d\sigma(s, k_1^2, k_2^2) = \frac{1}{1 + \Pi(k_1^2)} \cdot \frac{1}{1 + \Pi(k_2^2)} \tag{5.10}$$

$$\cdot \frac{\pi^2}{2\,s} \int \frac{d^3k_1}{2k_{10}(2\pi)^3} \int \frac{d^3k_2}{2k_{20}(2\pi)^3} \left| A_{\mu\nu}\varepsilon^\mu(\lambda_1)\varepsilon^\nu(\lambda_1) \right|^2 \, \delta(p_+ - p_- - k_1 - k_2)$$

$$\Gamma(k^2) = \frac{1}{1 + \Pi(k^2)} \frac{1}{2\pi} \sum_{decays} \frac{1}{2\sqrt{k_1^2}} \int dLIPS \, \left| B^\rho \, \varepsilon_\rho(\lambda) \right|^2 \, \delta\left(k_1 - \sum q_i\right) . \tag{5.11}$$

The quantities occuring here have to be calculated including radiative corrections. The numerical evaluation of these formulas can be simplified keeping the desired accuracy by expanding those terms which have a smooth k^2 dependence around $k^2 = M_Z^2$:

$$\Gamma(k^2) \approx \Gamma(M^2) \cdot \frac{\sqrt{k^2}}{M} \tag{5.12}$$

$$d\sigma(s, k_1^2, k_2^2) = d\sigma_B(s, k_1^2, k_2^2) \cdot \underbrace{\frac{d\sigma(s, k_1^2, k_2^2)}{d\sigma_B(s, k_1^2, k_2^2)}}_{smooth}$$

$$\approx d\sigma_B(s, k_1^2, k_2^2) \frac{d\sigma(s, M_Z^2, M_Z^2)}{d\sigma_B(s, M_Z^2, M_Z^2)}. \tag{5.13}$$

This approximation contains the off-shell Born cross section $d\sigma_B$ which has for $s = (k_1 + k_2)^2 \approx 4\,M_Z^2$ just from phase space a strong k_1^2, k_2^2 dependence since $k_{1,2}^2 \approx M_Z^2$. All other quantities i. e. $\Gamma, d\sigma$ are taken on-shell. A direct comparison has shown that the relative error introduced by this approximation is smaller than 10^{-3}.

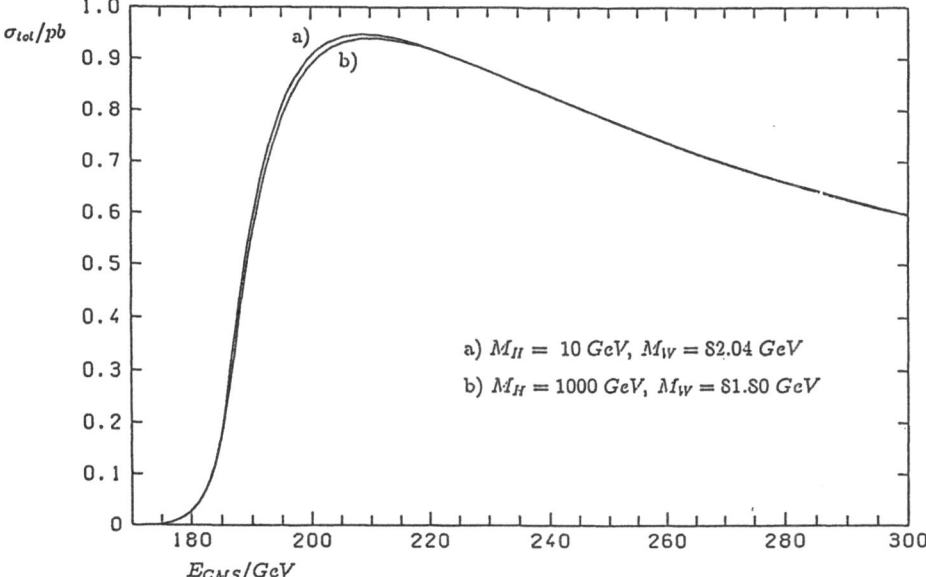

Dependence of the total cross section on M_H with M_Z and G_F kept fixed.

Figure 5.5:

Taking $M_Z = 93\,GeV, G_\mu, M_H = 100\,GeV, m_t = 60\,GeV$ as numerical input and the predicted value $\Gamma_Z = 2.5\,GeV$ as well as a photon cutoff $\Delta E = 0.1 E_{beam}$ we obtain the results shown in fig. 5.4. There are shown the results without and with radiative corrections for zero width bosons and for bosons with physical width with no cutoffs for the invariant mass of the intermediate Z and a cutoff of $M_Z \pm 10\,GeV$. We observe that in the *LEP II* energy region the finite width effects are as important as the radiative corrections. The figure also demonstrates that an accurate determination of the mass — which will be of importance for the W boson case — needs a very careful study of both the radiative correction and finite width effects.

Finally we present the M_H and m_t dependence of the threshold cross section in figs. 5.5. and 5.6. We find a weak sensitivity on the Higgs mass but a strong one on the top mass when this exceeds 100 GeV.

6. Conclusion

$e^+ e^-$ annihilation into pairs of gauge bosons opens the possibilities for decisive tests of the standard model. At *SLC/LEP I* two-photon annihilation is still a pure QED process allowing to test QED at $\sqrt{s} = 100\,GeV$. The required calculations including all the $O(\alpha)$ radiative corrections exist and are free from ambiguities of presently unknown parameters of the SM or the possible existence of further fermion generations and their masses.

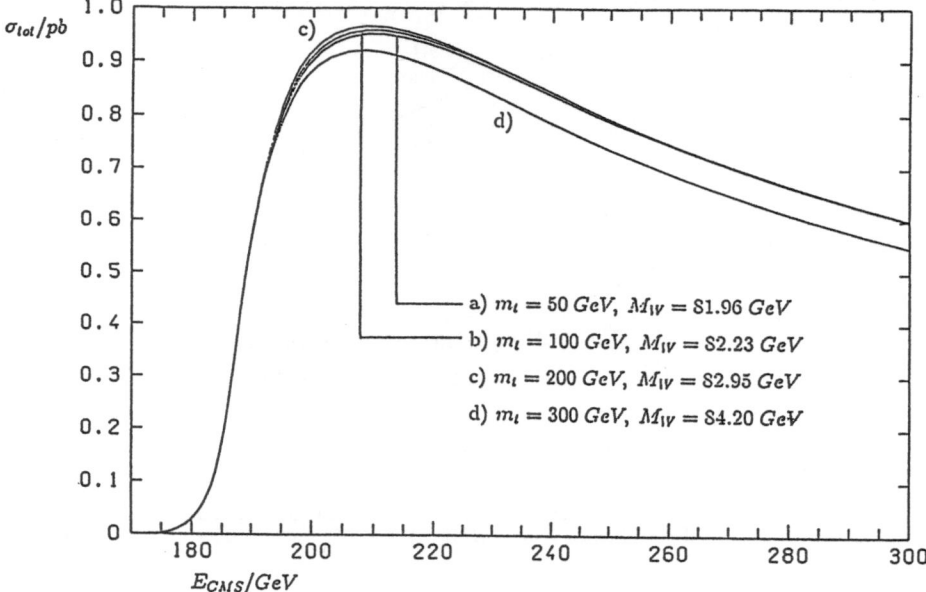

Dependence of the total cross section on m_t with M_Z and G_F kept fixed.

Figure 5.6:

$O(\alpha)$ calculations including 1-loop electroweak virtual corrections and soft bremsstrahlung with analytic results also exist for $e^+e^- \rightarrow \gamma Z, ZZ, WW$. What is missing in this case are Monte Carlos for hard photon bremsstrahlung and the finite width effects. The latter have been discussed in the case of Z-pair production for simple cuts. When these MC programs are written it will be possible to compare theory and experiment at the level of accuracy (1-3%) which we finally expect for *LEP II* experiments. Then we will have the possibility to measure such important parts of the SM as the W-mass, the non-abelian coupling, providing also the means to look for new physics.

Acknowledgement:
I would like to thank Hans Kühn for the invitation and for the very pleasent stay at Schloß Ringberg.

References

[1]   M. Böhm, W. Hollik, H. Spiesberger: Progr. Phys. **34**, 687 (1986)

[2]   A. Sirlin: Phys. Rev. **D22**, 2695 (1980)

[3]   F. Berends, R. Gastmans: Nucl. Phys. **B61**, 414 (1973)

[4]   M. Böhm, T. Sack: Z. Phys. **C33**, 157 (1986)

[5]   M. Böhm, T. Sack: Z. Phys. **C35**, 119 (1987)

[6]   A. Denner, T. Sack: Nucl. Phys. **B305**, 221 (1988)

[7]   A. Denner, T. Sack: Nucl. Phys. **B**, in print (1988)

[8]   M. Böhm, A. Denner, T. Sack, W. Beenakker, F. Berends, H. Kuijf: Nucl. Phys. **B304**, 463 (1988).

[9]   J. Fleischer, F. Jegerlehner, M. Zralek: Z. Phys. **C42**, 409 (1989)

[10]   M. Capdequi-Peyranere, G. Grunberg, F.M. Renard, M. Talon: Nucl. Phys. **B149**, 243 (1979)

[11]   P. Mevy, M. Perrotet, F.M. Renard: CERN-TH 4741/87

[12]   F. Berends, G. Burgers, W. van Neerven: Leiden 1989

[13]   L. Brown, R.P. Feynman: Phys. Rev. **85**, 231 (1952)

[14]   I. Harris, L. Brown: Phys. Rev. **105**, 1656 (1957)

[15]   Y. Tsai: Phys. Rev. **137**, B3730 (1965)

[16]   See the other contributions of these proceedings

[17]   G Passarino, M. Veltman: Nucl. Phys. **B160**, 151 (1979)

[18]   G. 't Hooft, M. Veltman: Nucl. Phys. **B153**, 365 (1979)

[19]   R. Mertig: Application of symbolic manipulation programs for Feynman diagrams in non-Abelian gauge theories, Diploma Thesis Würzburg, 1989

Radiative Corrections for e^+e^- Collisions   Editor: J.H. Kühn
© Springer-Verlag Berlin, Heidelberg 1989

Radiative Corrections to W-Pair Production at LEP

F. JEGERLEHNER

Paul Scherrer Institute, CH-5232 Villigen PSI, Switzerland

June 23, 1989

1. Introduction

The most interesting aspect of W-pair production at LEP2 will be the direct test of gauge boson self-interaction by the measurement of the production cross section $\sigma(e^+e^- \rightarrow W^+W^-)$. One will be able to determine with good precision the so far experimentally unverified triple gauge couplings γW^+W^- and ZW^+W^- which are predicted by the Standard Model [1,2]. Their s-channel contributions greatly reduce the t-channel ν-exchange cross-section by the gauge cancellation mechanism [3]. The main topics which can be investigated are the following [2]

- Determination of the W-mass by measurement of σ_{tot} near the W-pair production threshold ($E_b \simeq M_W$). The W-mass measurement is equivalent to a direct determination of Δr, given by [4]

$$\Delta r = 1 - \frac{\pi \alpha}{\sqrt{2} G_\mu} \frac{1}{M_W^2 (1 - \frac{M_W^2}{M_Z^2})}. \tag{1}$$

  This measurement thus provides an important test of quantum corrections, since Δr vanishes at the tree level. Δr is the most sensitive quantity to virtual heavy particle effects, [5] and also depends (logarithmically) on the unknown Higgs mass.
  The low cross-section requires long running times for the experiments. By collecting 500 pb^{-1} at 5 different energies, a precision of $\delta M_W \simeq 100\ MeV$ for the W-mass measurement is supposed to be achievable.

- The determination of the angular distribution $\frac{d\sigma}{d\cos\theta}$ and of the ratio $\frac{\sigma_L}{\sigma_T}$ of cross-sections for production of longitudinal (σ_L) and transversal (σ_T) W's will provide limits on (possibly anomalous) triple gauge couplings of the W.

- A careful analysis of the W-decays $W \rightarrow q\bar{q}, l\nu_l$ will provide not only precise values for the total- and partial widths, but also a drastic improvement of the determination of the heavy flavor Kobayashi-Maskawa mixing matrix elements.

- From the investigations of the invariant mass distribution and the endpoint of the decay spectrum, independent determinations of the W-mass are possible. The precision of the mass measurements from the decays is supposed to be comparable to the one which can be reached from the investigation of the $e^+e^- \to W^+W^-$ threshold.

Of course, the process actually observed in experiments is $e^+e^- \to W^+W^- \to f_1\bar{f}_2f_3\bar{f}_4$. The W's must be reconstructed from their decays. The main event topologies are

- 4 jet final states (about 50 % of the total). The detection efficiency for a W-pair from these final states is supposed to reach 70 %.

- 2 jet plus isolated lepton plus missing energy. These will allow W-pair reconstruction with 55 % efficiency.

- Two isolated leptons plus missing energy. The detection efficiency is about the same as in the previous case.

In all cases the background has been estimated to lie below 1 %. The maximum total cross-section is $\sigma_{max} \simeq 18pb$ ($E_b \simeq 96GeV$) and about 10^4 events per experiment are expected (within two years of operation).

Typically, the precision expected is better than 1 % (0.1 % for the W-mass measurement). Radiative corrections therefore must be take into account to better than 1 % , requiring a full one-loop calculation of this process.

One-loop calculations including soft photon Bremsstrahlung have been presented in Refs. [6-9]. The results of Refs.[8] and [9] agree to better than 0.3 % in the most critical cases (large energies, large scattering angles). In this contribution we review the main results on radiative corrections to $e^+e^- \to W^+W^-, W^+W^-(\gamma)$ relevant for LEP2 experiments. Results for related processes $e^+e^- \to \gamma\gamma, \gamma Z, ZZ$ and some properties of $e^+e^- \to W^+W^-$ have been discussed by M. Böhm [10] in these proceedings.

2. The W-pair production cross-section

The Born cross-section for W-pair production in e^+e^--annihilation

$$e^+(p_+, \bar{\sigma}) + e^-(p_-, \sigma) \to W^+(p_{W+}, \bar{\lambda}) + W^-(p_{W-}, \lambda) \tag{2}$$

has been discussed extensively in the literature [11]. For the discussion of radiative corrections we need the general form of the cross-section. We denote by σ and λ the helicities of the incoming and outgoing particles, respectively. Kinematical variables used are

$$
\begin{aligned}
s &= (p_+ + p_-)^2 = 4E_b^2 \\
t &= (p_+ - p_{W+})^2 = \frac{1}{2}(2m_e^2 + 2M_W^2 - s + s\beta_e\beta_W \cos\theta)
\end{aligned}
\tag{3}
$$

with $\beta_e = \sqrt{1 - 4m_e^2/s}$ and $\beta_W = \sqrt{1 - 4M_W^2/s}$. In the c.m. frame, we denoted by E_b the beam energy and by θ the scattering angle between the electron and the W^-. In the approximation of vanishing electron mass ($m_e \ll M_W, E_b$) two initial states only, $e_L^- e_R^+$ and $e_R^- e_L^+$ (L=lefthanded, R=righthanded), yield nonvanishing cross-sections. Hence the helicity of the positron is fixed for given electron helicity $h_e = \sigma$ ($= \Delta\sigma = \frac{1}{2}(\sigma - \bar{\sigma})$), and we can label the initial state by h_e, simply. The transition matrixelement has the form

$$M(h_e; \lambda\bar{\lambda}) = \bar{v}(p_+, \bar{\sigma}) \mathbf{F}_{h_e}^{\mu\nu} \mathbf{P}_{h_e} u(p_-, \sigma) \varepsilon_\mu^*(p_{W+}, \bar{\lambda}) \varepsilon_\nu^*(p_{W-}, \lambda) \tag{4}$$

where v and u denote the spinors of the positron and electron, respectively, and ε_μ and ε_ν are the W polarization vectors. $\mathbf{P}_{h_e} = \frac{1}{2}(1 + h_e\gamma_5)$ is the chiral projector onto fixed electron helicity and $\mathbf{F}$ is a 4×4 matrix the form of which is given below.

At one loop order CP is conserved for the process $e^+ e^- \rightarrow \tilde{W}^+ W^-$ within the Standard Model. CP invariance implies

$$M(\sigma, \bar{\sigma}; \lambda, \bar{\lambda}) = M(-\bar{\sigma}, -\sigma; -\bar{\lambda}, -\lambda)$$

which, for the (in the approximation $m_e \simeq 0$) nonvanishing amplitudes, reads

$$M(h_e; \lambda, \bar{\lambda}) = M(h_e; -\bar{\lambda}, -\lambda).$$

This means that final states with the same $\Delta\lambda = (\lambda - \bar{\lambda})$ have identical cross-sections if $|\lambda| + |\bar{\lambda}| \neq 0$. Thus, for each h_e there are 6 independent cross-sections which can be measured and, correspondingly, there are 6 independent physical amplitudes. Thus the matrix $\mathbf{F}$ can be decomposed into 6 CP-invariant amplitudes for each electron helicity. It is convenient however to use the following overcomplete decomposition in terms of 8 amplitudes:

$$
\begin{aligned}
\mathbf{F}_{h_e}^{\mu\nu} &= \sum_{i=1}^{8} X_i^{(h_e)} \mathbf{O}_{X_i}^{\mu\nu} = \sum_{i=1}^{4} S_i^{(h_e)} \mathbf{O}_{S_i}^{\mu\nu} + \sum_{i=1}^{2} T_i^{(h_e)} \mathbf{O}_{T_i}^{\mu\nu} + \sum_{i=7}^{8} X_i^{(h_e)} \mathbf{O}_{X_i}^{\mu\nu} \\
&= S_1^{(h_e)}[\not{P} g^{\mu\nu} + 2(\gamma^\mu P^\nu - \gamma^\nu P^\mu)] + S_2^{(h_e)}(\gamma^\mu P^\nu - \gamma^\nu P^\mu) \\
&\quad + S_3^{(h_e)} \not{P} \frac{P^\mu P^\nu}{P^2} + S_4^{(h_e)} i\varepsilon^{\mu\nu\alpha\beta} \gamma_\alpha r_\beta \\
&\quad + T_1^{(h_e)} \gamma^\mu(\not{p}_- - \not{p}_{W-})\gamma^\nu + T_2^{(h_e)}(\gamma^\mu p_-^\nu - \gamma^\nu p_+^\mu) \\
&\quad + X_7^{(h_e)} \not{P} \frac{Q^\mu Q^\nu}{P^2} + X_8^{(h_e)} \not{P} \frac{P^\mu Q^\nu - P^\nu Q^\mu}{P^2}.
\end{aligned}
\tag{5}
$$

The momentum variables used in this expression are $P = p_+ + p_-$, $Q = p_+ - p_-$ and $r = p_{W+} - p_{W-}$. The decomposition is chosen in such a way that the individual amplitudes are crossing symmetric and CP-even. The first four amplitudes $X_{i=1,\cdots,4}$ have been denoted by S_i in order to exhibit that s-channel diagrams only contribute to these S-amplitudes. Similarly, t-channel diagrams only contribute to the T-amplitudes $T_1 = X_5$ and $T_2 = X_6$. The invariant amplitudes S_i, T_i and X_i are functions of s,t and the particle masses.

Two dependent operators can be eliminated using the on-shell relations (Chisholm identities)

$$\mathbf{O}_{X_8} = -(h_e \mathbf{O}_{S_4} + \beta c \mathbf{O}_{S_2})$$
$$\mathbf{O}_{S_4} = h_e[2(\mathbf{O}_{T_1} - \mathbf{O}_{T_2}) - (\mathbf{O}_{S_1} - \mathbf{O}_{S_2})]$$

where $c = \cos\theta$ and $\beta = \beta_W$. The differential cross-section then reads

$$\frac{d\sigma(h_e; \lambda\bar{\lambda})}{d\cos\theta} = \frac{\beta}{32\pi s}|\sum_{i=1}^{8} X_i^{(h_e)}\mathbf{M}_{X_i}(h_e; \lambda\bar{\lambda})|^2 \tag{6}$$

where

$$M_{X_i}(h_e; \lambda\bar{\lambda}) = \bar{v}(p_+, \bar{\sigma})\mathbf{O}_{X_i}^{\mu\nu}\mathbf{P}_{h_e}u(p_-, \sigma)\varepsilon_\mu^*(p_{W+}, \bar{\lambda})\varepsilon_\nu^*(p_{W-}, \lambda) \tag{7}$$

are the helicity transition matrix elements associated with the operators $\mathbf{O}_{X_i}$. The reduced matrix elements $\tilde{M}_{X_i}(h_e; \lambda\bar{\lambda})$, defined by

$$M_{X_i}(h_e; \lambda\bar{\lambda}) = -2\sqrt{2}M_W^2\gamma^2 d_{\Delta\sigma\Delta\lambda}^{J_0}(\theta)\tilde{M}_{X_i}(h_e; \lambda\bar{\lambda}) \tag{8}$$

are listed in Table 1. γ is the Lorentz factor defined by $\gamma^2 = s/(4M_W^2)$. In the Born approximation, only the three amplitudes $S_1^{(\pm)}$ and $T_1^{(-)}$ are non-zero. They are given by

$$S_{10}^{(-)} = \frac{g^2}{s - M_Z^2}(v - a) - \frac{e^2}{s} = -\frac{e^2}{s}(1 - \frac{1}{1 - M_Z^2/s}) - \frac{g^2}{2s}\frac{1}{1 - M_Z^2/s}$$

$$S_{10}^{(+)} = \frac{g^2}{s - M_Z^2}(v + a) - \frac{e^2}{s} = -\frac{e^2}{s}(1 - \frac{1}{1 - M_Z^2/s}) \tag{9}$$

$$T_{10}^{(-)} = -\frac{g^2}{2t} = \frac{g^2}{s}\frac{1}{\frac{1+\beta^2}{2} - \beta\cos\theta}$$

with $v = \sin^2\Theta_W - \frac{1}{4}$ and $a = \frac{1}{4}$, the vector- and axial-vector neutral-current amplitudes (Zee-vertex), $e^2 = 4\pi\alpha$ and $g^2 = 4M_W^2\sqrt{2}G_\mu$. In Eq. (9) the second form of the Born amplitudes is given in order to better visualize the gauge cancellation at work. The lowest order cross-section is then given by the simple form

$$\frac{d\sigma_0(\pm; \lambda\bar{\lambda})}{d\cos\theta} = \frac{\beta}{32\pi s}|\mathbf{M}_0(\pm; \lambda\bar{\lambda})|^2 \tag{10}$$

with

$$\mathbf{M}_0(-; \lambda\bar{\lambda}) = S_{10}^{(-)}\mathbf{M}_{S_1}(-; \lambda\bar{\lambda}) + T_{10}^{(-)}\mathbf{M}_{T_1}(-; \lambda\bar{\lambda})$$
$$\mathbf{M}_0(+; \lambda\bar{\lambda}) = S_{10}^{(+)}\mathbf{M}_{S_1}(+; \lambda\bar{\lambda})$$

The Born cross-section determines the main features of W-pair production:

- The dominant production mode is $e_L^- e^+ \to W^- W^+$. $e_R^- e^+ \to W^- W^+$ is suppressed by two orders of magnitude mainly because there is no ν-exchange contribution in this case. The s-channel γ- and Z-exchange terms cancel more effectively than the t-channel ν-exchange against the s-channel Z-exchange contributions.

  Longitudinally polarized beams would allow to study the pure triple gauge coupling process by using right-handed electrons.

Table 1: Reduced matrix elements $\bar{M}_{X_i}(h_e; \lambda\bar{\lambda})$ ($c = \cos\theta$, $s = \sin\theta$).

amplitude state	$d^{J_0}_{\Delta\sigma\Delta\lambda}(\theta)$	S_1	S_2	S_3	S_4
$(\pm;00)$	$\mp s/\sqrt{2}$	$2\gamma^2\beta(3-\beta^2)$	$4\gamma^2\beta$	$-2\gamma^2\beta^3$	0
$(\pm;--),(\pm;++)$	$\mp s/\sqrt{2}$	2β	0	0	0
$(+;-0)(+;0+)$	$(1-c)/2$	$4\gamma\beta$	$2\gamma\beta$	0	$2\gamma\beta^2$
$(-;0-)(-;+0)$	$(1-c)/2$	$4\gamma\beta$	$2\gamma\beta$	0	$-2\gamma\beta^2$
$(+;0-)(+;+0)$	$(1+c)/2$	$4\gamma\beta$	$2\gamma\beta$	0	$-2\gamma\beta^2$
$(-;-0)(-;0+)$	$(1+c)/2$	$4\gamma\beta$	$2\gamma\beta$	0	$2\gamma\beta^2$
$(\pm;\mp\pm)$	$\mp s(1-c)/2$	0	0	0	0
$(\pm;\pm\mp)$	$\pm s(1+c)/2$	0	0	0	0

T_1	T_2	X_7	X_8
$\gamma^2(\beta(3-\beta^2)-2c)$	$2\gamma^2(\beta-c)$	$2\gamma^2\beta c^2$	$-4\gamma^2\beta^2 c$
$(\beta-c)$	$-c$	$-\beta(1-c^2)$	0
$-\gamma(1-2\beta-\beta^2+2c)$	$-\gamma(1-\beta+2c)$	$2\gamma\beta c\,(1+c)$	$-2\gamma\beta^2(1+c)$
$-\gamma(1-2\beta-\beta^2+2c)$	$-\gamma(1-\beta+2c)$	$2\gamma\beta c\,(1+c)$	$-2\gamma\beta^2(1+c)$
$\gamma(1+2\beta-\beta^2-2c)$	$\gamma(1+\beta-2c)$	$-2\gamma\beta c\,(1-c)$	$2\gamma\beta^2(1-c)$
$\gamma(1+2\beta-\beta^2-2c)$	$\gamma(1+\beta-2c)$	$-2\gamma\beta c\,(1-c)$	$2\gamma\beta^2(1-c)$
$-\sqrt{2}$	$-\sqrt{2}$	$2\sqrt{2}\beta\,(1+c)$	0
$-\sqrt{2}$	$-\sqrt{2}$	$-2\sqrt{2}\beta\,(1-c)$	0

- Longitudinal W's contribute substantially to the total cross-section. In the forward-backward direction W_L's are produced as frequently as the transversal W_T's. This arises from angular momentum conservation, and because $|\Delta\sigma| = 1$, which implies $J_z = \pm 1$ and thus $|\Delta\lambda| = 1$.

- Deviations from the Standard Model gauge couplings in general lead to substantially larger cross-sections, due to lack of gauge cancellations. For example [12], for composite W's, like for the ρ meson in hadron physics, there is no reason why they should have $SU(2)_L$ gauge couplings. Obviously, they would have standard electric coupling and an anomalous magnetic moment $\mu = \frac{e}{2M_W}(1 + \kappa)$. In the Standard Model we have $\kappa_{SM} = 1$. The absence of a $SU(2)_L$ coupling amounts to a change of the ZWW coupling to $-tan^2\Theta_W$ times the Standard Model value, and an anomalous magnetic moment implies a non-vanishing S_2 amplitude at the tree level

$$S_2^{(\pm)} = (\kappa - 1)\, S_1^{(\pm)}.$$

The effects of such a change are shown in Fig. 1.

3. Radiative corrections

Virtual corrections to $e^+e^- \to W^+W^-$ are due to ν-, γ- and Z- self-energies, $e\nu_e W$, $e^+e^-\gamma$, e^+e^-Z, $W^+W^-\gamma$, W^+W^-Z form factors and box diagrams as indicated in Fig. 2. The self-energies may be written in the form:

$$
\begin{aligned}
\text{``}\nu_e\nu_e\text{''} &= i(1 - \gamma_5)\not{p}_\nu \Sigma_1^{\nu\nu}(t) \qquad\qquad\qquad (11)\\
\text{``}\gamma\gamma\text{''} &= i(ev)^2 g^{\mu\nu} A_1^{\gamma\gamma}(s) + \cdots\\
\text{``}\gamma Z\text{''} &= i(ev)M_Z g^{\mu\nu} A_1^{\gamma Z}(s) + \cdots\\
\text{``}ZZ\text{''} &= iM_Z^2 g^{\mu\nu} A_1^{ZZ}(s) + \cdots,
\end{aligned}
$$

where the dots stand for amplitudes which do not contribute to the cross-section in our approximation. Similarly, the form factors may be written in the form (external $e^\pm$ and $W^\pm$ lines taken on-shell):

$$
\begin{aligned}
\text{``}We\nu_e\text{''} &= i\frac{M_W}{\sqrt{2}v}\{\gamma^\mu A_1^{We\nu}(t) + \frac{(p_e + p_\nu)^\mu}{M_W^2}\not{p}_\nu A_3^{We\nu}(t) + \cdots\}(1 - \gamma_5) \quad (12)\\
\text{``}\gamma ee\text{''} &= ie\{\gamma^\mu A_1^{\gamma ee}(s) + \gamma^\mu\gamma_5 A_2^{\gamma ee}(s) + \cdots\}\\
\text{``}Zee\text{''} &= i\frac{2M_Z}{v}\{\gamma^\mu A_1^{Zee}(s) + \gamma^\mu\gamma_5 A_2^{Zee}(s) + \cdots\}\\
\text{``}VWW\text{''} &= i\ C_V\{(r^\mu g^{\rho\sigma} + 2(g^{\mu\rho}P^\sigma - g^{\mu\sigma}P^\rho))A_1^{VWW}(s)\\
&\quad + (g^{\mu\rho}P^\sigma - g^{\mu\sigma}P^\rho)A_2^{VWW}(s)\\
&\quad + r^\mu \frac{PP^{\rho\sigma}}{P^2}A_3^{VWW}(s)\\
&\quad + i\varepsilon^{\rho\sigma\mu\alpha}r_\alpha A_4^{VWW}(s)\}
\end{aligned}
$$

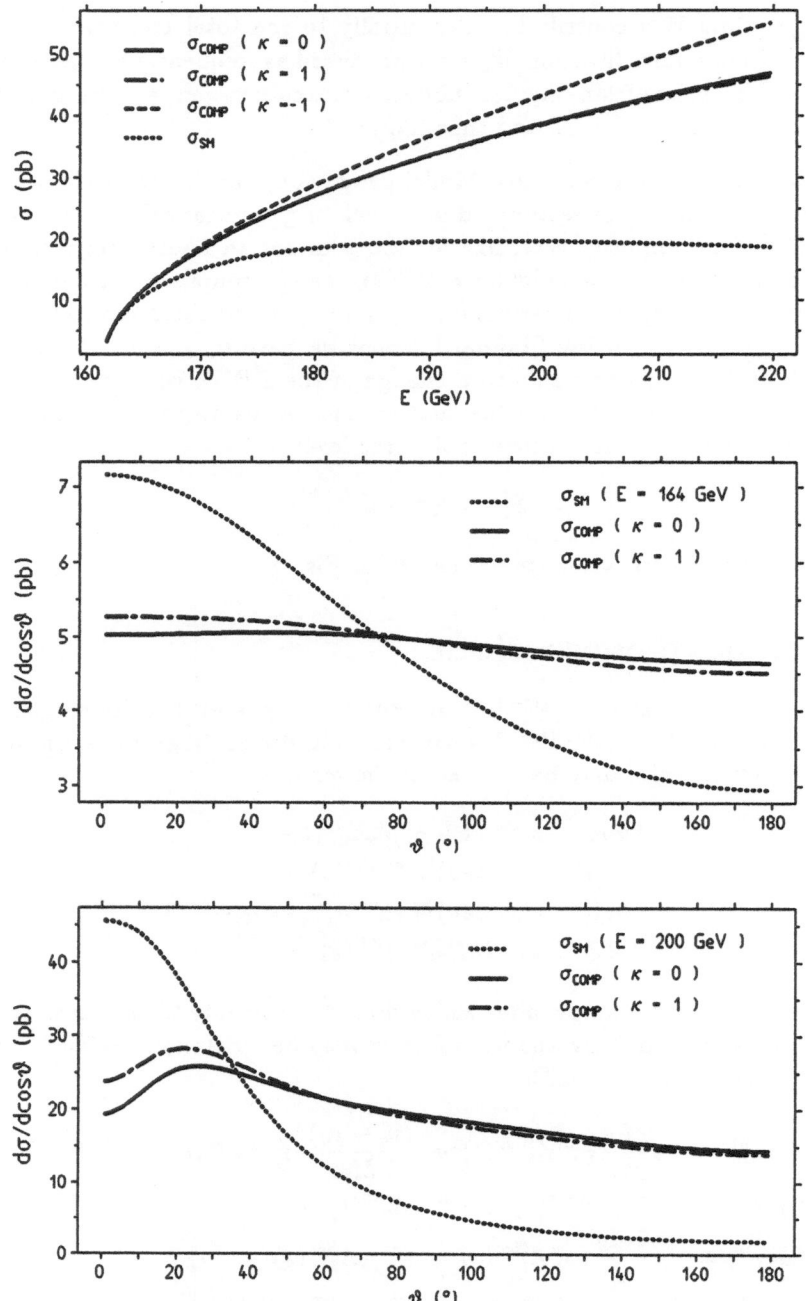

Figure 1: $e^+e^- \rightarrow W^+W^-$ lowest order cross-sections for anomalous couplings

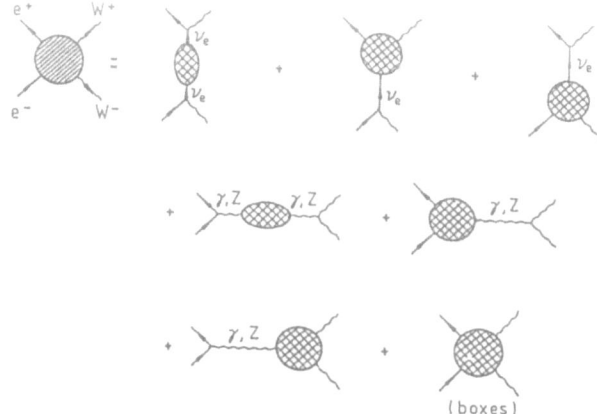

Figure 2: Virtual corrections for the process $e^+e^- \to W^+W^-$

with $C_\gamma = e$ and $C_Z = -\frac{2M_W^2}{vM_Z}$.

We use the on-shell renormalization scheme of Ref. [13]. The renormalized amplitudes are obtained by adding the appropriate counter terms as follows: For each external (amputated) field i, a wavefunction renormalization factor $(1 + \frac{1}{2}\delta Z_i)$ must be applied. For each physical field, the wavefunction factor must be determined by the residue of the propagator pole in order to get gauge invariant renormalized on-shell amplitudes. The wavefunction renormalizations for the fermions have terms proportional to γ_5. For the electron we write $\delta Z_e = z_v + z_a \gamma_5$, for the neutrino $\delta Z_\nu = z_\nu(1 - \gamma_5)$. The parameter renormalizations follow from the substitutions

$$M_V^2 \to M_V^2 \left(1 + \frac{\delta M_V^2}{M_V^2}\right), \quad v^{-1} \to v^{-1} \left(1 + \frac{\delta v^{-1}}{v^{-1}}\right)$$

and

$$e \to e \left(1 + \frac{\delta e}{e}\right)$$

for the parameters in the Born terms.

Throughout we use the physical particle masses, M_Z and M_W in particular, as input parameters and consequently the definition

$$\sin^2 \Theta_W = 1 - \frac{M_W^2}{M_Z^2} \tag{13}$$

for the weak mixing parameter. Then, $\sin^2 \Theta_W$ is renormalized by

$$\frac{\delta \sin^2 \Theta_W}{\sin^2 \Theta_W} = \frac{\cos^2 \Theta_W}{\sin^2 \Theta_W} \left(\frac{\delta M_Z^2}{M_Z^2} - \frac{\delta M_W^2}{M_W^2}\right).$$

v^{-1} is given by either

$$v^{-2} = \sqrt{2}G_\mu \quad or \quad v^{-2} = \frac{\pi\alpha}{M_W^2 \sin^2 \Theta_W} \tag{14}$$

depending on whether the Fermi constant G_μ or the fine structure constant $\alpha = e^2/4\pi$ is chosen as an input parameter in addition to the masses. Accordingly, in the first case

$$\frac{\delta v^{-1}}{v^{-1}} = \frac{1}{2}\frac{\delta G_\mu}{G_\mu}$$

$$\frac{\delta e}{e} = \frac{1}{2}\left(\frac{\delta G_\mu}{G_\mu} + \frac{1}{M_Z^2 - M_W^2}\{(M_Z^2 - 2M_W^2)\frac{\delta M_W^2}{M_W^2} + M_W^2\frac{\delta M_Z^2}{M_Z^2}\}\right)$$

with $\frac{\delta G_\mu}{G_\mu}$ determined from μ-decay. In the second case

$$\frac{\delta e}{e} = \frac{1}{2}\frac{\delta \alpha}{\alpha}$$

$$\frac{\delta v^{-1}}{v^{-1}} = \frac{1}{2}\left(\frac{\delta \alpha}{\alpha} - \frac{1}{M_Z^2 - M_W^2}\{(M_Z^2 - 2M_W^2)\frac{\delta M_W^2}{M_W^2} + M_W^2\frac{\delta M_Z^2}{M_Z^2}\}\right)$$

with $\frac{\delta \alpha}{\alpha}$ determined by electric charge renormalization. The explicit expressions for the counter-terms may by found in Ref. [14] (see also [13,9]). The invariant amplitudes defined in Eq. (5) may now be written in terms of the renormalized self-energies and form factors as follows:

$$S_1^{(\mp)} = \sqrt{2}G_\mu\{([(v \mp a) + (v \mp a)A_{1r}^{ZZ}(s)]\frac{M_Z^2}{s - M_Z^2 + iM_Z\Gamma_Z} \tag{15}$$
$$-A_{1r}^{\gamma Z}(s)\chi_\gamma(s) + (A_{1r}^{Zee}(s) \mp A_{2r}^{Zee}(s))) + (v \mp a)A_{1r}^{ZWW}(s)]\chi_Z(s)$$
$$-[1 + A_{1r}^{\gamma\gamma}(s)\chi_\gamma(s) - \frac{v \mp a}{2\cos^2\Theta_W}A_{1r}^{\gamma Z}(s)$$
$$+(A_{1r}^{\gamma ee}(s) \mp A_{2r}^{\gamma ee}(s)) + A_{1r}^{\gamma WW}(s)]\chi_\gamma(s)\} + S_{1,box}^{(\mp)}$$
$$S_i^{(\mp)} = \sqrt{2}G_\mu\{(v \mp a)A_i^{ZWW}(s)\chi_Z(s) - A_i^{\gamma WW}(s)\chi_\gamma(s)\} + S_{i,box}^{(\mp)} \ ; \ i = 2,3,4$$
$$T_1^{(-)} = \sqrt{2}G_\mu\{-(1 + 2\Sigma_r^{\nu_e\nu_e}(t) + 2A_{1r}^{We\nu_e}(t))\chi_\nu(t)\} + T_{1,box}^{(-)}$$
$$T_2^{(-)} = \sqrt{2}G_\mu\{-A_3^{We\nu_e}(t)\} + T_{2,box}^{(-)}$$

$$\tag{16}$$

Only the amplitudes $S_1^{(\mp)}$ and $T_1^{(-)}$, which are non-zero at the tree level, are renormalized at one loop order. The amplitudes $T_{1,2}^{(+)}$ and $X_{7,8}^{(\mp)}$ get contributions from boxes only and $S_{4,box}^{(\mp)} = 0$. The pole factors are given by

$$\chi_Z = \frac{4M_W^2}{s - M_Z^2 + iM_Z\Gamma_Z}; \ \chi_\gamma = \frac{4M_W^2\sin^2\Theta_W}{s}; \ \chi_\nu = \frac{4M_W^2}{t}$$

The explicit formal expressions for the renormalized amplitudes can be found in Ref. [9]. Notice that by our choice of amplitudes, each s-channel amplitude picks up exactly one ZWW form factor and one γWW form factor.

We have calculated the invariant amplitudes in the dimensional regularization scheme [15]. An anticommuting γ_5 has been used in order to avoid spurious anomalies [16]. Vector current conservation has been imposed in order to get a unique result for the individual fermion triangle contributions. In any case, anomalies cancel familywise due to lepton-quark duality.

Calculations have been performed in the 't Hooft gauge with an arbitrary gauge parameter. This allows to test gauge invariance for all renormalized on-shell amplitudes. On-shell infrared singularities have been regularized using an infinitesimal photon mass m_γ. The analytic part of the calculation has been done with the help of SCHOONSCHIP.

4. Bremsstrahlung

Infrared finite cross-sections can be obtained only if real photon emission of the charged states is taken into account. The Bremsstrahlung diagrams are depicted in Fig. 3. Here we only include soft photons for which we have the usual factorization

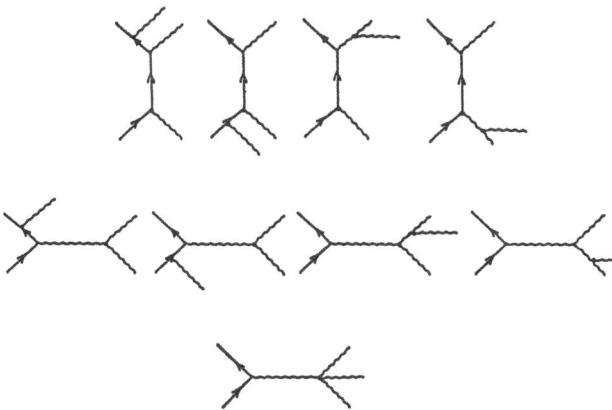

Figure 3: Bremsstrahlung diagrams for $e^+e^- \rightarrow W^+W^-$

of the Bremsstrahlung cross-section

$$d\sigma_B = C_{Br} d\sigma_0 \qquad (17)$$

into a Bremsstrahlung integral and the Born cross-section.

The Bremsstrahlung integral is given by

$$C_{Br} = -\frac{e^2}{(2\pi)^3} \int_{|\vec{k}|\leq\omega} \frac{d^3k}{2\omega_k} (\hat{p}_i - \hat{p}_f)^2; \quad \omega_k = \sqrt{m\gamma^2 + \vec{k}^2} \qquad (18)$$

where

$$\hat{p}_i = \left(\frac{p_+}{kp_+} - \frac{p_-}{kp_-}\right) \quad and \quad \hat{p}_f = \left(\frac{p_{W^+}}{kp_{W^+}} - \frac{p_{W^-}}{kp_{W^-}}\right)$$

are the radiation factors from initial and final state particles, respectively. The result of the integration may be written as a sum of four terms

$$C_{Br} = C(s, m_e^2) + C(s, M_W^2) + C_{int}(t, s, m_e^2, M_W^2) - C_{int}(u, s, m_e^2, M_W^2) \qquad (19)$$

which correspond to initial and final state radiation and t- and u- channel interference terms. The functions are given by

$$C(s, m_i^2) = \frac{e^2}{\pi^2}\left\{(1 - \frac{y_i}{2})(F_1(y_i)\ln\frac{2\omega}{m_\gamma} - F_2(y_i)) - \frac{1}{2}(\ln\frac{2\omega}{m_\gamma} - F_1(y_i))\right\} \quad (20)$$

$$C(s, m_e^2) \simeq \frac{e^2}{2\pi^2}\left\{(\ln\frac{s}{m_e^2} - 1)\ln\frac{2\omega}{m_\gamma} - \frac{1}{4}\ln^2\frac{m_e^2}{s} + \frac{1}{2}\ln\frac{s}{m_e^2} - \frac{\pi^2}{6}\right\}_{m_e \to 0}$$

with

$$F_1(y_i) = -\frac{1}{2\beta_i}\ln\xi_i$$

$$F_2(y_i) = \frac{1}{2\beta_i}\left\{\frac{\pi^2}{6} - Sp(\xi_i) + \frac{1}{4}\ln^2(\xi_i) - \ln(\xi_i)\ln(1 - \xi_i)\right\}$$

where

$$y_i = \frac{4m_i^2}{s}, \quad \beta_i = \sqrt{1 - y_i}, \quad \xi_i = \frac{1 - \beta_i}{1 + \beta_i}.$$

For the interference term we find

$$C_{int}(t, s, m_e^2, M_W^2) = \frac{e^2}{2\pi^2}\left\{\ln\frac{(M_W^2 - t)^2}{M_W^2 m_e^2}\ln\frac{2\omega}{m_\gamma}\right. \quad (21)$$

$$+ Sp(1 + \frac{st}{M_W^2(M_W^2 - t)}\frac{1 + \beta_W}{2})$$

$$+ Sp(1 + \frac{st}{M_W^2(M_W^2 - t)}\frac{1 - \beta_W}{2})$$

$$\left. - Sp(1 + \frac{st}{(M_W^2 - t)^2}) + \cdots\right\}$$

where the dots stand for terms which drop out in the difference between the t- and u- channel contributions. $Sp(x) = -\int_0^1(dt/t)\ln(1 - xt)$ is the Spence function. The infrared singular part of C_{Br} is given by

$$C_{Br}^{sing} = \frac{e^2}{\pi^2}\left\{-\ln\frac{2\omega}{m_\gamma}\left[1 + \frac{1}{2}\ln\frac{m_e^2}{s} + \left(1 - \frac{y_W}{2}\right)\frac{1}{2\beta_W}\ln\xi_W + \ln\frac{M_W^2 - u}{M_W^2 - t}\right]\right\}. \quad (22)$$

The singular terms cancel against similar terms appearing in the virtual corrections of the amplitudes $S_1^{(\mp)}$ and $T_1^{(-)}$ such that

$$\frac{2Re\delta S_1^{(\pm)}}{ReS_{10}^{(\pm)}} + C_{Br}, \quad \frac{2Re\delta T_1^{(-)}}{ReT_{10}^{(-)}} + C_{Br},$$

$$\frac{Re\delta S_1^{(-)}}{ReS_{10}^{(-)}} + \frac{Re\delta T_1^{(-)}}{ReT_{10}^{(-)}} + C_{Br} \quad (23)$$

are infrared finite and free of large double-logarithms. The sums Eq. (23) contain terms which depend on the soft-photon cut off $\omega \ll E_b$

$$C_{IR}(\omega) = \frac{e^2}{\pi^2}\left\{-\frac{1}{2}\left(\ln\frac{2\omega}{m_e} + \ln\frac{2\omega}{M_W}\right)\right. \quad (24)$$

$$\left. - \ln\frac{\omega}{E_b}\left[\frac{1}{2}\ln\frac{m_e^2}{s} + \left(1 - \frac{y_W}{2}\right)\frac{1}{2\beta_W}\ln\xi_W + \ln\frac{M_W^2 - u}{M_W^2 - t}\right]\right\}.$$

For small ω the perturbation expansion breaks down and multi soft photon effects must be summed by exponentiation [17]

$$1 + C_{IR}(\omega) + \cdots = \exp C_{IR}(\omega) \qquad (25)$$

and the corrected cross-section takes the form

$$\frac{d\sigma^*}{d\cos\theta} = \frac{d\sigma_0}{d\cos\theta}(\exp C_{IR}(\omega) + C_{Br} + 2ReC - C_{IR}(\omega)) \qquad (26)$$

where $2ReC$ denotes the virtual corrections. The possible values for the photon energy cut off are restricted by phase space to

$$\omega = x_r E_b \left\{ \frac{2}{1 + x_r + \sqrt{(1 - x_r)^2 - y_W}} \right\} \qquad (27)$$

which goes to $\omega \simeq x_r E_b$ for $E_b \ll M_W$. x_r is the fraction of W-energy which is hidden in soft photons.

5. Numerical results

For the numerical calculation of the cross-section we have used the following parameters: We choose $M_W = 81$ GeV and $\sin^2 \Theta_W = 0.2281$ [18] i.e. $M_Z = 92.18$ GeV. The hadronic contributions to the photon vacuum polarization are evaluated using the effective quark masses

$$m_u = 53 \text{ MeV}, \quad m_d = 71 \text{ MeV}$$
$$m_s = 174 \text{ MeV}, \quad m_c = 1.5 \text{ GeV}$$
$$m_b = 4.5 \text{ GeV}$$

which, in the range 50 GeV $\leq \sqrt{s} \leq$ 200 GeV, give a perfect fit for the result which has been obtained by evaluating the e^+e^--data with the help of dispersion relations [19].

In the numerical evaluation of the radiative corrections we have to choose a soft photon cut off x_r. As we can see from Eq. (27) we cannot keep x_r fixed when we approach threshold ($y_W \to 1$). In the results given below we choose

$$x_r = min \left\{ 1 - \frac{M_W}{E_b}, 0.1 \right\}$$

i. e. $x_r = 0.1$ if this is kinematically allowed and $x_r = 1 - M_W/E_b$ if E_b lower than 90 GeV (for $M_W = 81$ GeV). In Fig. 4 the total cross-section is shown as a function of the c.m. energy $\sqrt{s}$ for both the G_μ and the α scheme. Fig. 5 shows the same corrections for the angular distribution for $\sqrt{s} = 250$ GeV. Numerical values are presented in Tabs. 2 and 3.

At the Born level the cross-sections differ by $\delta\sigma = \frac{\sigma(\alpha) - \sigma(G_\mu)}{\sigma(G_\mu)} \simeq -13.7$ % in the two schemes. This difference, as expected, becomes substantially smaller after inclusion

Table 2: σ_{tot}(pb) for some values of the beam energy E_b. Input parameters: $M_Z = 92.18$ GeV, $M_W = 81$ GeV, $m_t = 60$ GeV and $m_H = 100$ GeV .

	E_b (GeV)	σ_0	σ	$C_B(\%)$	$C_F(\%)$	$C_{tot}(\%)$
$\sigma = \sigma(G_\mu)$	165	10.044	8.543	-16.60	1.66	-14.95
	170	14.830	13.292	-12.04	1.66	-10.37
	175	17.220	15.888	-9.41	1.67	-7.73
	180	18.554	17.440	-7.68	1.68	-6.00
	185	19.294	17.855	-9.14	1.69	-7.46
	190	19.666	18.013	-10.10	1.70	-8.40
	195	19.798	18.000	-10.79	1.70	-9.08
	200	19.771	17.871	-11.32	1.71	-9.61
	205	19.635	17.665	-11.75	1.72	-10.03
	210	19.426	17.409	-12.11	1.73	-10.38
	225	18.557	16.491	-12.88	1.74	-11.14
	250	16.882	14.874	-13.65	1.76	-11.89
	275	15.292	13.401	-14.13	1.77	-12.36
	300	13.884	12.121	-14.48	1.78	-12.70
	400	9.889	8.547	-15.35	1.78	-13.57
	500	7.493	6.430	-15.94	1.76	-14.18
$\sigma = \sigma(\alpha)$	165	8.668	8.537	-16.56	15.04	-1.52
	170	12.799	13.193	-11.97	15.05	3.08
	175	14.861	15.714	-9.31	15.06	5.74
	180	16.012	17.210	-7.58	15.06	7.48
	185	16.650	17.653	-9.05	15.07	6.02
	190	16.971	17.836	-9.98	15.08	5.10
	195	17.086	17.843	-10.65	15.09	4.43
	200	17.062	17.731	-11.18	15.09	3.92
	205	16.945	17.539	-11.59	15.10	3.51
	210	16.764	17.295	-11.94	15.11	3.17
	225	16.015	16.403	-12.70	15.12	2.42
	250	14.569	14.814	-13.46	15.14	1.68
	275	13.196	13.357	-13.93	15.15	1.22
	300	11.982	12.088	-14.27	15.16	0.88
	400	8.534	8.535	-15.14	15.15	0.02
	500	6.466	6.428	-15.72	15.14	-0.58

Table 3: $d\sigma/d(\cos\theta)$ (pb) for different values of the scatering angle θ. Input parameters: $M_Z = 92.18$ GeV, $M_W = 81$ GeV, $m_t = 35$ GeV and $m_H = 100$ GeV

θ (deg)	σ_0	$\sigma(G_\mu)$	$C_B(\%)$	$C_F(\%)$	$C_{tot}(\%)$
1	41.673	38.362	-9.97	2.03	-7.95
10	40.198	37.005	-9.99	2.05	-7.94
20	35.471	32.650	-10.05	2.10	-7.95
30	28.743	26.444	-10.14	2.15	-8.00
40	22.157	20.367	-10.28	2.20	-8.08
50	16.787	15.411	-10.45	2.26	-8.20
60	12.751	11.687	-10.67	2.32	-8.35
70	9.811	8.976	-10.91	2.39	-8.51
80	7.687	7.018	-11.18	2.48	-8.70
90	6.144	5.598	-11.47	2.57	-8.89
100	5.011	4.556	-11.77	2.68	-9.09
110	4.166	3.779	-12.08	2.79	-9.28
120	3.525	3.191	-12.39	2.92	-9.47
130	3.035	2.742	-12.69	3.05	-9.64
140	2.660	2.400	-12.97	3.18	-9.80
150	2.382	2.145	-13.22	3.30	-9.92
160	2.187	1.969	-13.42	3.41	-10.01
170	2.072	1.864	-13.54	3.48	-10.06
179	2.035	1.830	-13.58	3.51	-10.08

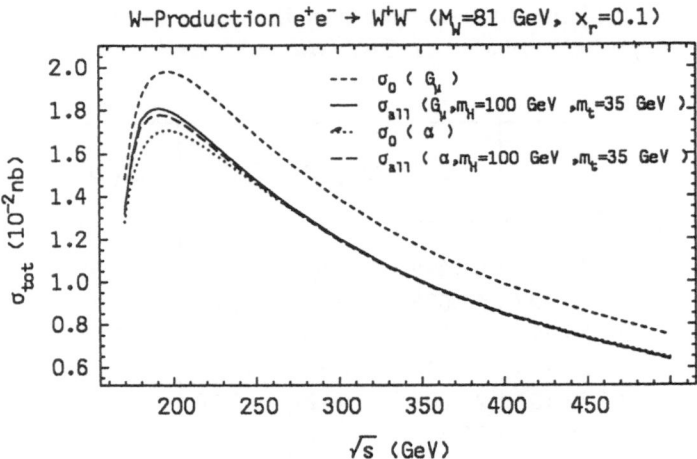

Figure 4: Total cross-section in lowest order(σ_0) and including radiative corrections (σ_{all}) in the G_μ- and α-scheme, respectively

of the one-loop corrections. Nevertheless there remains a nonnegligible difference which signals missing higher order effects. The difference is largest, $\simeq$ -1.7%, at about 180 GeV ($\simeq$ 12 GeV below the peak), -1.1% at 170 GeV and -1.2% at 200 GeV. For a required precision of better than 1% the missing leading higher order effects must be included. Since we know that the parameters of the two schemes are related by

$$\sqrt{2}G_\mu = \frac{\pi\alpha}{M_W^2 \sin^2\Theta_W} \frac{1}{1 - \Delta r},$$

we have to linear order ($O(\alpha)$) in the cross-section

$$\sigma_0(G_\mu) \simeq \sigma_0(\alpha)(1 + 2\Delta r).$$

This is what is usually included at $O(\alpha)$ in the α-scheme. To second order we get

$$\sigma_0(G_\mu) \simeq \sigma_0(\alpha)(1 + 2\Delta r + 3\Delta r^2).$$

and the additional term $3\Delta r^2$ is the missing next leading term. Typically for $\Delta r \simeq$ 7% we get $3\Delta r^2 \simeq$ 1.5% which essentially accounts for the above mentioned difference.

The tables also include the percentage corrections from fermion loops C_F and the remaining bosonic correction C_B. The latter is large and negative mainly due to QED effects, which lower the cross-section because of the soft photon cut applied here. The difference between the G_μ and α scheme mainly shows up in the fermionic corrections C_F which are large ($\simeq$ 15.4%) in the α-scheme, but small ($\simeq$ 2.3%) in the G_μ-scheme. In the latter case, the leading light fermion contributions originating from the photon vacuum polarization are absent.

For further results we refer to Refs. [8] and [9]. In [8] results are given for the α-scheme, in [9] for the G_μ-scheme. Using the same scheme and parameters we found

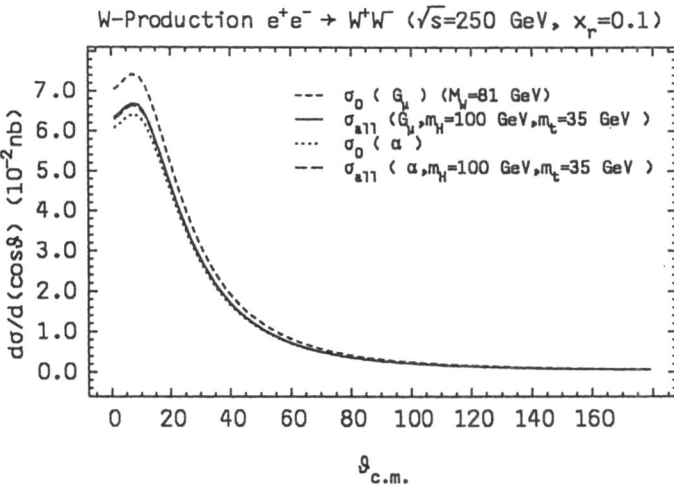

Figure 5: Angular distribution in the G_μ- and α-scheme

perfect agreement of the two independent calculations [9]. The approach used in Ref. [8] has been outlined by M. Böhm in the previous talk at this meeting.

6. Higgs- and top-mass dependence

The sources of m_H- and m_t- dependence are:

- contributions from γ, Z and W self-energy diagrams
- contributions from ZWW and γWW form factors

Using the G_μ-scheme, with G_μ, M_Z and M_W as input parameters, the dependence of the cross-section on m_H and m_t is very weak, as can be seen from Table 4. Of particular importance is the possible top-mass dependence of the W-mass measurement [20]. A crucial fact is that the threshold region is dominated by the t-channel exchange terms, where only the renormalization effects (counter terms) depend on unknown physics mainly showing up in the vector boson self-energies. Only the amplitude $T_1^{(-)}$ exhibits such terms. According to Eq. (15)

$$T_1^{(-)} = -\sqrt{2}G_\mu \frac{2M_W^2}{t}\{1 + 2\Sigma_r^{\nu_e\nu_e}(t) + 2A_{1r}^{We\nu_e}(t)\} + \cdots \qquad (28)$$

$$= -\sqrt{2}G_\mu \frac{2M_W^2}{t}\{1 + \Delta C^W + \cdots\} + \cdots$$

where the m_H and m_t dependence of ΔC^W has been analyzed in Ref. [21]. Formally,

$$\Delta C^W = \frac{Re\Pi^{WW}(M_W^2)}{M_W^2} - \frac{\Pi^{WW}(0)}{M_W^2} - Re\frac{d\Pi^{WW}}{dq^2}(M_W^2)$$

$$= -M_W^2 \pi^{WW}(M_W^2)$$

Table 4: σ_{tot} (pb) for some values of m_t and m_H. Energies and masses in GeV.

	(m_H, m_t)	$E_{c.m.}=$ 163	170	180	200
G_μ-scheme	(100,30)	5.042	13.392	17.566	18.005
	(100,200)	4.998	13.297	17.461	17.908
	(100,300)	5.001	13.314	17.496	17.965
	variation	0.85 %	0.71 %	0.60 %	0.54 %
α-scheme	(100,30)	5.039	13.240	17.268	17.793
	(100,200)	4.489	11.908	15.613	16.041
	(100,300)	3.938	10.576	13.959	14.293

	$(E_{c.m.}, m_t)$	$m_H=$ 10	100	500	1000
G_μ-scheme	(195,60)	17.861	18.000	17.988	17.986

is determined solely by the (twice subtracted) W self-energy function

$$\Pi^{WW}(q^2) = \Pi^{WW}(0) + q^2 \pi^{WW}(q^2).$$

One finds for the top contribution $\Delta C^{Wt} = -2K$ for $m_t \gg M_W$ and $\Delta C^{Wt} = 2K$ for $m_t \ll M_W$ with $K = \frac{\sqrt{2}G_\mu M_W^2}{16\pi^2}$, i.e. not even a logarithmic dependence on m_t is present!

Similarly, for the Higgs dependence, we find $\Delta C^{WH} = 0$ for $m_H \gg M_W$ and $\Delta C^{WH} = 2K(\ln\frac{m_H^2}{M_W^2} + \frac{47}{12})$ for $m_H \ll M_W$. The potentially interesting infrared log for a light Higgs disappears if Higgs Bremsstrahlung off the final state W's is taken into account. Hence there is a very weak dependence on both m_t and m_H only near threshold. The situation is quite different if we use α, M_Z and M_W as input parameters. Since

$$\sqrt{2}G_\mu = \frac{\pi\alpha}{M_W^2 \sin^2\Theta_W} \frac{1}{1 - \Delta r}$$

we get

$$T_1^{(-)} = -\frac{\pi\alpha}{M_W^2 \sin^2\Theta_W} \frac{2M_W^2}{t}\{1 + \Delta r + \Delta C^W + \cdots\} + \cdots \qquad (29)$$

instead of Eq. (28), the following quadratic m_t- and logarithmic m_H-dependence results [21]

$$\Delta r^{top} = \frac{\sqrt{2}G_\mu M_W^2}{16\pi^2}\left\{-3\frac{c_W^2}{s_W^2}\frac{m_t^2}{M_W^2} + 2\left(\frac{c_W^2}{s_W^2} - \frac{1}{3}\right)\ln\frac{m_t^2}{M_W^2} + \cdots\right\}$$

$$\Delta r^{Higgs} \simeq \frac{\sqrt{2}G_\mu M_W^2}{16\pi^2}\left\{\frac{11}{3}(\ln\frac{m_H^2}{M_W^2} - \frac{5}{6})\right\} \quad (m_H \gg M_W).$$

(see also Ref. [20] for plots of results which include finite widths effects).

The α-scheme is therefore completely inadequate for a model independent determination of the W-mass since the cross-section for given α and M_Z depends in an essential way on *two parameters*, M_W and m_t. In contrast for given G_μ and M_Z the cross-section is a function of M_W only to high accuracy!

7. Summary

The investigation of W production at LEP 200 will lead to crucial tests of the gauge structure of the vector bosons. Deviations of the vector boson couplings from the values predicted by the local $SU(2)_L \otimes U(1)_Y$ gauge group are expected to lead to substantial deviations from the Standard Model predictions , increasing the cross-sections at larger values of E_b.

The inclusion of the full one-loop radiative corrections for the analysis and interpretation of the experimental data is mandatory since these effects are far larger than the experimental precision which will be reached. Leading two-loop effects will have to be included as well.

In addition to the effects included so far, the incorporation of the finite width effects from the decaying W's and of the hard-photon Bremsstrahlung is indispensible. Furthermore, there are indications [22] that background processes like $e^+e^- \rightarrow W^+e^-\nu_e$ (where the $e^-\nu_e$ pair is not a decayed W^-) may not be completely ignored in a careful analysis of the threshold behavior (small $e^+e^- \rightarrow W^+W^-$ cross-section).

"New physics effects" will be tested via the precision measurement of the W-mass, which is equivalent to the precision measurement of Δr. Also, flavor tagging of hadronic W-decays will yield much improvement in our knowledge of some Kobayashi-Maskawa matrix elements.

The sensitivity of the W-production cross-section to new physics effects coming from particles like the Higgs, the top, squarks, sleptons e.t.c. seem to be beyond the experimental possibilities.

A final remark concerns to status of the full one-loop calculations for $e^+e^- \rightarrow W^+W^-$ including soft photon emission. Using the same input parameter scheme, the same input parameters and soft photon cuts we obtain complete agreement with Ref. [8]. It may be worthwile to mention that the bookkeeping and the numerical programs wich have been used by the two groups of Refs. [8] and [9] are quite different such that this part of the calculation may be considered as fully established.

Acknowledgements

I wish to thank Johann Kühn for the invitation to this interesting meeting and for the very kind hospitality at Schloss Ringberg. I am very grateful to Jochem Fleischer for help in preparing the numerical tables and to Chris Fasano for reading the manuskript.

References:

1. S. Glashow, Nucl. Phys. B 22 (1961) 579;
   S. Weinberg, Phys. Rev. Lett. 19 (1967) 1264;
   S. Glashow, I. Illiopoulos, L. Maiani, Phys. Rev. D 2 (1970) 1285

2. "Physics with LEP", CERN 86-02, (1986) eds. J. Ellis and R. Peccei;
   ECFA Workshop on LEP 200, CERN/ECFA 87-08 (1987),
   eds. A. Böhm and W. Hoogland

3. C. H. Llewellyn Smith, Phys. Lett. 46 B (1973) 233;
   J. M. Cornwall, D. N. Levin, G. Tiktopoulos, Phys. Rev. D 10 (1974) 1145

4. A. Sirlin, Phys. Rev. D 22 (1980) 971

5. M. Veltman, Nucl. Phys. B 123 (1977) 89;
   M. Consoli, S. Lo Presti, L. Maiani, Nucl. Phys. B 223 (1983) 474;
   J. Fleischer, F. Jegerlehner, Nucl. Phys. B 228 (1983) 1

6. M. Lemoine, M. Veltman, Nucl. Phys. B 164 (1980) 445

7. R. Philippe, Phys. Rev. D 26 (1982) 1588

8. M. Böhm et al., Nucl. Phys. B 304 (1988) 463

9. J. Fleischer, F. Jegerlehner, Z. Phys. C 42 (1989) 409

10. M. Böhm, these proceedings

11. O. P. Sushkov, V. V. Flambaum, I. B. Khriplovick,
    Sov. J. Nucl. Phys. 20 (1975) 537;
    W. Alles, Ch. Boyer, A. J. Buras, Nucl. Phys. B 119 (1977) 125
    and Ref. [2]

12. H. Schlereth, "On Composite Weak Bosons",
    Preprint NBI-HE-89, Niels Bohr Institute, 1989
    and references therein.

13. J. Fleischer, F. Jegerlehner, Phys. Rev. D 23 (1982) 2001

14. F. Jegerlehner, in "Testing of the Standard Model ",
    eds. M. Zrałek, R. Mańka, World Scientific Publ., Singapore, 1988

15. G. 't Hooft, M. Veltman, Nucl. Phys. B 44 (1972) 189

16. W. A. Bardeen, R. Gastmans, B. Lautrup, Nucl. Phys. B 46 (1972) 319

17. D. R. Yennie, S. C. Frautschi, H. Suura, Ann. Phys. (N.Y.) 13 (1961) 379

18. U. Amaldi et al., Phys. Rev. D 36 (1987) 1385;
    G. Costa et al., Nucl. Phys. B 297 (1988) 244

19. F. Jegerlehner, Z. Phys. C 32 (1986) 195;
    H. Burkhardt et al., in "Polarization at LEP", CERN 88-06 (1988),
    eds. G. Alexander at al.

20. B. Grządkowski, Z. Hioki, Phys. Lett. 197 (1987) 213;
    B. Grządkowski, Z. Hioki, H. J. Kühn, Phys. Lett. 205 (1988) 388;
    Z. Hioki, Preprint TOKUSHIMA 88-01, Tokushima (1988)

21. F. Jegerlehner, Z. Phys. C 32 (1986) 425

22. J. C. Romão, P. Nogueira, Preprint IFM 8/88, Lisboa (1988)

Part III

Beyond the Standard Model

Radiative Corrections for e⁺e⁻ Collisions Editor: J.H. Kühn
© Springer-Verlag Berlin, Heidelberg 1989

SEARCHES FOR A NEW GAUGE BOSON AT DIFFERENT LEP PHASES

C. Verzegnassi[*]

Theoretical Physics Division, CERN
1211 Geneva 23, Switzerland

1. INTRODUCTION

In this seminar, I will show that one of the several remarkable features of LEP would be that of allowing a search for the existence of one extra vector boson of extended gauge origin and mass up to a few hundred GeV in a very systematic and exhaustive way. This search would proceed in different steps, corresponding to three various phases of the machine that I will call phases 1, 1a and 2, respectively. They would roughly correspond to a first stage with total energy $E \lesssim 110$ GeV and $\int L \, dt \simeq 100 \, pb^{-1}$, an intermediate stage with either initial beam longitudinal polarization or very high luminosity ($\sim 10^8$ Zs) available, and a final state with $E_{max} \simeq 200$ GeV, $\int L \, dt \lesssim 500 \, pb^{-1}$. Each of these phases would be able to reveal and to identify different signals of the new Z' generated by independent parameters. Thus, a negative search at LEP would, at least, set limits on all the parameters of the involved models that would be, in general, competitive with those achievable in future pp, $p\bar{p}$ collider experiments.

Since the theoretical framework involving a new gauge boson has already been exhaustively discussed in several excellent papers, I will be limited to the essential notations and parametrizations. More systematic treatments can be found elsewhere [1].

2. NEW GAUGE BOSON PARAMETERS

A proper theoretical discussion requires now some preliminary definitions. I will consider the case in which an extra boson exists, generated by an extended gauge symmetry. In particular, I will concentrate on two general classes of models: those with just one extra U(1) symmetry and those with a previous left-right symmetry. In the first case, one is left with a gauge group

$$G_1 = SU(2)_L \times U(1)_Y \times U(1)_{Y'} \tag{1}$$

such that

$$Q = I_{3L} + \tfrac{1}{2}Y \tag{2}$$

i.e., the new gauge boson Z'_0 is simply associated to a neutral generator that does not contribute to the electric charge. In the second case, one will have eventually a gauge group

$$G_{LR} = SU(2)_L \times SU(2)_R \times U(1)_{B-L} \tag{3}$$

with

$$Q = I_{3L} + I_{3R} + (B-L)/2 \tag{4}$$

In both cases, one will be able to write down the following expression for the neutral current Lagrangian:

---

[*] On leave of absence from Department of Theoretical Physics, University of Trieste, and INFN, Sezione di Trieste, Italy.

$$- L_N = J_\mu^{em} \cdot A^\mu + J_\mu^{Z_0} Z_0^\mu + J_\mu^{Z'_0} Z_0'^\mu \tag{5}$$

where A^μ is the photon field and Z_0, Z'_0 are the "mathematical" gauge bosons. The first one is coupled to the current

$$J^{Z_0} = \frac{g_L^2}{\sqrt{g_L^2 - e^2}} \; [J_{3L} - \frac{e^2}{g_L^2} \; J^{em}] \tag{6}$$

while the second one would be coupled, in case (1), to the current

$$J_{(Y')}^{Z'_0} = g_{Y'} \; J_{Y'} \tag{7}$$

and in case (3) to the current

$$J_{(LR)}^{Z'_0} = g_Y \; [\alpha_{LR} \, J_{3R} - 1/(2\alpha_{LR}) \, J_{B-L}] \tag{8}$$

with

$$\alpha_{LR} = \sqrt{g_R^2/g_Y^2 \; - 1} \tag{9}$$

($g_{L,Y,R}$ are the various gauge couplings).

The physical gauge bosons Z, Z' are the mass eigenstates. If the mathematical Z_0, Z'_0 mass matrix is written as

$$M^2{}_0 \equiv \begin{vmatrix} A & B \\ B & C \end{vmatrix} \tag{10}$$

then

$$\begin{vmatrix} Z \\ Z' \end{vmatrix} = \begin{vmatrix} \cos \theta_M & \sin \theta_M \\ -\sin \theta_M & \cos \theta_M \end{vmatrix} \begin{vmatrix} Z_0 \\ Z'_0 \end{vmatrix} \equiv D \begin{vmatrix} Z_0 \\ Z'_0 \end{vmatrix} \tag{11}$$

Since the various theoretical frameworks fix the _mathematical_ Z_0, Z'_0 couplings, one sees that to specify the effect of the physical states one needs to know, besides their masses M_Z, $M_{Z'}$, the mixing angle θ_M as well. Note that the latter is a model-dependent quantity. In fact, from the properties of the diagonalizing matrix D, one finds the exact relation:

$$TG^2 \; \theta_M = y - 1 - y\sqrt{1 - 2/y} \tag{12}$$

where $y = a^2/x^2$ and

$$a = 1/\varepsilon - 1 \quad ; \quad \varepsilon = \frac{M_Z^2}{M_{Z'}^2} \tag{13}$$

$$x = \sqrt{2} \; B/M_Z^2 \tag{14}$$

showing that θ_M is related to M_Z and $M_{Z'}$ via the off-diagonal element B, fixed by the unknown Higgs structure of the different models. Thus, a perfectly reasonable choice of parameters would be that of M_Z, M'_Z and x; for the latter, the properties of the matrix D, Eq. (11) fix the upper bound

$$x \leqslant a/\sqrt{2} \tag{15}$$

Alternatively, a popular choice is that of M_Z, ε, θ_M. I shall follow this notation, since the existing experimental bounds [2] are given for θ_M, e, rather than for x, ε.

Quite generally, the effects of a non-directly produced Z' on a certain physical observable measurable in e^+e^- collisions at LEP will be of two kinds:

(i) virtual effects, produced by the exchange of a virtual Z';

(ii) mixing effects, corresponding to the shift of the properties of the physical Z from those of the mathematical Z_0, which is involved in the existing theoretical formulae. The shift affects both the couplings of Z_0 to fermions and the Z_0 mass that appears in the various definitions of $\sin^2\theta \equiv s^2$. This second effect, actually, can be reabsorbed in a redefinition of s^2, identical to the one which we would obtain by allowing a modification at tree level of the ρ parameter [3]. I will define the effect of this shift as

$$\delta\rho_{Z'} = \frac{M_{Z_0}^2 - M_Z^2}{M_Z^2} \equiv \frac{A - M_Z^2}{M_Z^2} = a \sin^2\theta_M > 0 \qquad (16)$$

A general property of this term is that, on Z resonance, it will always appear in the various theoretical predictions multiplied by the same coefficient that multiplies the one-loop correction term $\Delta\rho(0)$, defined as [4]

$$\Delta\rho(0) \equiv -\frac{\Pi_W(0)}{M_W^2} + \frac{\Pi_Z(0)}{M_Z^2} \qquad (17)$$

whose (known) property is to be quadratic in the mass of a heavy top and/or of heavy (split) new particles (or sparticles). Thus, if a certain prescription is given that eliminates $\Delta\rho(0)$ from a certain observable, the same prescription will automatically eliminate the mass-shift effect $\delta\rho_{Z'}$ [3].

Away from Z resonance, and for small values of the mixing angle (consistent with the existing experimental limits [2]), the virtual contribution coming from the Z' exchange is generally dominating. However, on top of the Z resonance, this contribution vanishes exactly and one is left with mixing effects only. I will concentrate now on the shift of the Z couplings to fermions. In other words, I will write the neutral current Lagrangian as:

$$- L_N = J_\mu^{e.m.} A^\mu + J_\mu^Z Z^\mu + J_\mu^{Z'} Z'^\mu \qquad (18)$$

where

$$J_\mu^Z = \cos\theta_M J_\mu^{Z0} + \sin\theta_M J_\mu^{Z'0} \qquad (19)$$

Defining the fermionic couplings as

$$J_\mu^Z(f) = \bar{f} [\gamma_\mu v_f^Z + \gamma_\mu\gamma_5 a_f^Z] f \qquad (20)$$

one sees that

$$v,a|^Z = \cos\theta_M \ v,a|^{Z0} + \sin\theta_M \ v,a|^{Z'0} = v,a|^{Z0} + \theta_M \ v,a|^{Z'0} + O(\theta_M^2) \quad (21)$$

Thus, a measurement of the physical Z couplings can isolate the pure θ_M effect. In general, though, a physical observable on Z resonance will also depend on the mass shift term $\delta\rho_{Z'}$, Eq. (16), that introduces an extra dependence on $M_{Z'}^2$.

To conclude this introduction, I have still to specify the class of models of type (7) [one extra $U(1)_{Y'}$] that I will consider. Following the common philosophy, I will assume that the extra $U(1)_{Y'}$ can be of the most general E_6 origin. This leads to an extra Z' of the following nature [5]:

$$Z'_0 = \cos\beta \ Z'_\chi + \sin\beta \ Z'_\psi \qquad (22)$$

where $Z'_{\chi,\psi}$ correspond to different symmetry breaking patterns down to the final low energy symmetry group of Eq. (1). In the following, I will assume $g_\chi = g_\psi = g_Y$, but allow $\cos\beta$ to vary in its full range. On the contrary, α_{LR}, Eq. (9), will be allowed to vary in its meaningful range [6]

$$\sqrt{2/3} < \alpha_{LR} \lesssim \sqrt{3/2} \qquad (23)$$

Note that the L-R symmetric model with the value $\alpha_{LR} = \sqrt{2/3}$ produces the same results as the extra $U(1)_{Y'}$ model with $\cos\beta = 1$. This known [6] property will provide a useful check of the relevant theoretical expressions.

The experimental information on the parameter θ_M is shown in Fig. 1 at variable $\cos\beta$. The corresponding limits on $M_{Z'}$ can also be derived [2,7].

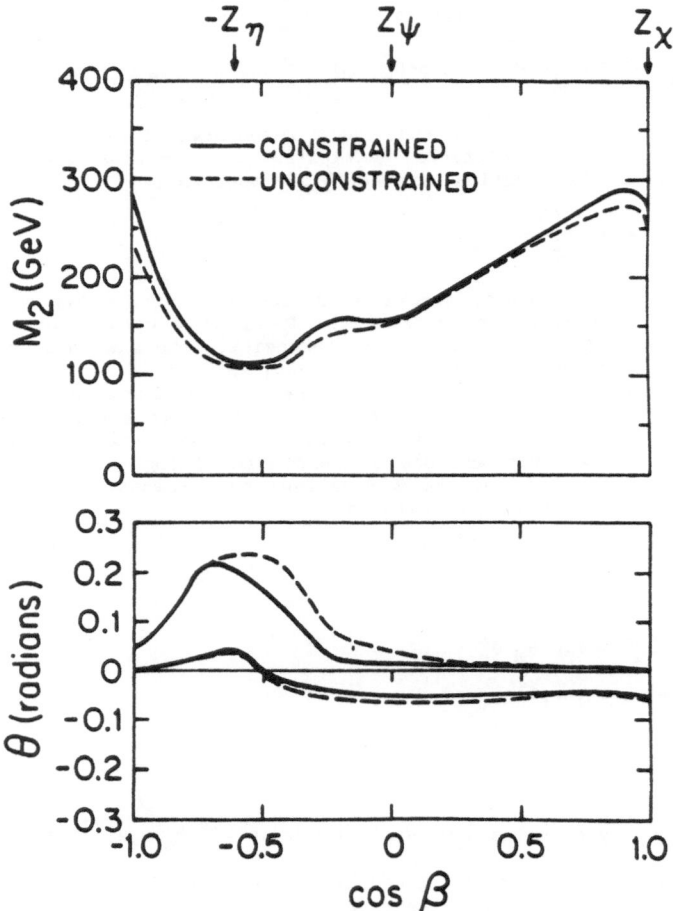

Fig. 1 Lower limits on $M_{Z'} \equiv M_2$ and allowed $\theta_M \equiv \theta$ range for an E_6 gauge boson, taken from Ref. [2] (constrained $\equiv \rho_{tree} = 1$).

In the following, I will assume the (rather conservative) limit

$$M_{Z'} \gtrsim 130 \text{ GeV} \tag{24}$$

coming from low-energy neutral current data [2]. Alternatively, a value of ~150 GeV could be taken [7]. This would not alter the next discussion appreciably.

3. SEARCH FOR MIXING EFFECTS: PHASE 1

I assume a preliminary measurement of M_Z to have been performed to an accuracy of ~±50 MeV. I also assume a number of Zs of order 10^6 to have been produced.

A very relevant quantity for Z' searches would then be the Z leptonic width [8]. From the theoretical expression at tree level,

$$\Gamma_{Z \to f\bar{f}} \Big|_{M_Z} = \frac{N}{12\pi} M_Z [v_f^2 + a_f^2] \tag{25}$$

[where $N = 1$ for leptons and $\sim 3(1+\alpha_s/\pi)$ for quarks] it is relatively easy to derive the mixing effect due to the extra Z'. In fact, it is sufficient to compute this contribution at tree level, ignoring Z' effects on radiative corrections (this is rigorously licit for values of θ_M not vanishingly small, which will be relevant for this discussion; for $\theta_M \to 0$, one has to be more careful, as illustrated in Sirlin's talk).

Consider for simplicity the deviation of, e.g., the muonic width from its value computed in the Standard Model (SM) (the latter value can be found tabulated in previous works [9]). This reads, in the two classes of models, Eqs. (22), (8):

$$\frac{\delta\Gamma_{Z \to \mu\bar{\mu}}}{M_Z} = \frac{\Gamma_{Z \to \mu\bar{\mu}} \text{ (SM)}}{M_Z} \left[\theta_M \left(\frac{-2\cos\beta}{\sqrt{6}} - \frac{\sqrt{10}}{3}\sin\beta\right) + \delta\rho_{Z'}\right] \tag{26}$$

$$\frac{\delta\Gamma_{Z \to \mu\mu}^{(LR)}}{M_Z} = \frac{\Gamma_{Z \to \mu\bar{\mu}} \text{ (SM)}}{M_Z} \left[\theta_M (-\alpha_{LR}) + \delta\rho_{Z'}\right] \tag{27}$$

To illustrate the effect of the shift at variable $\cos\beta$, it is better to consider the two mixing effects separately. In fact, the mass shift $\delta\rho_{Z'}$ term is always positive, and of a typical size of a few per cent. The coupling shift, conversely, can be of both signs depending on $\cos\beta$ in the extra U(1) case, Eq. (26), while it remains positive in the L-R symmetric model, Eq. (27). From Fig. 1 and Ref. [2], it is possible to draw the allowed ranges of the coupling shifts at variable $\cos\beta$ and α_{LR}. As one sees from Fig. 2, this can become strongly negative, up to values

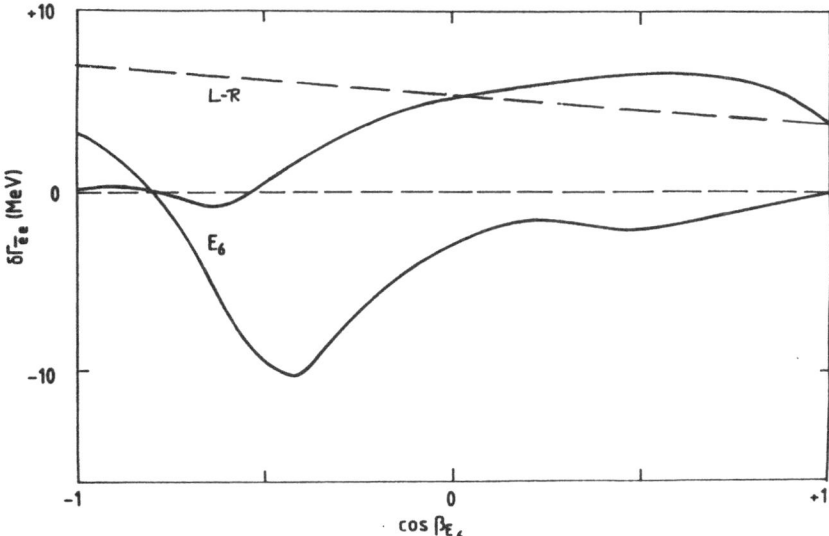

Fig. 2 Allowed range of variation of the coupling shift effect on the muonic Z width for the two models, Eq. (22) full lines, and Eq. (8) dashed lines, taken from Ref. [12].

of ~−10 MeV, for certain values of $\cos\beta \simeq -0.5$. Even adding a few per cent contribution from the positive $\delta\rho_Z$, term, this would leave a characteristic signature (a width <u>smaller</u> than that predicted by the SM) that, to my knowledge, no other types of <u>extra physics</u> could generate. In alternative, the positive shift that might be produced by Eq. (27) in the case of a Z'_0 of left−right symmetry origin or by Eq. (26) for values of $\cos\beta \neq -0.5$ would also be large and hardly explainable by other known mechanisms. Note that the maximum size of the expected overall theoretical and experimental error on the width (obtained by including the possible uncertainty coming from the top mass) would be of ~±2 MeV, well below the maximum possible Z' effects.

If a signal of clear deviation from the SM prediction, consistent with that produced by a new Z', were observed in the leptonic width, the next goal would be that of identifying the parameters of the model. In practice, each of the two considered cases is defined by (at least) three parameters, i.e., θ_M, $\cos\beta$ (or α_{LR}) and $\delta\rho_Z$, ~ $a\theta_M^2$. Clearly, this program requires the availability of other suitable measurable quantities. An immediate possibility [8] would be that of considering the ratio

$$R' \equiv 3/59 \frac{\Gamma_5}{\Gamma_{\mu\bar\mu}} \quad ; \quad \Gamma_5 \equiv \Gamma_{Z\to u,d,s,c,b} \tag{28}$$

In fact, the theoretical expression of R' in the two cases, Eqs. (8) and (22), is such that one can isolate the $\delta\rho_Z$, parameter in a certain linear combination of R' and $\Gamma_{Z\to\mu\bar\mu}$. More precisely, one finds [8]:

$$R' + \frac{50}{59} \left(\frac{\Gamma_{Z\to\mu\bar\mu}}{M_Z} \frac{9}{\alpha(M_Z)}\right) = [R' + \frac{50}{59} \left(\frac{9}{\alpha(M_Z)} \frac{\Gamma_{Z\to\mu\bar\mu}}{M_Z}\right)]^{SM} + \frac{4}{3}\delta\rho_Z, \tag{29}$$

from which, taking into account the experimental and theoretical uncertainties, one should be able to derive $\delta\rho_Z$, to about 1 %. Then, to fix the remaining parameters, another quantity would be required.

A convenient possibility would be provided by the final τ polarization A_τ, to be measured in a not too distant future still belonging to phase 1 [10]. If a measurement of A_τ were available, from the three observables $\gamma \equiv 9/\alpha(M_Z) \Gamma_{Z\to\mu\bar\mu}/M_Z$, R' and A_τ it would then become possible to define two linear combinations that would be free of $\delta\rho_Z$, and, therefore, of the one-loop term $\Delta r(0)$ as well, according to the previous discussion [3]. For these quantities, the residual theoretical error from the SM would be therefore practically negligible. In particular, one could consider the two observables [11]:

$$R' - \tfrac{1}{2}\gamma \equiv [R' - \tfrac{1}{2}\gamma]^{SM} + n \tag{30}$$

$$R' - \frac{10}{59} A_\tau \equiv [R' - \frac{10}{59} A_\tau]^{SM} + q \tag{31}$$

The theoretical expression of the variations n, q produced by the two kinds of considered Z' has been computed elsewhere [12]. From that, one can derive the region in the plane of the variations (n,q) belonging to the two models shown in Fig. 3. The evaluation of $(\theta_M, \cos\beta)$ or (θ_M, α_{LR}) can then proceed in a standard way [12]. In particular, it would be possible to derive the new limits on θ_M stemming from a negative result of the previous searches at the expected experimental accuracies of phase 1 [11]. These are shown in Fig. 4, again at variable $\cos\beta$, together with the previous limits from Ref. [2]. As one sees, a much more uniform limit would now be achieved, i.e.,

$$|\theta_M| \lesssim 0.025-0.030 \tag{32}$$

over the full $\cos\beta$ range.

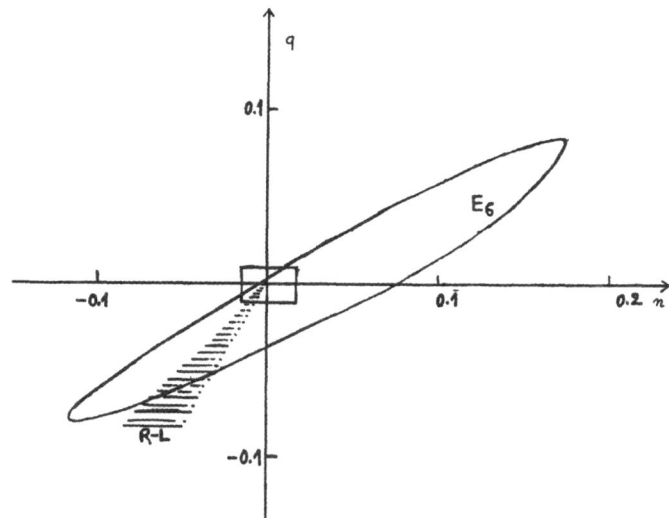

Fig. 3 Allowed region in the plane of the variables (n,q), Eqs. (30) and (31), for the two models of Eqs. (22) and (8), taken from Ref. [12]. The rectangle around the origin represents the overal uncertainty.

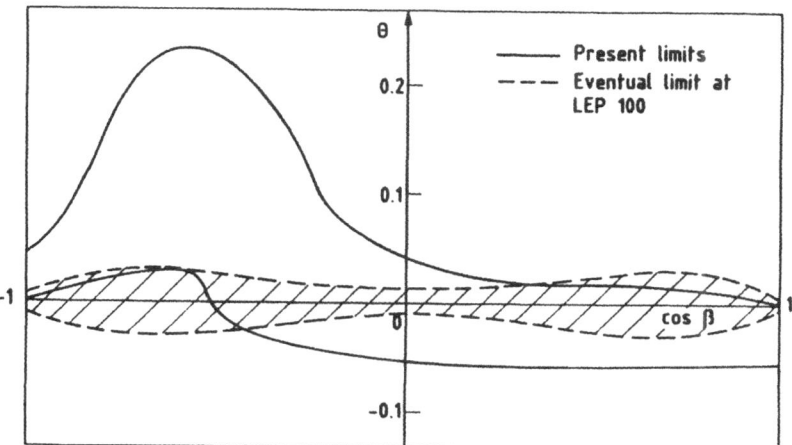

Fig. 4 Limits on $|\theta_M|$ stemming from a negative Z' search at LEP1, from Ref. [11].

The previous discussion has assumed that the extra Z' was of some extended gauge nature. However, other possible models exist that predict the existence of a new Z' from rather different theoretical origins. In particular, the new Z' could arise from models that do not require the existence of any Higgs boson. Examples of such a situation are either composite models of (W,Z) bosons [13] or models with a strongly interacting electroweak sector [14]. If a signal of deviation from the SM prediction were seen in the (n,q) plane, it would become important to isolate the theoretical model that could produce it. In fact, as shown in Fig. 5, this possibility would not be automatically provided at this stage, since several regions of possible confusion between the various models (Y, Y_L Z* correspond to three different versions of composite (W,Z) models, while model Z_V corresponds to that of Ref. [14]) would still remain [12]. This residual confusion could only be eliminated by proper measurements belonging to phase 1a, as shown in the next section.

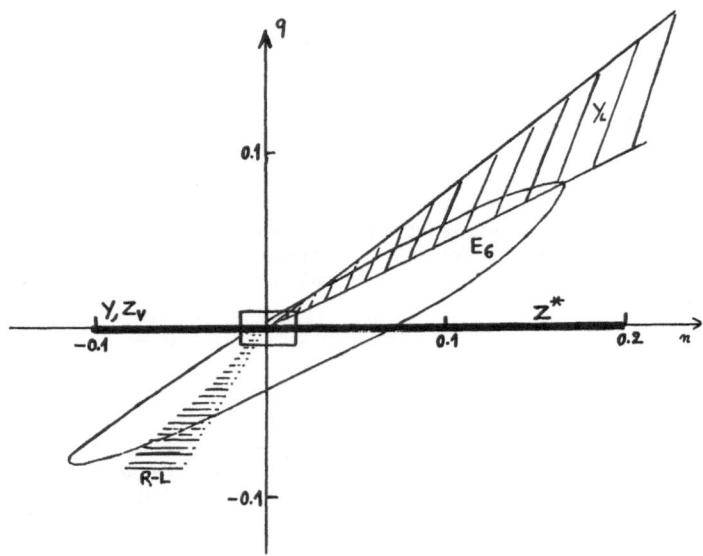

Fig. 5 The same as in Fig. 3, but including the allowed regions of models (Y,Y_L,Z*,Z_V), from Ref. [12].

4. FULL IDENTIFICATION OF THE THEORETICAL MODEL: PHASE 1a

The possibility that longitudinally polarized initial beams become available at LEP and the theoretical motivations for this effort have been the subject of an entire yellow book [9], and I will not discuss them now. Here I will only show one of the (several) outcomes of this fascinating eventuality. If longitudinal polarization became available, it would be possible to measure three specific polarized asymmetries, i.e., the quantities:

$$A_{LR}^{(h)} \equiv A = \frac{\sigma(e^+e_L^- \to \text{had}) - \sigma(e^+e_R^- \to \text{had})}{\sigma(e^+e_L^- \to \text{had}) - \sigma(e^+e_R^- \to \text{had})} \tag{33}$$

$$A^{b,c}_{FB(pol)} \equiv A^{b,c} \quad \equiv \frac{1}{P} \frac{(N^{b,c}_{F,P} - N^{b,c}_{F,-P}) - (N^{b,c}_{B,P} - N^{b,c}_{B,-P})}{(N^{b,c}_{F,P} - N^{b,c}_{F,-P}) + (N^{b,c}_{B,P} - N^{b,c}_{B,-P})} \qquad (34)$$

The theoretical properties of these observables have been exhaustively discussed in the literature [4,15,16]. Their main feature is that, on Z resonance, they would provide independent measurements of the initial lepton (A) and of the final quarks ($A^{b,c}$) couplings to the Z. From these three quantities, one could generate two linear combinations [17]

$$\tilde{A}_b = A^b - \frac{1}{15} A = \tilde{A}^{SM}_b + x \qquad (35)$$

$$\tilde{A}_c = A^c - \frac{9}{25} A = \tilde{A}^{SM}_c + y \qquad (36)$$

that would be free of oblique EW radiative corrections. These quantities would feel the effect of a new Z' of extended gauge origin. However, from their properties, they would <u>not</u> feel the effect of a new Z' of (Y, Y_L, Z^*, Z_V) origin [3]. Therefore, they would be a unique filter for observation (or exclusion) of the models of Eqs. (22) and (8) [18], as shown by Fig. 6.

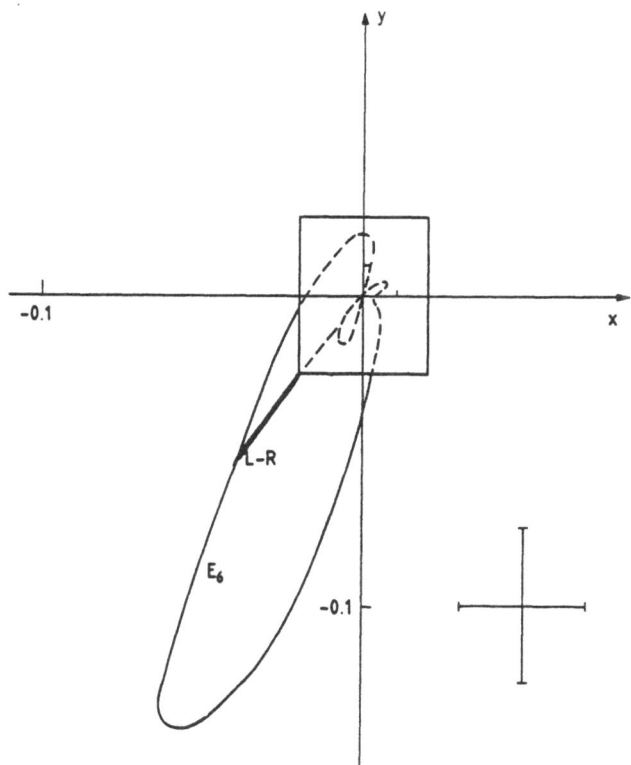

Fig. 6 Allowed domain in the plane of the variables (x,y), Eqs. (35) and (36) for the two models, Eqs. (22) and (8), from Ref. [18]. The rectangle corresponds to the overall uncertainties (left-right symmetric models are confined on a segment in the figure).

One should stress that an alternative and completely equivalent possibility would be provided if the set $(A, A^{b,c})$ were replaced by the set $(A_\tau, A_{FB}^{b,c})$ including the previously mentioned final τ polarization and the <u>unpolarized</u> forward-backward asymmetries for b, c quarks (but also s, c would do). In fact, one could then replace A by A_τ and $\overline{A}^{b,c}$ by [19]

$$\overline{A}_{b,c} \equiv \frac{A_{FB}^{b,c}}{A_\tau} - \left\{ \begin{matrix} 1/15 \\ 9/25 \end{matrix} \right. A_\tau \tag{37}$$

With a very high luminosity ($\sim 10^8$ Zs) [20] the experimental error on $A_{FB}^{b,c}$ [21] and A_τ [10] would then become such as to allow for a measurement of Eqs. (37) to nearly the same accuracy as that of Eqs. (35) and (36), and an analogous identification of a new Z' signal could be performed (for M_Z not much smaller than ~ 92 GeV [16]).

In conclusion, I have shown that at the end of phases 1 and 1a, one would be able to draw definite conclusions about a possible new Z'. These would not give, though, any precise indication about its mass. The only possible statement that one might derive from the new bound of Eq. (32), by combining it with the existing experimental limits [2], would be that

$$M_{Z'} \gtrsim 150 \text{ GeV} \tag{38}$$

implying that this particle could in principle be directly produced at LEP2, phase 2. Even if this were not the case, it would still be possible, though, to obtain very interesting information in this phase by considering those virtual effects (vanishing on top of Z) that would now be dominant. This will be discussed in the final section.

5. SEARCH OF VIRTUAL Z' EFFECTS AT LEP2

A detailed analysis of virtual effects of a new Z' on a number of quantities that should be measured at LEP2 has been recently performed [22]. The result of the investigation is that, under realistic experimental conditions, the best quantity in this respect is the ratio R' previously defined by Eq. (28), with the understanding that an analogous quantity might be used if the top were sufficiently light to be produced in that phase. Figure 7 shows the effect on R' of a specific couple of representative models of the two classes, Eqs. (22) and (8) at variable $\sqrt{s}$ for $M_{Z'} = 250$ GeV, showing that the effect would be large and positive. As thoroughly discussed in Ref. [22], this would be due to a positive interference of an increase in the hadronic numerator of R' with a decrease of its leptonic denominator.

Figures 8 and 9 show the effect of a new Z' of the most general kind, Eqs. (22) and (8) at $\sqrt{s} = 200$ GeV and variable $M_{Z'}$. Again, as one sees, the effect would be very large over a sizeable fraction of the possible values of ($\cos\beta$, α_{LR}) up to rather large masses. This would lead to an identification of the signal, e.g., for certain particular values of $\cos\beta$, up to values of $M_{Z'} \simeq 400$ GeV (for the same values of $\cos\beta$, the direct production at tevatron would fail for $M_{Z'} \gtrsim 300$ GeV [7]). Moreover, from a combined measurement of R' <u>and</u> of the hadronic longitudinal polarization asymmetry A_{LR}^h (assuming that the latter can be actually measured at LEP2) a clear identification of the theoretical model would also be possible in the relevant mass range, as discussed in Ref. [22].

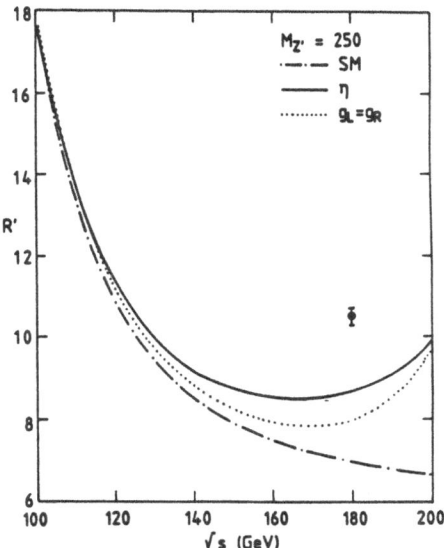

Fig. 7 Effects on the ratio R', Eq. (28), of a couple of representative models of
the classes, Eqs. (8) and (22) at variable √s and $M_{Z'}$ = 250 GeV, from
Ref. [22].

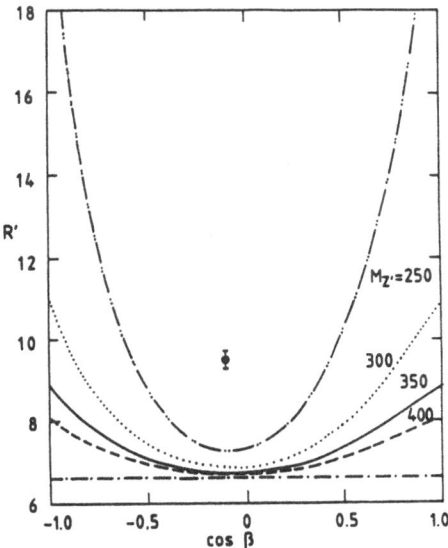

Fig. 8 Effects of a Z' of variable mass on the ratio R' for √s = 200 GeV in the
model of Eq. (22), from Ref. [22].

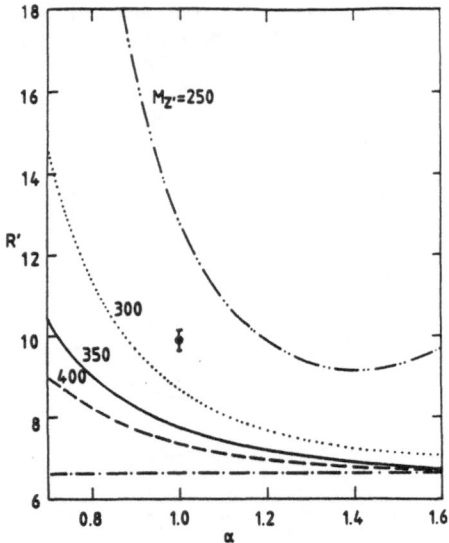

Fig. 9 Same as Fig. 7 for the model of Eq. (8), from Ref. [22].

6. CONCLUSIONS

In conclusion, from a combination of experiments at three different LEP phases, it would be possible to search in a systematic and consistent way for the existence of a new Z' of extended gauge origin down to values of the mixing angle $|\theta_M| \gtrsim 0.025$ (phase 1) and up to values of the mass $M_{Z'} \lesssim 400$ GeV (phase 2). These limits would be competitive with those obtainable at future collider experiments, particularly those for θ_M. If no evidence of a new Z' were brought up by these searches, the surviving range of allowed parameters for such a hypothetical particle might begin to appear, to say the least, discouraging.

REFERENCES

1. See, for an updated discussion and list of references:
   R. Casalbuoni, D. Dominici, F. Feruglio and R. Gatto - "New Gauge Vector
   Bosons at LEP1", LEP Physics Workshop 1989, in publication.

2. U. Amaldi, A. Böhm, L.S. Durkin, P. Langacker, A.K. Mann, W.J. Marciano,
   A. Sirlin and H.H. Williams - Phys.Rev. D36 (1987) 1385.

3. F. Boudjema, F.M. Renard and C. Verzegnassi - Nucl.Phys. B314 (1989) 301.

4. I follow the notations of B.W. Lynn, M.E. Peskin and R.G. Stuart in "Physics
   at LEP", CERN 86-02, (J. Ellis and R. Peccei editors), p. 90.

5. P, Langacker, R.W. Robinett and J.L. Rosner - Phys.Rev. D30 (1984) 1470;
   D. London and J.L. Rosner - Phys.Rev. D34 (1986) 1530.

6. R.W. Robinett and J.L. Rosner - Phys.Rev. D25 (1982) 3036.

7. J. Ellis, P.J. Franzini and F. Zwirner - Phys.Lett. B202 (1988) 417.

8. F.M. Renard and C. Verzegnassi - CERN Preprint TH. 5372 (1989), to appear in
   Phys.Lett. B.

9. See, e.g.: G. Burgers - in "Polarization at LEP", CERN Report 88-06 (1988)
   (G. Altarelli, A. Blondel, G. Coignet, E. Keil, D.E. Plane and D. Treille
   editors), p. 121.

10. See, e.g., the discussion given by Z. Was - "Polarization at LEP" (previous
    reference), p. 250.

11. F.M. Renard and C. Verzegnassi - Phys.Lett. B221 (1989) 197.

12. F. Boudjema and F.M. Renard - LEP Physics Workshop 1989, Ref. 1.

13. H. Fritzsch, R. Kögerler and D. Schildknecht - Phys.Lett. 114B (1982) 157;
    M. Kuroda, D. Schildknecht and K.H. Schwarzer - Phys.Rev. D35 (1987) 297.

14. R. Casalbuoni, S. de Curtis, D. Dominici and R. Gatto - Nucl.Phys. B282
    (1987) 234;
    R. Casalbuoni, P. Chiappetta, D. Dominici, F. Feruglio and R. Gatto -
    Nucl.Phys. B310 (1988) 181.

15. B.W. Lynn and C. Verzegnassi - Phys.Rev. D35 (1987) 3326.

16. A. Blondel, B.W. Lynn, F.M. Renard and C. Verzegnassi - Nucl.Phys. B304
    (1988) 438;
    See also: A. Blondel, F.M. Renard and C. Verzegnassi - Ref. 9, p. 197.

17. M. Cvetic and B.W. Lynn - Phys.Rev. D35 (1987) 51.

18. F.M. Renard and C. Verzegnassi - Phys.Lett. B217 (1989) 199.

19. F. Boudjema, F.M. Renard and C. Verzegnassi - Phys.Lett. B214 (1988) 151.

20. See the discussion given by D. Treille - These Proceedings.

21. See the discussion given by J. Drees - These Proceedings.

22. A. Blondel, F.M. Renard, P. Taxil and C. Verzegnassi - CERN Preprint TH. 5314
    (1989), submitted to Nucl.Phys.

Radiative Corrections for e⁺e⁻ Collisions Editor: J.H. Kühn
© Springer-Verlag Berlin, Heidelberg 1989

A Simple Renormalization Framework
for $SU(2)_L \times U(1)_Y \times \widetilde{U}(1)$ Models

Giuseppe Degrassi and Alberto Sirlin
Department of Physics, New York University, New York, NY 10003

1. Introduction

$SU(2)_L \times U(1)_Y \times \widetilde{U}(1)$ models have received considerable attention during the last several years [1]. They often arise as the low energy limit of interesting GUT and superstring theories. It is understood that gluons and quarks and leptons of the three generations have the same transformation properties under $SU(2)_L \times U(1)_Y$ as in the standard model (SM). There may exist additional exotic fields. In the cases of interest there exists an orthogonality constraint between the $\widetilde{U}(1)$ and $U(1)_Y$ quantum numbers: this can be traced to the fact that such models are descendants of gauge theories with simple Lie groups.

We consider a class of these models with Higgs bosons transforming as doublets or singlets of $SU(2)_L$; they are referred to as constrained models.

As is well known, because of the mixing, m_{Z_1} is smaller than m_Z in the SM and, as a consequence, $\rho \equiv m_W^2 / m_{Z_1}^2 \cos^2 \theta_W > 1$. At the tree level an important relation holds true:

$$tan^2\varphi = \frac{m_W^2 / \cos^2 \theta_W - m_{Z_1}^2}{m_{Z_2}^2 - m_W^2 / \cos^2 \theta_W} \tag{1}$$

or

$$\rho - 1 = sin^2\varphi \left[\frac{m_{Z_2}^2}{m_{Z_1}^2} - 1 \right]$$

where φ is the mixing angle relating the mass eigenstates Z_1 and Z_2 to the Z and C bosons of $SU(2)_L \times U(1)_Y$ and $\widetilde{U}(1)$. The gauge sector of the constrained models involves e, m_W, m_{Z_1}, m_{Z_2}, $sin^2 \theta_W$ and the $\widetilde{U}(1)$ gauge coupling $\tilde{g}$; φ is a dependent parameter specified by Eq.(1). In the cases of interest, renormalization group arguments relate the value of $\tilde{g}$ to $g = e/ \sin \theta_W$ so that effectively the gauge sector involves five parameters.

When radiative corrections are included [2] there are two possible strategies: i) keep Eq.(1) as an exact relation or ii) allow for the possibility that (1) is corrected to $O(\alpha)$. We adopt i). This has obvious advantages: the dependence of φ on the other parameters is simple and transparent; recent analyses [1] use (1) as exact; Eq.(1) follows from the identification of φ with the angle that diagonalizes the renormalized mass matrix.

It is also covenient to identify m_W, m_{Z_1}, m_{Z_2} in (1) with the physical masses. As present experiments are compatible with $\varphi = 0$, we should allow 0 in the range of variability of φ. On the other hand, in some models values of m_{Z_2} as low as $\simeq 110$ GeV are allowed. This leads to an important constraint: in order (1) to hold for finite m_{Z_2} and arbitrary small φ, $\cos^2\theta_W$ must be defined so that

$$\lim_{\varphi \to 0} \cos^2\theta_W = m_W^2/m_{Z_1}^2. \tag{2}$$

We will see that if the definition of $\cos^2\theta_W$ is not consistent with (2), but differs from (2) by $O(\alpha)$, the response of the radiative corrections is catastrophic: terms of $O(\alpha/\varphi)$ arise and one loses the validity of the perturbation expansion in the neighborhood of $\varphi = 0$. Note that Eq.(2) is a consistency condition that follows from our insistence that (1) be exact in the presence of radiative corrections and our identification of m_W and m_{Z_1} with the physical masses. In the present analysis, the occurence of the potential terms of $O(\alpha/\varphi)$ is prevented and one is led to a definition of $\sin^2\theta_W$ in terms of G_μ, α and m_W^2 which is very close to the corresponding $SU(2)_L \times U(1)_Y$ expression (with a slightly modified Δr). A strategy to incorporate approximately the $O(\alpha)$ terms in the neutral currents of these theories is outlined. The discussion identifies in a simple way the mathematical origin of the potential non-analytic terms, and emphasizes the role that the m_t dependence of the radiative corrections may have in determining the tenability of these theories.

2. Mass Matrix of the Vector Bosons

After spontaneous symmetry breaking the mass terms for the vector bosons have the structure

$$\mathcal{L}_M^{VB} = m_{0_W}^2 W_\mu^\dagger W^\mu + \frac{1}{2} m_{0_Z}^2 Z_\mu Z^\mu + \frac{1}{2} m_{0_C}^2 C_\mu C^\mu + m_{0_{ZC}}^2 Z_\mu C^\mu$$

where

$$Z_\mu = c_0 W_\mu^3 - s_0 B_\mu$$
$$A_\mu = s_0 W_\mu^3 + c_0 B_\mu$$

where W_μ^3, B_μ and C_μ are neutral gauge fields associated with $SU(2)_L$, $U(1)_Y$ and $\overline{U}(1)$, respectively; A_μ and Z_μ are then identified with the photon and the neutral $SU(2)_L$ vector boson fields, respectively. ($c_0 \equiv \cos\theta_{0_W}$; $\tan\theta_{0_W} \equiv g_0'/g_0$).

In constrained models, we have $m_{0_Z}^2 = m_{0_W}^2/c_0^2$. In these expressions all the masses and coupling constants are unrenormalized quantities. To generate counterterms we write:

$$m_{0_W}^2 = m_W^2 - \delta m_W^2, \ \ m_{0_C}^2 = m_C^2 - \delta m_C^2, \ \ m_{0_{ZC}}^2 = m_{ZC}^2 - \delta m_{ZC}^2, \ \ c_0^2 = c^2 - \delta c^2.$$

(the general strategy is analogous to the treatment of $SU(2)_L \times U(1)_Y$ in Ref. [3].) This leads to

$$
\begin{aligned}
\mathcal{L}_M^{VB} = \; & m_W^2 W_\mu^\dagger W^\mu + \frac{1}{2}\frac{m_W^2}{c^2} Z_\mu Z^\mu + \frac{1}{2}m_C^2 C_\mu C^\mu + m_{ZC}^2 Z_\mu C^\mu \\
& - \delta m_W^2 W_\mu^\dagger W^\mu - \frac{1}{2}\frac{m_W^2}{c^2}\left(\frac{\delta m_W^2}{m_W^2} - \frac{\delta c^2}{c^2}\right) Z_\mu Z^\mu \\
& - \frac{1}{2}\delta m_C^2 C_\mu C^\mu - \delta m_{ZC}^2 Z_\mu C^\mu \dots
\end{aligned}
$$

where the ellipses stand for higher order terms. We see that the renormalized mass matrix is of the form

$$
\frac{1}{2}\left(Z^\mu \; C^\mu \right)\begin{pmatrix} \frac{m_W^2}{c^2} & m_{ZC}^2 \\ m_{ZC}^2 & m_C^2 \end{pmatrix}\begin{pmatrix} Z_\mu \\ C_\mu \end{pmatrix}.
$$

It can be diagonalized by an orthogonal transformation

$$
Z^\mu = \chi Z_1^\mu - \sigma Z_2^\mu; \quad C^\mu = \sigma Z_1^\mu + \chi Z_2^\mu
$$

where $\chi \equiv cos\varphi$, $\sigma \equiv sin\varphi$ and Z_1^μ and Z_2^μ are the mass eigenfields. In terms of this new fields we have

$$
\begin{aligned}
\mathcal{L}_M^{VB} = \; & m_W^2 W_\mu^\dagger W^\mu + \frac{1}{2}m_{Z_1}^2 Z_{1\mu} Z_1^\mu + \frac{1}{2}m_{Z_2}^2 Z_{2\mu} Z_2^\mu \\
& - \delta m_W^2 W_\mu^\dagger W^\mu - \frac{1}{2}\delta m_{Z_1}^2 Z_{1\mu} Z_1^\mu - \frac{1}{2}\delta m_{Z_2}^2 Z_{2\mu} Z_2^\mu - \delta m_{Z_1 Z_2}^2 Z_{1\mu} Z_2^\mu
\end{aligned}
$$

where

$$
\begin{aligned}
\delta m_{Z_1}^2 &= \chi^2 \delta m_Z^2 + \sigma^2 \delta m_C^2 + 2\chi\sigma\delta m_{ZC}^2 \\
\delta m_{Z_2}^2 &= \sigma^2 \delta m_Z^2 + \chi^2 \delta m_C^2 - 2\chi\sigma\delta m_{ZC}^2 \\
\delta m_{Z_1 Z_2}^2 &= -\chi\sigma\delta m_Z^2 + \chi\sigma\delta m_C^2 + (\chi^2 - \sigma^2)\delta m_{ZC}^2
\end{aligned}
$$

$$
\delta m_Z^2 \equiv \frac{m_W^2}{c^2}\left(\frac{\delta m_W^2}{m_W^2} - \frac{\delta c^2}{c^2}\right)
$$

Eliminating δm_{ZC}^2 and δm_C^2

$$
\delta m_{Z_1 Z_2}^2 = \frac{\chi}{2\sigma}\delta m_{Z_1}^2 + \frac{\sigma}{2\chi}\delta m_{Z_2}^2 - \frac{1}{2\chi\sigma}\delta m_Z^2.
$$

To identify m_W, m_{Z_1} and m_{Z_2} with the physical masses we set

$$
\begin{aligned}
\delta m_W^2 &= ReA_{WW}(m_W^2) + t_{ww} \\
\delta m_{Z_1}^2 &= ReA_{Z_1 Z_1}(m_{Z_1}^2) + t_{z_1 z_1} \\
\delta m_{Z_2}^2 &= ReA_{Z_2 Z_2}(m_{Z_2}^2) + t_{z_2 z_2}
\end{aligned}
$$

where the t's stand for the contribution of the tadpole and tadpole counterterms and the A's are the coefficients of $g^{\mu\nu}$ in the corresponding self-energies:

$$\Pi_{ij}^{\mu\nu}(q) \;=\; A_{ij}(q^2)g^{\mu\nu} + B_{ij}(q^2)q^\mu q^\nu.$$

Inserting δm_Z^2, δm_W^2, $\delta m_{Z_1}^2$, $\delta m_{Z_2}^2$ in $\delta m_{Z_1 Z_2}^2$:

$$\delta m_{Z_1 Z_2}^2 \;=\; \frac{\chi}{2\sigma}\left[ReA_{Z_1 Z_1}(m_{Z_1}^2) + t_{z_1 z_1}\right] + \frac{\sigma}{2\chi}\left[ReA_{Z_2 Z_2}(m_{Z_2}^2) + t_{z_2 z_2}\right]$$
$$-\frac{1}{2\sigma\chi c^2}\left[ReA_{WW}(m_W^2) + t_{ww} - \frac{m_W^2}{c^2}\delta c^2\right].$$

It is important to note that $\delta m_{Z_1 Z_2}^2$ contains terms that potentially diverge as $\varphi \to 0$ (or equivalently $\sigma \to 0$). To prevent this we choose

$$\frac{\delta c^2}{c^2} = \frac{ReA_{WW}(m_W^2) + t_{ww}}{m_W^2} - \frac{c^2}{m_W^2}\left(ReA_{Z_1 Z_1}(m_{Z_1}^2) + t_{z_1 z_1}\right) + f(\sigma) \qquad (3)$$

with $f(\sigma) \sim \sigma$ as $\sigma \to 0$ ($f(\sigma)$ will be determined later) Inserting $\delta c^2/c^2$ into $\delta m_{Z_1 Z_2}^2$ and using the relation

$$t_{z_1 z_2} = \frac{\chi}{2\sigma}t_{z_1 z_1} + \frac{\sigma}{2\chi}t_{z_2 z_2} - \frac{1}{2\chi\sigma}(t_{ww}/c^2)$$

one finds

$$\delta m_{Z_1 Z_2}^2 \;=\; \frac{\sigma}{2\chi}\left[ReA_{Z_2 Z_2}(m_{Z_2}^2) - ReA_{Z_1 Z_1}(m_{Z_1}^2)\right]$$
$$+ \frac{1}{2\chi\sigma}\left[\frac{t_{ww}}{c^2} - t_{z_1 z_1}\right] + \frac{m_W^2}{2\sigma\chi c^2}f(\sigma) + t_{z_1 z_2}$$

As $\sigma \to 0$, $t_{z_1 z_1} \to t_{zz} \to t_{ww}/c^2$. Each term is regular as $\sigma \to 0$, and so is the overall $Z_1 Z_2$ self-energy wich is proportional to $A_{Z_1 Z_2}(q^2) + t_{z_1 z_2} - \delta m_{Z_1 Z_2}^2$.

To determine completely $\delta c^2/c^2$ and $\delta m_{Z_1 Z_2}^2$ we must still choose $f(\sigma)$.

3. Interactions of Vector Mesons with the Matter Fields

The interactions of the vector mesons with the matter fields is described by

$$\mathcal{L}_{int} \;=\; -\frac{g_0}{\sqrt{2}}(W_\mu^\dagger J_W^\mu + h.c.) - \frac{g_0}{c_0}Z_\mu(J_Z^\mu)_0 - g_0 s_0 A_\mu J_\gamma^\mu - \tilde{g}_0 J_C^\mu C_\mu$$

where

$$(J_Z^\mu)_0 \;=\; \frac{1}{2}J_3^\mu - s_0^2 J_\gamma^\mu; \quad J_C^\mu = \bar{f}\gamma^\mu(\widetilde{Y}_L a_- + \widetilde{Y}_R a_+)f$$

$a_\mp \equiv (1 \mp \gamma_5)/2$. Setting $g_0 = g - \delta g$, $\tilde{g}_0 = \tilde{g} - \delta\tilde{g}$, etc. ..., we obtain

$$\mathcal{L}_{int} \;=\; \hat{\mathcal{L}}_{int} + \delta\mathcal{L}$$

$$\mathcal{L}_{int} = -\frac{g}{\sqrt{2}}(W_\mu^\dagger J_W^\mu + h.c.) - \frac{g}{c}Z_\mu J_Z^\mu - gsA_\mu J_\gamma^\mu - \tilde{g}J_C^\mu C_\mu$$

$$\delta\mathcal{L} = \frac{\delta g}{\sqrt{2}}(W_\mu^\dagger J_W^\mu + h.c.) + \frac{g}{c}\left(\frac{\delta g}{g} - \frac{\delta c}{c}\right)Z_\mu J_Z^\mu - \frac{g}{c}\delta s^2 J_\gamma^\mu Z_\mu$$
$$+ \delta e A_\mu J_\gamma^\mu + \delta\tilde{g}J_C^\mu C_\mu$$

$$\delta e \equiv s\delta g + g\delta s$$

In terms of Z_1^μ and Z_2^μ

$$\mathcal{L}_{int} = -\frac{g}{\sqrt{2}}(W_\mu^\dagger J_W^\mu + h.c.) - gsJ_\gamma^\mu A_\mu - \frac{g}{c}\left[J_{Z_1}^\mu Z_{1\mu} + J_{Z_2}^\mu Z_{2\mu}\right]$$

$$J_{Z_1}^\mu \equiv \chi J_Z^\mu + \frac{\tilde{g}c}{g}\sigma J_C^\mu; \quad J_{Z_2}^\mu \equiv -\sigma J_Z^\mu + \frac{\tilde{g}c}{g}\chi J_C^\mu.$$

Similarly,

$$\delta\mathcal{L} = \frac{\delta g}{\sqrt{2}}(W_\mu^\dagger J_W^\mu + h.c.) + \delta e J_\gamma^\mu A_\mu$$
$$+ \frac{g}{c}Z_{1\mu}\left[\delta p J_{Z_1}^\mu - \delta t J_{Z_2}^\mu - \chi\delta s^2 J_\gamma^\mu\right]$$
$$+ \frac{g}{c}Z_{2\mu}\left[-\delta t J_{Z_1}^\mu + \delta q J_{Z_2}^\mu + \sigma\delta s^2 J_\gamma^\mu\right]$$

(δp, δt, δq are linear combinations of δg, δc, $\delta\tilde{g}$). Identifying $e \equiv gs$ with the positron charge fixes δe:

$$\frac{2\delta e}{e} = -\Pi_{\gamma\gamma}^{(l)}(0) - \Pi_{\gamma\gamma}^{(h)}(0) - \frac{7e^2}{8\pi^2}\left(\frac{1}{n-4} + \gamma - \ln 4\pi + \ln\frac{m_W}{\mu} - \frac{1}{21}\right).$$

4. μ-Decay

At this stage in Ref. [3], all the basic counterterms of the gauge sector were determined so that one could happily proceed to calculate μ-Decay; now we use this process to define precisely g^2 or, equivalently, $\sin^2\theta_W$. So far we have fixed δe, δm_W^2, $\delta m_{Z_1}^2$, $\delta m_{Z_2}^2$. We still need δs^2 and $\delta\tilde{g}$. After separating the traditional photonic corrections of the V-A theory one finds

$$\mathcal{M} = \mathcal{M}^0/(1 - \widetilde{\Delta r})$$
$$\widetilde{\Delta r} = \frac{[ReA_{WW}(q^2) - ReA_{WW}(m_W^2)]}{q^2 - m_W^2} + V + B - \frac{2\delta g}{g}$$

where V and B represent vertex and box diagrams and

$$\frac{2\delta g}{g} = \frac{2\delta e}{e} - \frac{2\delta s}{s} = \frac{2\delta e}{e} + \frac{\delta c^2}{s^2}.$$

Recalling (3), we have

$$\frac{2\delta g}{g} = \frac{2\delta e}{e} + \frac{c^2}{s^2}\left[\frac{ReA_{WW}(m_W^2) + t_{ww}}{m_W^2} - \frac{c^2}{m_W^2}\left(ReA_{Z_1Z_1}(m_{Z_1}^2) + t_{z_1z_1}\right) + f(\sigma)\right].$$

Inserting this expression into $\widetilde{\Delta r}$ and setting $q^2 = 0$ (as it is appropriate in μ-Decay) leads to

$$\widetilde{\Delta r} = \Delta r_1(\sigma) - \frac{c^2}{s^2} f(\sigma)$$

$$\Delta r_1(\sigma) = \frac{Re A_{WW}(m_W^2) - A_{WW}(0)}{m_W^2} - \frac{2\,\delta e}{e} + V + B$$

$$+ \frac{c^2}{s^2} \left[c^2 \frac{Re A_{Z_1 Z_1}(m_{Z_1}^2) + t_{z_1 z_1}}{m_W^2} - \frac{Re A_{WW}(m_W^2) + t_{ww}}{m_W^2} \right]$$

In the limit $\varphi = \sigma = 0$ the self energies and $\delta e/e$ become identical with the corresponding contributions in the $SU(2)_L \times U(1)_Y$ correction Δr (one should include the contributions from additional particles). Z_2^μ becomes identical to C^μ; it does not contribute to V but it does contribute to B [4]:

$$(B)_{Z_2} = -\frac{3}{8\pi^2} \tilde{g}^2 (\widetilde{Y}_L)_l^2 \frac{ln\,x}{(x-1)}$$

where $(\widetilde{Y}_L)_l$ are the appropriate quantum numbers for the ordinary leptons and $x \equiv m_{Z_2}^2/m_W^2$. Thus:

$$\Delta r_1(0) = \Delta r + (B)_{Z_2}$$

The simplest strategy is to choose

$$\frac{c^2}{s^2} f(\sigma) = \Delta r_1(\sigma) - \Delta r_1(0)$$

which satisfies the requirement $f(0) = 0$. Then

$$\widetilde{\Delta r} = \Delta r_1(0) = \Delta r + (B)_{Z_2}$$

Recalling

$$\frac{G_\mu}{\sqrt{2}} = \frac{g^2}{8\,m_W^2\,(1 - \widetilde{\Delta r})}$$

and $g \equiv e/\sin\theta_W$, we find

$$\sin^2\theta_W = \frac{\pi\alpha/(G_\mu\sqrt{2})}{m_W^2\,(1 - \widetilde{\Delta r})}. \tag{4}$$

This is consistent with our constraint (2) because, as $\varphi = 0$, the radiative corrections to μ decay should become those of the $SU(2)_L \times U(1)_Y$ theory, evaluated with $\cos^2\theta_W \equiv m_W^2/m_{Z_1}^2$, plus the additional contribution of the new boson, $C \equiv Z_2$. But these are precisely Δr and $(B)_{Z_2}$, respectively.

In the models of interest

$$\widetilde{Y}_{L,R} = \cos\beta\, Q_{L,R}^\chi + \sin\beta\, Q_{L,R}^\psi$$

$$\tilde{g} = \sqrt{\frac{3}{8}}\, g\, tan\theta_W \sqrt{\lambda} \qquad (\lambda \simeq 1).$$

Using the quantum numbers for the leptons

$$(B)_{Z_2} = -\frac{9\alpha\lambda}{16\pi \cos^2\theta_W}\left(cos\beta + sin\beta\sqrt{\frac{5}{27}}\right)^2 \frac{ln\,x}{(x-1)}$$

Noting that $(cos\beta + sin\beta(5/27)^{\frac{1}{2}})^2 \le 1.185$ we obtain $|(B)_{Z_2}| \le 1.6 \cdot 10^{-3}$, $1.0 \cdot 10^{-3}$, $7 \cdot 10^{-4}$ for $m_{Z_2} = 100$, 150 and 200 GeV, and $\lambda = 1$. For the Z_η boson in $E_6 \rightarrow SU(3) \times SU(2)_L \times U(1)_Y \times U(1)_\eta$, $cos\beta = (3/8)^{\frac{1}{2}}$, $sin\beta = -(5/8)^{\frac{1}{2}}$ and $(B)_{Z_2} = -1.0 \cdot 10^{-4}$, $-6.3 \cdot 10^{-5}$, $-4 \cdot 10^{-5}$ for the same values of m_{Z_2} and λ. Thus, $\sin^2\theta_W$ is very close to the $SU(2)_L \times U(1)_Y$ expression in terms of G_μ, α and m_W^2.

5. Neutrino-Induced Neutral Current Phenomena

At the tree level these processes are described by the diagrams of Fig. 1.

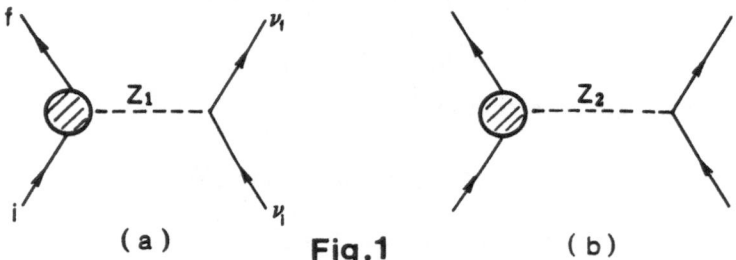

(a)   Fig.1   (b)

The $O(\alpha)$ electroweak corrections can be separated into loop corrections to Fig. (1a), to Fig. (1b) and mixing γZ_2 ad $Z_1 Z_2$ amplitudes. (By convention γZ_1 amplitudes are included in the loop corrections to Fig. (1a)).

We first consider the fermionic contribution to the γZ_2 amplitudes depicted in Fig. (2).

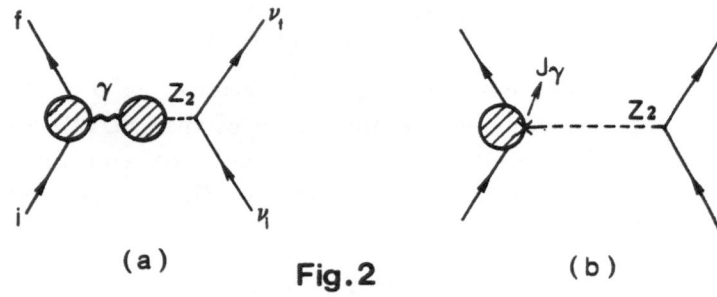

(a)   Fig.2   (b)

They are proportional to

$$\left(\frac{A_{\gamma Z_2}}{q^2} - \frac{\sigma\delta s^2}{sc}\right)_f = \frac{\tilde{g}e\chi}{12\pi^2}\sum_i Q^i_{el}(\widetilde{Y}^i_L + \widetilde{Y}^i_R)\left[\frac{1}{n-4} + Pf\right]$$

where the sum runs over all the fermions, Q^i_{el} is the electric charge of the ith fermion and $\mathbf{Pf}$ means "finite part". In the models of interest

$$\sum_i Q^i_{el}(\widetilde{Y}^i_L + \widetilde{Y}^i_R) = 0$$

In fact, $T_{3L}\, Y,\, \widetilde{Y}$ are proportional to neutral generators of the higher group and $Tr\,(T_i T_j) \propto \delta_{ij}$. Thus the above expression is convergent.

Next we consider the $Z_j\, Z_k\ (j,k=1,2)$ amplitudes depicted in Fig. 3.

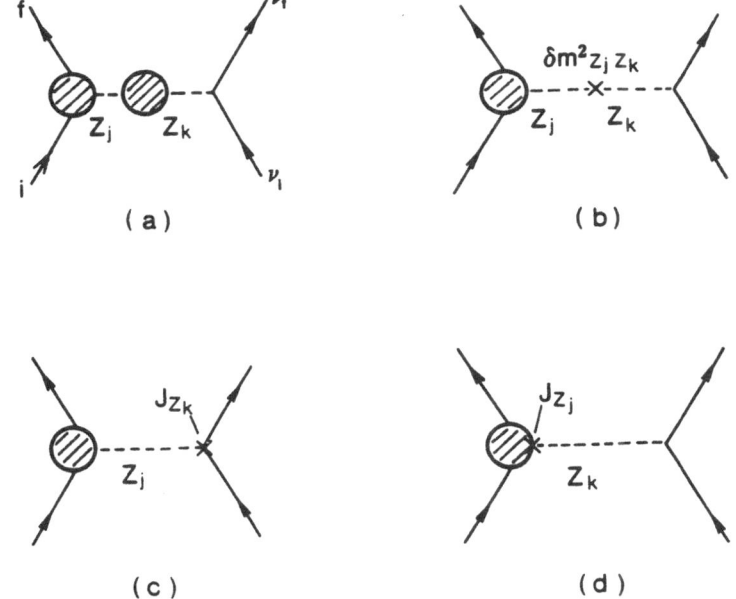

Fig.3

For $j=k$ all the counterterms have been previously determined except $\delta\tilde{g}$. The divergent part of the latter is fixed by requiring the cancellation of the remaining divergences for $j=k=1$ or 2. Next one considers $j=2,\ k=1$. The sum of the 4 diagrams becomes proportional to

$$\left[C_{Z_2 Z_1}(q^2) - \frac{m^2_W}{2\,c^2}\frac{f(\sigma)}{\sigma\chi}\right]$$

After using the orthogonality condition, evaluation of the fermionic contributions leads to

$$C_{Z_2 Z_1}(q^2) = -\frac{g^2}{4\pi c^2} \sum_i m_i^2 \left[-\frac{\chi\sigma}{4} + \frac{\tilde{g}c}{g}(\chi^2 - \sigma^2) T_{3i} (\widetilde{Y}_L^i - \widetilde{Y}_R^i) \right.$$
$$\left. + \chi\sigma\frac{\tilde{g}^2 c^2}{g^2}(\widetilde{Y}_L^i - \widetilde{Y}_R^i)^2 \right] \frac{1}{n-4} + Pf.$$

One finds that the divergent part is cancelled by the term proportional to $f(\sigma)/\sigma$. In particular the divergent part of $C_{Z_2 Z_1}$ survives as $\sigma \to 0$. However $f(\sigma)$ involves $\frac{c^2}{m_W^2} Re A_{Z_1 Z_1}(m_{Z_1}^2)$ and in turn this contains a divergent term proportional to σ and $\sum_i m_i^2 T_{3i}(\widetilde{Y}_L^i - \widetilde{Y}_R^i)$. It survives in $f(\sigma)/o$ as $\sigma \to 0$ and cancels the divergent part of $C_{Z_2 Z_1}$ in the same limit.

In conclusion, all the divergent parts in the one-loop corrections to ν-induced neutral current phenomena are cancelled when the orthogonality relation and the counterterms involving δe, δm_W^2, $\delta m_{Z_1}^2$, $\delta m_{Z_2}^2$, δs^2 and $\delta \tilde{g}$ are applied.

Regarding the finite parts of the corrections, we note that the experimental upper bounds on φ and the amplitude $\mathcal{M}_{Z_2}^0$ are quite small. (The latter is a consequence of either $m_{Z_2}^2 \gg m_{Z_1}^2$ and or the fact that the Z_2 couplings to the ordinary fermions are quite weak.) A reasonable approximation, at least at low energy, is to neglect the loop corrections to $\mathcal{M}_{Z_2}^0$ and the mixing $\gamma - Z_2$ and $Z_1 - Z_2$ contributions and evaluate the electroweak corrections to the dominant $\mathcal{M}_{Z_1}^0$ amplitude in the limit $\varphi = 0$. The latter become then the corrections of the $SU(2)_L \times U(1)_Y$ theory evaluated with $m_{Z_1}^2 = m_W^2/c^2$. This is essentially the approximation used in recent phenomenological studies [1] of the $SU(2)_L \times U(1)_Y \times \widetilde{U}(1)$ models, except for the small shift $\Delta r \to \widetilde{\Delta r}$ in the definition of $\sin^2 \theta_W$.

6. Observations

The above renormalization scheme is consistent with (1) even for arbitrarily small φ and any mass $m_{Z_2} > m_{Z_1}$. We were guided by the requirement that non-analytic terms of $O(\alpha/\varphi)$ should be absent in the radiative corrections . This was implemented by the constraint $f(\sigma) \sim \sigma$ as $\sigma \to 0$. This strategy led us in turn to a definition of $\sin^2 \theta_W$ which is very close to the SM expression in terms of G_μ, α and m_W^2 and Δr, namely (4). In principle, by altering our choice of the finite part of $f(\sigma)$ we could slightly shift $\widetilde{\Delta r}$ and correspondingly the definition of $\sin^2 \theta_W$ by terms of $O(\sigma)$. But suppose that an attempt is made to redefine $\sin^2 \theta_W$ so that it differs from the above by terms of $O(\alpha)$ but of zeroth order in σ, while maintaining (1) as exact. This requires $f(0) \neq 0$ and generates a potentially large radiative corrections in $\delta m_{Z_1 Z_2}^2$, namely $(m_W^2/2\sigma\chi c^2)f(0) = m_W^2 O(\alpha/\sigma)$.

What is the origin of such terms? The Born (tree-level) amplitudes contain, because of the mixing, terms linear in σ such as

$$\frac{ig\tilde{g}}{c}\sigma\, J^\mu_Z J_{C\mu}\left[\frac{1}{q^2-m^2_{Z_1}}-\frac{1}{q^2-m^2_{Z_2}}\right].$$

For small φ, (1) can be written as $\sigma^2\simeq(m^2_W/c^2-m^2_{Z_1})/(m^2_{Z_2}-m^2_W/c^2)$. Suppose we shift c^2 by $\Delta c^2=-\Delta s^2$. Then σ changes by

$$\Delta\sigma=\frac{m^2_W}{c^4}\frac{\Delta s^2}{2\,\sigma}\frac{1}{(m^2_{Z_2}-m^2_W/c^2)}$$

and the Born terms change by

$$\simeq\frac{-ig\tilde{g}}{c}J^Z_\mu J^{\mu C}\frac{m^2_W}{c^4}\frac{\Delta s^2}{2\,\sigma}\frac{1}{(q^2-m^2_{Z_1})(q^2-m^2_{Z_2})}.$$

To compensate, the radiative corrections generate an equal and opposite contribution in the mixing diagrams. The relative correction (with respect to the $\sigma J^Z_\mu J^{\mu C}$ Born terms) is

$$\simeq\frac{-m^2_W}{m^2_{Z_2}-m^2_{Z_1}}\frac{\Delta s^2}{2\,c^4\sigma^2}.$$

If $\sigma\simeq0.05$ and $m^2_{Z_2}\simeq150$ GeV, this is $\simeq-139\,\Delta s^2$. Thus $|\Delta s^2|$ must be $\ll0.007$ in order that the $O(\alpha)$ term be much smaller than the corresponding Born term. The situation becomes more sensitive for smaller σ and $m^2_{Z_2}$.

Because of (1), in these constrained models we must have $m^2_W/\cos^2\theta_W\geq m^2_{Z_1}$. It is interesting to note that, because of (4), this may be true only for some range of values of m_t. For example, let us assume that $m_W=81$ GeV and $m_{Z_1}=92$ GeV with great precision. Then $\cos^2\theta_W$ should be ≤0.775. Inserting $\Delta r=0.0713$ (the value corresponding to $m_t=45$ GeV and $m_H=100$ GeV) in (4) gives $sin^2\theta_W=0.228$ or $c^2=0.772$ which is consistent. But, if $m_t=150$ GeV, $\Delta r=0.0412$ and $\cos^2\theta_W=0.779$ which contradicts the above upper bound. In conclusion, with improved experimental precision, it is possible that the m_t dependence of Δr may play an interesting role in deciding the tenability of constrained $SU(2)_L\times U(1)_Y\times\widetilde{U}(1)$ models.

References

1. U. Amaldi et al., Phys. Rev. **D36**, 1385 (1987); G. Costa et al., Nucl. Phys. **B297**, 244 (1988) and articles cited therein.

2. G. Degrassi and A. Sirlin, "Renormalization of Constrained $SU(2)_L\times U(1)_Y\times\widetilde{U}(1)$ Models", NYU preprint (1989). The present talk is based on the analysis of this paper.

3. A. Sirlin, Phys. Rev. **D22**, 971 (1980).

4. W. J. Marciano and A. Sirlin, Phys. Rev. **D35**, 1672 (1987).

5. W. J. Marciano and A. Sirlin, Phys. Rev. **D22**, 2695 (1980).

Part IV

Experimental Strategies

Radiative Corrections for e⁺e⁻ Collisions   Editor: J.H. Kühn
© Springer-Verlag Berlin, Heidelberg 1989

Precision Measurement of Electroweak Parameters at the Stanford Linear Collider

Patricia Rankin

Department of Physics, University of Colorado, Boulder, Colorado, 80309-390.

1. Abstract

This paper discusses the precision measurement of the Z^0 line shape at the SLC, concentrating on systematic errors which differ from those affecting similar measurements at LEP. Some measurements designed to test the Standard Model are also discussed, including those of various asymmetries. The emphasis is again on the exploitation of particular features of the SLC such as the planned longitudinal polarization of the electron beam.[1]

2. Introduction

The Standard Model of Electroweak interactions has proven very successful, both at providing a structure which preserves unitarity, and at relating many apparently disparate phenomena. The theory has a $SU(2) \times U(1)$ gauge structure requiring three experimental measurements to renormalize the three bare parameters of the model (the $SU(2)$ coupling strength, the $U(1)$ coupling strength, and the vacuum expectation value of the Higgs field).[2] Once the defining parameters have been chosen, predictions can be made. As we enter the era of SLC/LEP physics, no deviations from the predictions of the Standard Model have been seen. However, as we make more and more exact predictions and more and more types of test measurements (at more and more energy scales), the greater the chance that the model will fail. It may fail "softly" in such a way that the discrepancies can be resolved by the addition of new particle generations, or more radical changes may be required.

We expect the mass of the Z^0 resonance to become one of the defining parameters of the Standard Model following accurate measurements at the SLC (Stanford Linear Collider) and at LEP (Large Electron Positron collider). This paper begins with a discussion of how the Z^0 mass will be measured at the SLC, and concentrates on those systematic errors which will differ from those which will limit experiments at LEP. It continues by discussing the real challenge for experiments at the SLC and LEP which is to test the predictions of the Standard Model. One way of extracting a value for $\sin^2\theta_w$ to high precision, at least in principle, is to measure forward-backward asymmetries using unpolarized beams. The difficulties associated with these measurements and their sensitivity will be reviewed. This paper closes with a discussion of polarization at the SLC and the measurement of the left-right polarization asymmetry.

3. Measurement of the Z^0 Mass

The mass of the Z^0 must be measured as accurately as possible since the error on this measurement will determine how stringent the future tests of the Standard Model can be. Measurements of the Z^0 mass and width will be amongst the first and the most important physics results coming from the SLC. The Z^0 mass is the third and last parameter of the three parameters, α, G_F, and M_{Z^0}, which completely specify the electroweak sector of the basic Standard Model (which presumes the existence of a single Higgs doublet). The other two parameters are known to one part in a million and one part in fifty-thousand respectively. The aim is to measure the Z^0 mass to an accuracy of around one part in two-thousand (about a 50 MeV error). This will allow all of the couplings in the theory to be calculated to high precision for comparison to values derived from other SLC measurements and from other experiments such as neutrino scattering and measurements of atomic parity violation. In the future we can expect that a correct prediction of the Z^0 mass will be a requirement imposed on any candidate theory of grand unification.

It is clear that the absolute value of the center-of-mass energy of the beams (before correction for energy losses due to initial state radiation) at the interaction point needs to be known as exactly as possible. Measurements of the line shape of the Z^0 resonance and many tests of the Standard Model also require high precision of relative beam energy measurements, often over long periods of time. The measurement of the forward-backward muon asymmetry, for example, is very sensitive to changes in the center-of-mass energy, a 40 MeV shift causes a 10% change in the value of A_{FB} at the Z^0 peak. The success of the SLC physics program will partly be determined by the precision which can be reached in measuring the collision energy.

The SLC is a single pass machine which means that there is only one chance for the electron and positron bunches to collide. The collision energy can vary from pulse to pulse, and the energy of the electron beam may not necessarily be equal to that of the positron beam.[3] The accurate determination of the available energy at the interaction point required the installation in each beam dump of apparatus specifically designed to measure beam energies. A schematic of these devices, termed energy spectrometers, is shown in figure 1.[4] Two small horizontal bend "wiggler" magnets are used primarily to produce swaths of

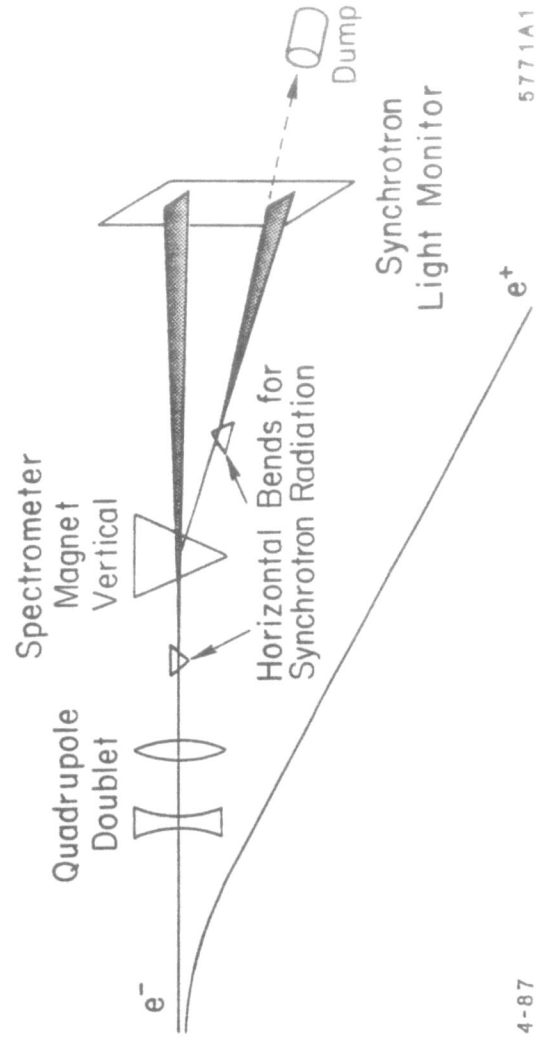

Figure 1: A conceptual drawing of an energy spectrometer.

single beam at the beam dump is 20 MeV. Relative energy measurements can be made to 10 MeV accuracy.

However, it is not completely trivial to combine the energies of the beams at the dumps to calculate the center-of-mass energy at the interaction point. An allowance must be made for energy lost when the beams are transported from the interaction point to the dumps (about 50 MeV) and also for losses due to beamstrahlung as the beams pass through each other (less than an MeV at turn on luminosities). A potentially more serious problem is that the beam conditions at the interaction point can influence the luminosity weighted center-of-mass energy. A residual momentum dispersion at the point of collision coupled with an offset in the crossing of the beams can shift the luminosity weighted average. This problem, however, lowers the luminosity achieved at the same time as it shifts the energy. An estimate of the size of this effect (see reference 4) gives a 30 MeV contribution to the error in the center-of-mass energy determination. Thus, the absolute energy at the interaction point is determined with **an overall systematic error of 40 MeV**.

The error on the center-of-mass energy measurement is expected to decrease as an understanding of SLC operation is gained since shifts in energy which depend on how the beams collide can be largely corrected for. For example, crossing errors at the IP can be measured even at low intensities to less than one beam sigma by looking at beam-beam deflections. If there are slow, coherent drifts in steering which cause a systematic bias in the energy measurement, they should be detectable by repeat scanning at given energy points. In any case, since the shifts in energy are signed, they will cancel stochastically when measurements are taken over reasonable intervals of time.

Several checks for unsuspected biases in the energy measurement will be made.The energy measured by the energy spectrometers will be read-out for event triggers, luminosity triggers, and for random (unbiased) triggers. The relevant estimate of the average center-of-mass energy for a scan point will be made using luminosity triggers. The energy estimate based on random triggers will tend to give the wrong energy if there is a correlation between luminosity and beam energy. The estimate based on the beam energy for Z^0 events will also be biased since it will be a cross-sectionally weighted average. The expected value of this bias can be calculated and checked using the pulse to pulse energy variance of the beams.

A systematic check of the energy measurement can also be made by combining measurements from only one spectrometer with information on the average acollinearity of wide angle Bhabha and muon pairs. The acollinearity distribution should center around zero for beams of equal energy. About 100 mu-pair decays need to be measured in the Mark II drift chamber (which has an angular synchrotron radiation. A large vertical bend magnet located between the two smaller magnets is used to analyse the beam momentum; it produces a deflection of about 18 mrad for a beam momentum of 50GeV/c. This angular deflection corresponds to a separation of about 27cm between the swaths of synchrotron radiation produced by the wiggler magnets when this radiation is detected by synchrotron light monitors further downstream.[5]

Table 1. Errors in single beam energy measurements

Error Source	Error
Magnetic Measurement	5 MeV
Survey Errors	5 MeV
Rotational errors in magnet alignment	16 MeV
Detector position resolution	10 MeV
Combined Systematic Error	20 MeV

The sources of error in the measurement of the energy of a single beam at the beam dump are summarized in Table 1. The first contribution to the error comes from the accuracy with which the magnetic field integrals of the magnets are known and monitored. The fields of the magnets were measured as accurately as possible using both a moving wire technique and a moving probe technique.[6] The magnetic fields are monitored with flip coils, nmr probes and by current recording. The position of the synchrotron radiation detectors with respect to the magnetic centers of the magnets is known to better than 1.5mm. This corresponds to a 5MeV error on the beam energy measurement since the system has a lever arm of at least 15m. The alignment of the vertical bend magnet contributes negligibly to the total error. Rotational misalignments of the horizontal magnets are potentially the source of the largest sytematic error on single beam energy measurements, the error quoted is for a measured misalignment of 2mrad and is therefore a conservative estimate since corrections can be applied. The final error quoted is the error on the measurement of the separation of the swaths of synchrotron radiation. The error quoted assumes that this measurement can be made to an accuracy of 80μ. This accuracy has already been achieved using phosphorus screens to detect the synchrotron radiation. A second synchrotron radiation detector - the WISRD (wire imaging synchrotron radiation detectors) is now coming into operation which is capable of greater accuracy. It will also provide for some redundancy in the system. The combined systematic error on the measurement of the absolute energy of a resolution of about 6.2 mrad) to measure any difference in the energies of the beams to better than 50 MeV.[7]

Finally, we expect that when the electron beam is polarized, it will be possible to calibrate its energy with an accuracy of about 25 MeV. Pulse to pulse measurements will not be made, but the different systematics which limit this measurement will allow another important systematic check on the measurement of the Z^0 mass. The upper half of Figure 2 shows how the expected error, on measurements of the position of the peak of the line shape, varies as a function of integrated luminosity.

The relationship between the peak position and that of the pole of the resonance is now well understood.[8] Many calculations have shown the need to incorporate higher order corrections to describe the radiation of initial state photons to sufficient accuracy. Once this is done, the seperation between the pole and the peak is known to better than 5 MeV. The cuts on the data used in the line shape measurement will be kept as small as is reasonable, so it should be possible to use analytic expressions of the form of the line-shape for fitting purposes and to use event generating Monte Carloes only to make small acceptance corrections.

Since the intention is to use all hadronic final states, it is necessary to consider the possibility that the differences in the interference between the Z^0 and the photon will subtly change the peak shape and hence the apparent position of the pole. These effects have been studied with the EXPOSTAR Monte Carlo[9] and the shifts are estimated to be less than 10 MeV.

4. Measurements of the Z^0 width, and of particle couplings

Once the Z^0 mass has been accurately measured, a prediction can be made for the resonances width, based on the known particles it couples to. A larger width than expected[10] is an indication of the existence of other particles to which the Z^0 can couple, or perhaps that the top quark mass is higher than the value assumed in calculating $\sin^2\theta_w$. For example, a fourth neutrino species contributes around 160 MeV/c^2 to the Z^0 width, while changing the mass of the top quark from 60 GeV to 180 GeV increases the width by about 40 MeV/c^2. The measurement of the width will ultimately be limited by errors in measuring the relative energies of points taken during a scan of the Z^0 line shape. Since any scan will extend over significant time intervals, the relative energy measurement needs to be stable during these periods or the error on the width will be increased. The relative error on center-of-mass beam energy measurements using the energy spectrometers is 30 MeV. The short term pulse

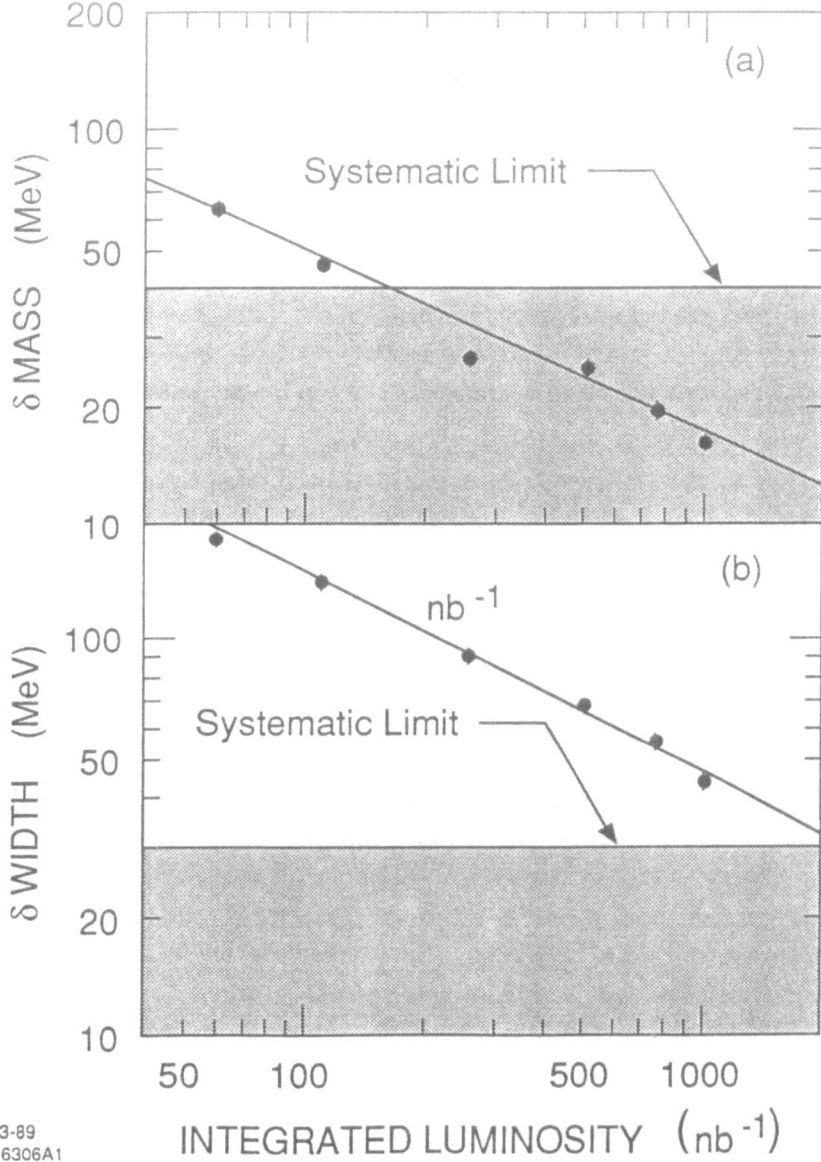

Figure 2: The error on the Z^0 mass and width measurements shown as a function of integrated luminosity. The systematic limits due to measurements of energy at the interaction point are indicated

to pulse energy resolution for both spectrometers is measured to be better than 5 MeV. This is an upper limit since it is obtained by studying the variation in the beam energy for consecutive triggers and attributing all fluctuations to the energy spectrometers rather than to the beams themselves. The lower half of Figure 2 shows the expected error on measurements of the width as a function of integrated luminosity.

The width of the Z^0 resonance is also broadened by QCD corrections to the partial widths of hadronic final states. These corrections amount to about a four percent enhancement for light quarks. The bottom quark needs to be treated as a special case (as would top if found); the partial width of the bottom quark has been shown to decrease as the top quark mass is raised.[11]

The value extracted from the line shape for the width of the resonance is sensitive to the absolute scale or normalization of the cross-section. It requires an understanding of the absolute luminosity, not just of the relative point to point luminosities measured during a scan. The luminosity monitors usually measure the cross-section due to small angle Bhabha's in regions dominated by t-channel exchange and are insensitive to the s-channel Z^0 resonance.

Two devices will be used to measure luminosities for the Mark II. The first of these (the small angle monitor or SAM) has an estimated energy resolution of about 5% for 50 GeV electrons and covers the angular range 50mrad $\leq$ $\theta \leq$ 140mrad. Over this angular range the contribution to the cross section from s-channel Z^0 interactions compared to that coming from t-channel photon interactions is around 1%, the overall counting rate in the detector being slightly higher than the peak Z^0 rate (about 42 nb cross section). The SAM has an estimated angular resolution of 0.2 mrad if the interaction point is known, decreasing to 3 mrad if it is not. The second device (mini-SAM) has a similar energy resolution but works at smaller angles (about 15 - 25 mrad), so the Z^0 contribution is less than 0.1% and the count rate is significantly higher than that in the SAM (around 250 nb).[12] Since the count rates are substantially higher in the mini-SAM, the statistical error on luminosity measurements is correspondingly smaller.

The Bhabha cross-section varies very rapidly at small angles, to lowest order it is given by

$$\sigma_0 \propto \left(\frac{1}{\Theta_{\min}^2} - \frac{1}{\Theta_{\max}^2} \right).$$

A one milli-rad change in the angular acceptance of the mini-SAM can change the observed cross-section by 15%. A 1% systematic error in a measurement made by the mini-SAM results from a 100μ uncertainty in its inner aperture radius or a 420μ uncertainty in the outer radius.

The mini-SAM is also very sensitive to variations in the machine performance, such as the effects of unequal beam energies (which mimic an acollinearity in the electron positron pair due to photon radiation and which will give about a 0.75mrad contribution to the acollinearity angle in the worst case) and misalignments of the beam axis with respect to the detector (again around a 0.7mrad angular contribution at worst). To decrease the sensitivity of both detectors to such effects they are designed so that the two sides of the detectors have slightly different angular acceptances.

Errors in the experimental measurements may be compounded by inaccuracies in the conversion of the count-rates to luminosities. This conversion depends on using Bhabha pair Monte-Carloes which may not be sufficiently accurate for SLC/LEP use. The first order Monte Carloes which exist have a similar "k_0 problem " to those found in first order (or in principle any finite order) muon pair generators. Initial state photon radiation is divided into two classes depending on whether the photon is "hard" or "soft". Hard photons have an energy greater than some cut-off energy (k_0) so three body final states are generated; soft photons have less than this energy and so are not generated. k_0 should be chosen to be below the photon energy detection threshold of the detector; if the energy of the photon is high enough to be detected, then this should be allowed for when generating the final state. Even if the photon is not directly detectable, it may still result in the final state fermion pair being measurably acollinear and the choice of k_0 should also allow for this possibility.

However, for first order Monte Carloes, k_0 cannot be chosen to be below a certain minimum value. This minimum value is determined by the need to generate only events with positive weights (and therefore consider only positive contributions to the cross section). The soft photon contribution to the cross section is included with the other two body contributions with which it is degenerate (corresponding to the emission and re-absorption of virtual photons) when the cross section is calculated. A theorem due originally to Kinoshita, Lee, and Nauenberg states that the singularity associated with the emission of extremely low energy real photons (which can be artificially controlled by giving the photon a fictitious mass) is cancelled by infra-red divergences associated with the virtual photons. The possible disadvantage of this technique is that the contribution from a pseudo-two body final state is balancing a genuinely two body contribution, and so care must be taken to calculate to sufficient accuracy for the balancing to work. If k_0 is too small, this cancellation will not work. The limit is set approximately by the value of k_0 which satisfies

$$0 = \frac{d\sigma_0}{d\Omega}\left[1 + \frac{4\alpha}{\pi}\left(\ln\frac{s}{m_e^2} - 1 + 2\ln\left[\tan\frac{\theta}{2}\right]\right)\right]\ln\frac{k_0}{E}$$

were θ is the angle of the pair with respect to the beam axis.

The value of the cross section or of any physical observable calculated should not depend on k_0, however, the cross-section predicted by the Monte Carlo is usually sensitive to this cut. This problem can be dealt with by tuning the k_0 cut to give the correct prediction, assuming of course that the correct value of the cross-section is known. This requires the existence of a higher order Monte Carlo or calculation.[13] The asymmetric angular acceptances of the detectors also make them more susceptible to higher order effects which change angular distributions. The first order Monte Carloes used at PEP/PETRA did not give good predictions of the observed acollinearity distribution at small angles.

It can be argued that it is difficult to get the experimental error on luminosity measurements to better than a percent, and that if this is true theoretical errors may not dominate. It can also be argued that SLC/LEP are machines for precision measurements and that the rewards for accurate absolute luminosity measurements are greater than before. Consider, for example, the peak cross section for muon pairs which is related to the total width by

$$\sigma_{\mu\mu}(s = M_{Z^0}^2) \propto \frac{\Gamma_{ee}\Gamma_{\mu\mu}}{\Gamma_{tot}^2}.$$

Assuming the theoretical values for the leptonic partial widths (these are completely dominated by the axial couplings which do not depend on the Z^0 mass), one can make an independent measurement of the total width. In this case, a 3% cross section measurement is equivalent to a precision of 40 MeV/c^2 in measuring the width (or about a quarter of a neutrino generation). Work is continuing on higher order Monte Carloes and these should allow the theoretical error on the measurements to be reduced to the fraction of a percent level.[14]

A related measurement to that of the muon cross section is that of the quantity R$'$ measured at the resonance peak, which is given by

$$R' = \frac{\Gamma(Z \rightarrow hadrons)}{\Gamma(Z \rightarrow \mu\mu)}.$$

This quantity is sensitive to interference between the photon and the Z^0. The muon cross section increases as a fraction of the total cross section as one moves down in energy away from the peak. As a result, the muon signal is enhanced relative to the hadronic one and R$'$ is decreased about 3% compared to the lowest order prediction.

The vector coupling constants can be extracted from the forward-backward asymmetry for unpolarized beams for various final state fermions. At the Z^0 peak the asymmetries are given by

$$A_{FB}^{f\bar{f}}(M_{Z^0}) = 3\left(\frac{a_e v_e}{(a_e^2 + v_e^2)}\right)\left(\frac{a_f v_f}{(a_f^2 + v_f^2)}\right).$$

The term involving the electron couplings is essentially that measured by the left-right polarization asymmetry. The second term which involves the couplings of the final state fermions effectively acts as a suppression factor. There are several problems associated with measuring these asymmetries. They are strong functions of energy (especially for the muon final state; this is less true if bottom quarks are being studied), and so the energy at the interaction point must be monitored very accurately. They are also extremely sensitive to the effects of initial state radiation. The measurements are exclusive; for example, strong cuts may be applied on the acollinearity of the muon pairs accepted. As a result, higher order Monte Carloes are essential to extract the particle couplings from the data, and may contribute significantly to the final error.[15] Since only specific final states are used, they require relatively higher statistics than measurements (such as the left-right asymmetry) without this restriction.

After 10 inverse picobarns of data taking (about three hundred thousand visible Z^0 decays), the error on the forward-backward asymmetry for the muon final state is expected to be about 1%, (the systematic error on this measurement due to uncertainties in the center-of-mass energy measurement is about 0.35%). The forward-backward asymmetry to the $b\bar{b}$ final state which has the advantage of a greater sensitivity to $\sin^2\theta_w$ and a larger cross-section will be measured to about 3%, the advantages being offset by the need to use an electron tag and larger systematic errors.

The major drawback to using the forward-backward asymmetries to derive a value for $\sin^2\theta_w$,however, is that the sensitivity to $\sin^2\theta_w$ depends on its value. If $\sin^2\theta_w$ has the value of 0.22, the ten inverse picobarn measurements correspond to an error on $\sin^2\theta_w$ of 0.0042 for muons and 0.0052 for bottom quarks. However, the sensitivity falls off rapidly with increasing values of $\sin^2\theta_w$. For example, if $\sin^2\theta_w$ has the value of 0.23, the corresponding error for muons is 0.0057 and for 0.24 it rises to 0.01.

Although the forward-backward asymmetries will be difficult to measure, and may not be as sensitive to the value of $\sin^2\theta_w$ as the left-right asymmetry discussed below, they will provide interesting constraints on the Standard Model. They are effectively sensitive to C-violating effects. When they are combined with the left-right asymmetry measurement the model will be tested more stringently than it is by either measurement alone.

5. Polarization

Linear accelerators do not disturb the polarization of their beams which makes polarizing the electron beam at the SLC comparatively straight-forward. A polarized electron source will be installed (A GaAs crystal) as will three spin rotation magnets (to allow maximal longitudinal polarization of the beam at the interaction point for any center-of-mass energy). The project is progressing smoothly;[16] the spin rotation magnets have been constructed and tested and are awaiting installation. The planned polarization of the electron beam is 45%. Three polarimeters will be installed; two (one at the end of the Linac, the other in the electron beam extraction line) will use Möller scattering and measure all three components of the polarization to about 5%, and the third, which is intended to be used for precise measurements of the longitudinal polarization (to 1%), will use Compton scattering (and will also be installed in the extraction line).[17]

Once the polarization facility begins working, we can measure A_{LR}, the asymmetry in the total cross section for left-handed versus right-handed electrons. This asymmetry measures only the electron couplings, at the Z^0 pole it is given by

$$A_{LR}(M_{Z^0}) = \frac{-2Pa_e v_e}{a_e^2 + v_e^2},$$

where P is the amount of beam polarization (positive for right-handed electrons). There is no dependence on the final state for measurements at the resonance pole.[18]. The insensitivity to final state effects is a direct result of the fact that this asymmetry measures the initial-state parity violating coupling of the e^- to the Z^0. Since it measures this coupling, it is sensitive to anything which modifies the coupling such as corrections to the production vertex, or corrections to the Z^0 self energy. It is therefore an extremely good probe of these types of electroweak corrections. The reader is reminded that since these corrections are influenced by virtual particles, they are sensitive to the effects of physics above the Z^0 energy scale.

As one moves away from the Z^0 pole, the effects of photon Z^0 interference introduce a slight final state dependence which can be corrected. A major advantage of this asymmetry is that the effects of initial state radiation largely cancel. It does not vary strongly with energy and so makes much looser requirements on the monitoring of the beam energy. It has the further advantage that

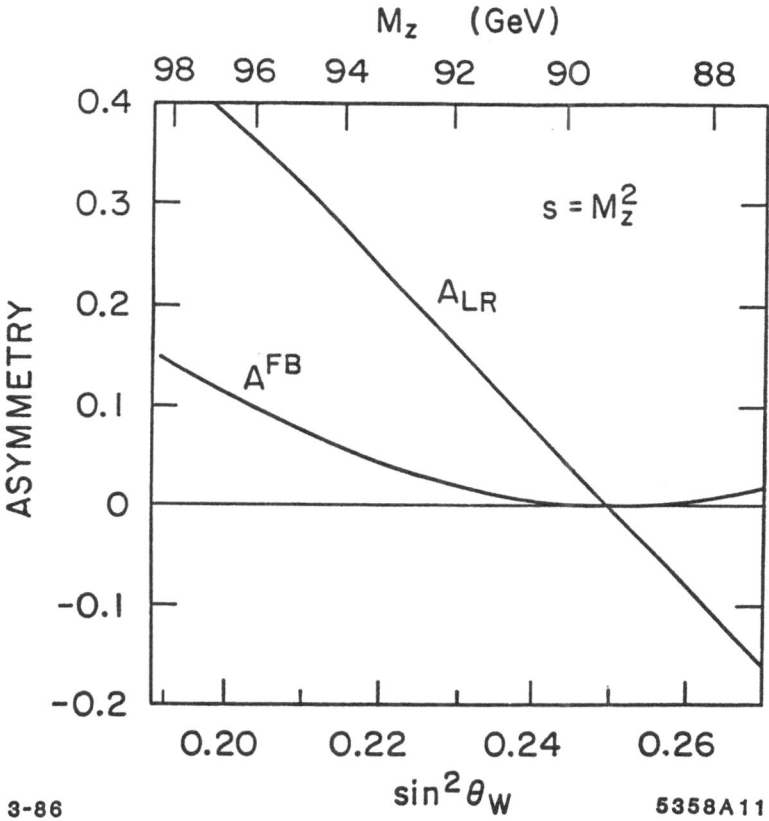

Figure 3: The values of the forward-backward asymmetry and of the left-right asymmetry plotted as a function of $\sin^2\theta_w$. The relationship between $\sin^2\theta_w$ and the Z^0 mass assumes a top quark mass of $50\,\mathrm{GeV}/c^2$

its sensitivity to $\sin^2\theta_w$ is independent of $\sin^2\theta_w$. Figure 3 illustrates the different dependencies on $\sin^2\theta_w$ of the forward-backward and left-right asymmetries (note that the relationship between M_z and $\sin^2\theta_w$ assumes a top quark mass of around $50\text{GeV}/c^2$). The error on $\sin^2\theta_w$ is simply one eighth of the error on the asymmetry measurement which is given by

$$\delta A_{LR} = \sqrt{\left(\frac{\delta P}{P} \times A_{LR}\right)^2 + \frac{(2.34 \times 10^{-4})}{(integrated\ luminosity\ in\ pb)}}.$$

For example, a ten inverse-picobarn sample corresponds to an error of 0.001 in $\sin^2\theta_w$ if the polarization is known to 3%, and decreases to 0.0007 for a 1% measurement of the polarization. This precision in the measurement of $\sin^2\theta_w$ is similar to the accuracy with which $\sin^2\theta_w$ can be derived from Z^0 mass measurements.

The only disadvantage to the left-right asymmetry is that it does not give any information about final state couplings. These may be of interest, especially if new particles are found at the SLC. However a modification of the forward-backward asymmetry has been proposed for use with polarized beams[19], which by taking particular combinations of forward and backward cross-sections gives

$$A_{FB}^{\tilde{f}} = \frac{3}{4}P\left(\frac{-2a_f v_f}{a_f^2 + v_f^2}\right).$$

Polarization even improves the measurement of b-quark asymmetries by increasing the tendency of the quark to align with the direction of the beam electrons.[20] When polarization is coupled with the CCD vertex detector of the SLD detector (which can approach within 15mm of the interaction point) it becomes feasible to study $b\bar{b}$ mixing at the SLC.[21]

6. Summary and Acknowledgements

The SLC is a very different machine from LEP and thus will allow some measurements to be made with very different systematic errors. This redundancy should be very useful given the importance of precise measurements of the Z^0 line shape. In addition, polarization at the SLC will allow a precise probe of the electroweak corrections.

I would like to thank Johann Kühn for inviting me to the conference, and for being such a pleasant host. I would also like to thank my colleagues at the University of Colorado (Professors Nauenberg and Franklin) who took over my teaching duties. I am grateful to the U.S. Department of Energy for support as a junior investigator. This work was supported by Department of Energy contract DEAC-0286ER-40253.

7. References

1. This paper is based on a talk which was given before any Z^0 decays had been observed at the SLC. Material which has been made obsolete by events will not be presented in this report. At the time of writing, the analysis of the early data from the SLC is too preliminary for inclusion.

2. If no assumption is made about the structure of the Higg's fields, a fourth measurements is needed to fix the value of what is usually termed the ρ parameter. The simplest models, with an isospin doublet structure, fix the ρ parameter to be one. Generalizations of the Standard Model allow this to be a fourth free parameter, measurement of which constrains the form of the Higgs sector. The renormalization scheme proposed by Lynn and Kennedy for example (SLAC-PUB-4039), specifies the running with q^2 of $\alpha, G_f, \sin^2\theta_w,$ and ρ. The contribution of Giampiero Passarino to these proceedings discusses this "fourth" parameter in more detail.

3. Actually the energy of the electron and positron beams may differ at the interaction point for conventional machines too, due to asymmetric placement of rf cavities with respect to detectors.

4. J. Kent at al., Precision Measurements of the SLC Beam Energy, SLAC-PUB-4922, LBL-26977.

5. M. Levi et al., Precision Synchrotron Radiation Detectors, SLAC-PUB-4921, LBL-26976.

6. M. Levi, J. Nash, S. Watson, Precision Measurements of the SLC Spectrometer Magnets, SLAC-PUB-4654 .

7. The effects of initial state radiation can be allowed for; they should not in any case produce significant systematic shifts in the observed energy difference.

8. See, for example, the contribution of Frits Berends to these proceedings.

9. A discussion of this Monte Carlo can be found in Kennedy et al, Electroweak cross-sections and asymmetries at the Z^0 , SLAC-PUB-4128.

10. Significantly smaller widths are very difficult to explain, and almost certainly imply the existence of very spectacular new physics. Increasing the Higgs mass substantially is the only way I know of to get such an effect within the Standard Model. However, the calculations of the effects of such a Higgs are almost certainly unreliable due to the breakdown of perturbative calculations as the Higgs becomes strongly interacting. One interesting possibility for new physics is the existence of a new E_6 gauge boson; this is discussed in a paper by F.M. Renard and C. Verzegnassi (CERN-TH.5372/89).

11. This is due to the fact that there is no automatic decoupling of the effects of heavy top quarks. This was discussed at this conference by Kühn and shown in tables presented by Berends.

12. The SAM is discussed in more detail in Mark II/SLC note:164. Information on the mini-SAM can be found in Mark II/SLC notes 150,152.

13. These comments reflect a conversation I had with Bennie Ward on the necessity for higher order Monte Carloes for calculating the Bhabha cross section. People comparing the results of Monte Carloes also need to be aware that comparisons should use the same value of k_0.

14. For example, the BHLUMI Monte Carlo of Jadach and Ward allows generation of two hard and multiple soft photons. This program is discussed in S. Jadach and B.F.L. Ward, Multiphoton Monte Carlo for Bhabha scattering at small angles, UTHEP-88-01.

15. This is a much harder problem than the prediction of the Z^0 line-shape. Not only must multiple photons be included, but how they are included becomes important. Not only must the amount of radiated energy be correct, but so must the overall change in momentum.

16. Its schedule has been determined by overall progress on the SLC rather than internal problems.

17. The reader who is interested in more details is referred to M.L.Swartz, Polarization at the SLC, SLAC-PUB-4689.

18. Many people have pointed out the particular sensitivity of the left-right asymmetry to corrections to the Z^0 self energy and the many desirable features of this quantity; for example, B.W.Lynn and C. Verzegnassi, Longitudinal e^- beam polarization asymmetry for $e^-e^+ \rightarrow$ hadrons, SLAC-PUB-3967.

19. A.Blondel,B.Lynn, F.Renard,C.Verzegnassi, Precision measurements of final state weak couplings from polarized electron annihilations, Montpellier PM/87-14.

20. P.G.Weismann, Polarization as a tool for studying particle properties, SLAC-PUB-4555.

21. W.T.Atwood, B meson physics with polarized electron beams at the SLC, SLAC-PUB-4668.

Radiative Corrections for e⁺e⁻ Collisions   Editor: J.H. Kühn
© Springer-Verlag Berlin, Heidelberg 1989

What Can We Learn from the Z Line Shape?
An Experimentalist's Survey

D. Schaile

Physikalisches Institut der Universität Freiburg

1. Introduction

For the preparation of LEP/SLC physics a big effort has been invested in the calculation of radiative corrections. One of the first experimental tests will be the measurement of the Z^0 line shape. Though the LEP/SLC era eventually will provide us with measurements which are much more sensitive to the details of radiative corrections as for example fermion asymmetries, the Z^0 line shape will probe one of the axioms these calculations are based on, namely the Standard Model. This paper first gives an overview on the experimental aspects of the Z^0 line shape like the choice of the event sample, backgrounds and systematic errors. We will put no emphasis on a specific detector as the measurement of the Z^0 line shape is not a challenge for any particular feature of the LEP/SLC detectors. The discussion of machine related problems however will be restricted to the still fairly homogeneous expectations for the LEP ring as the predictions on the SLC performance have acquired a wide span between pessimism and optimism in view of the initial start-up. Based on a realistic estimate of the experimental precision we then discuss the compromise between the accuracy of line shape parameters and the luminosity and give some prospects for the physics potential of the line shape measurement for the near future and beyond.

2. General considerations concerning the event sample

The line shape measurement should be available for all known fermions separately but not all channels will give small systematic errors. For a long time only the muon channel has been considered as candidate for the ultimate precision measurement of the Z^0 line shape (c.f. [1]) and the hadronic final state has been attributed the role of an interim solution for low luminosities. This opinion is no longer undisputed because of arguments summarised below. Around the Z^0 peak the ratio of hadronic events to μ pairs increases by a factor of five as compared to PEP/PETRA energies. As there is no comparable energy dependence of two photon events or machine backgrounds the signal to background ratio is shifted in favor of hadronic events. We also have to face the paradoxon that the cross-section $e^+e^- \to \mu^+\mu^-$ depends much stronger on the precision of the strong coupling constant α_s than the hadronic cross-section, because for the latter the α_s contribution to the total width and the final state will cancel in the vicinity of the resonance. At Z^0 energies it is easy to trigger on hadrons with full efficiency whereas all LEP experiments have sophisticated provisions for muons. An argument in favor of muons is the model independent determination of the acceptance which for hadrons depends on the fragmentation process. However the LEP/SLC era will provide us with a rich statistics of hadronic events for many distributions which will allow us to constrain the model significantly. In the end the limiting accuracy might not be set by acceptance calculation for either muons or hadrons but by the luminosity measurement or by the precision to which we determine the beam energy.

Having found no clear evidence that the systematic errors of the line shape measurement will be smaller for muons than for hadrons, we are left with the undisputed argument of statistics and will therefore concentrate on hadronic final states.

For the selection of hadronic events we have the following criteria:

i)   A small acceptance correction in general has small errors.
ii)   The background must be reduced to a level that uncertainties in its subtraction are negligible compared to other systematic errors.
iii)   The edges of acceptance should be chosen such that the acceptance does not depend strongly on the modelling of the detector resolution.

The criteria ii) and iii) require a compromise with respect to i). The final choice of criteria has to be based on a detailed study of the acceptance and the background rejection as function of detector resolution parameters. This analysis has been performed by each experiment and is subject to constant refinement. We will not reproduce this study here and just give a typical set of selection criteria in Table 1. Note that there is an implicit restriction of solid angle entering via the acceptance region for charged tracks and clusters (typically $|\cos\Theta| \leq .93$ or $.95$), because the requirement to have a visible energy of at least $0.5\sqrt{s}$ discards nearly all hadronic events for which the event axis points to the region where no tracks and clusters are accepted. With the cuts in Table 1 the acceptance of hadronic events is between $90-95\%$.

Table 1: Hadronic event selection criteria

Number of charged tracks	≥ 5
$E_{vis}/\sqrt{s}$	≥ 0.5
or	
$\Sigma p/\sqrt{s}$	$\geq .25$
z_{event}	$< 5-10$ cm

---

3. Backgrounds

At first sight the most important background are two-photon events. Most of these events however have a small invariant $\gamma\gamma$ mass which motivates a visible energy cut. After the cuts in Table 1 the two-photon contribution is negligible. At PEP/PETRA energies this background is estimated to be of the order of 1% with about equal uncertainties. Close to the Z^0 the uncertainty in its subtraction represents no problem as it will stay well below 0.1% within a region of $\sqrt{s} = m_z \pm 5$ GeV.

Also τ pairs are much less a problem at the Z^0 than at PEP/PETRA energies as the ratio $e^+e^- \to hadrons / e^+e^- \to \tau^+\tau^-$ increases from 4 to 20. The requirement to have at least 5 charged tracks removes the 1-1 and 1-3 prong events and will therefore discard $95-98$ % of all τ pairs, depending on the details of the track selection. Most of all Bhabha events will be clean 2 track events like μ pairs. Interactions in the detector material however give rise to electromagnetic showers. These events have been removed by scanning at PEP/PETRA energies, however an automatic procedure is required at LEP. For a typical LEP detector the event selection criteria in Table 1 are sufficient to reduce the accepted Bhabha background below the 0.1% level. An additional lepton rejection by about a factor of 4 can be obtained if one adds to the criteria in Table 1 an equivalent of the τ cut applied in the analysis of PEP/PETRA data which requires an invariant mass of more than 2 GeV in at least one hemisphere of the event.

The background which might turn out to be a problem are beam pipe interactions from off-momentum electrons. According to a generator of the LEP machine group we expect about $2 \ 10^3$ Hz of such interactions within a region of ± 3 m from the interaction point. Tracking these events through the full detector simulation only 2 Hz of those satisfy a loose 2-track trigger and none of those will pass the event cuts. But especially at the start of LEP operation this background might be worse and we have to await the scanning of the first thousand events to be sure.

Summing up all backgrounds we expect a background contribution of less than 0.5% over the energy range of the scan and the uncertainties in its subtraction will therefore be negligible.

4. Sources of systematic errors

As shown above the systematic error from the background subtraction can be neglected and we will therefore concentrate on the uncertainties of the acceptance, the luminosity measurement and the determination of the center of mass energy.

Most of the computing power spent by experimentalists is concerned with the determination of the acceptance. An important aspect is the precision of the detector simulation. PEP/PETRA experiments attribute an absolute error of 0.5% to their data due to uncertainties in the detector response. The high statistics expected at LEP might help to improve on the above error. The model dependence of the acceptance is induced by the fragmentation process and by radiative corrections and requires close collaboration with theoreticians. Figure 1 shows the energy dependence of the acceptance computed with the JETSET 7.1 Monte Carlo for the cuts given in Table 1. Without initial state radiation there is no energy dependence of the acceptance as indicated by the open symbols in Figure 1. An energy dependence is only introduced switching on QED corrections as indicated by the solid symbols in Figure 1. It is well known that JETSET 7.1 has only a first order treatment of initial state radiation which is unacceptable for the precision measurements expected at LEP/SLC. A big effort of comparing and understanding electroweak generators that treat photonic corrections in second order with exponentiation of soft photons has been started within the LEP physics workshop. First results of comparisons at the meeting in February '89 [2] looked rather alarming. The efficiency of accepting a μ pair, if one only imposes a cut on the invariant mass of the muons of 0.2 $\sqrt{s}$ agreed within 0.5% at the peak but only within 2% over an interval of $m_z \pm$ 4 GeV. In the mean time we have at least 3 generators (KORALZ [3], DYMU2 [4], MOE [5]) which agree with the analytical calculation of the line shape program ZBATCH [7] to better than 0.5% over the above energy range.

The fragmentation model dependence of the acceptance has two aspects: the present uncertainty of the fragmentation parameters and the extrapolation of models from $\sqrt{s} \approx 30$ GeV to $\sqrt{s} \approx 100$ GeV. Varying fragmentation parameters within their present limits results in a change of the acceptance of less than 0.5 % but the problem of extrapolation cannot be solved without a comparison to data. But as discussed above these uncertainties will mainly contribute to the overall scale and do not distort the line shape.

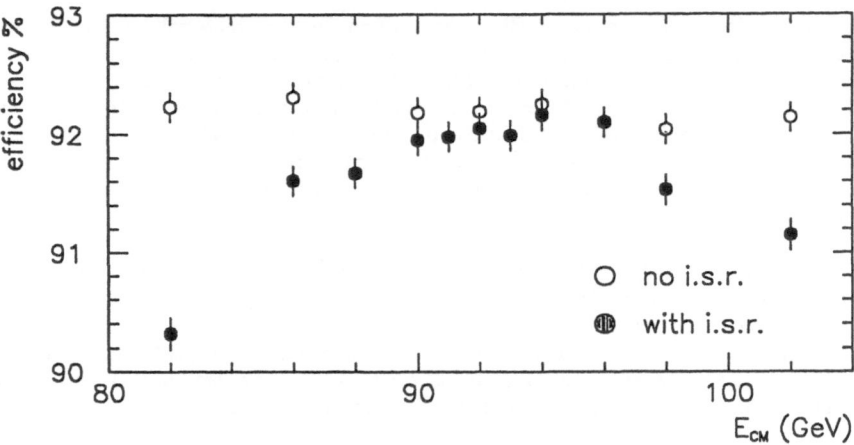

Figure 1:   Energy dependence of the acceptance for hadronic events generated by JETSET 7.1. The initial state radiation is switched off for points with open symbols and turned on for points with solid symbols.

A substantial source of systematic errors is the luminosity measurement. Most experiments have simple devices which cover the very small angle region (5−10 mrad) and a sophisticated luminosity monitor with tracking and calorimetry covering an angular range of about 50−120 mrad. The effective Bhabha cross-section in this detector is of the order of the hadronic cross-section and an order of magnitude higher in the far forward detector, which will however mainly be used for an online feedback for reasons of systematics. Experience from previous e^+e^- experiments shows that it is hard to get an absolute luminosity measurement to better than 2%, a realistic number for the first weeks of LEP operation is 5 %. The relative point to point normalisation depends on the understanding of radiative corrections, changes in the interaction position and the stability of background conditions. It is hoped that this error is well below 1%, but this has to be confirmed by measurements.

Finally a scan will also be affected by the absolute and relative precision of the machine energy determination. In the first phase of LEP the center of mass energy will be derived from the magnetic field measurement with an accuracy of $\Delta E_{CM} / E_{CM} = 5\ 10^{-4}$. During one fill the energy can be varied with a relative precision of $\Delta E_{CM} / E_{CM} = 5\ 10^{-5}$, limited by the accuracy of the power supply readings. Running down the dipoles for a new fill however reduces the energy reproducibility to $5\ 10^{-4}$ due to temperature fluctuations. If transverse polarisation becomes available the absolute energy scale could be measured to an accuracy of $2\ 10^{-5}$ by the method of depolarising resonances.

In Table 2 the expected measurement errors are summarised for the first 10^5 Z^0s together with an estimate of the ultimate precision of the hadronic cross-section. The diagram in Figure 2 gives the relative contribution of statistical and systematical uncertainties as function of $\sqrt{s}$ for an integrated luminosity of 200 nb^{-1} per point. In calculating the contribution of the point to point uncertainty introduced by the energy reproducibility we have assumed that there is no averaging over several fills which is probably pessimistic. From Figure 2 we conclude that the systematic errors become already sizable as compared to statistics at the integrated luminosity quoted.

Table 2: Contributions to the systematic errors for the line shape

	first 10⁵ Z⁰s		ultimate accuracy	
	normalisation	point to point	normalisation	point to point
detector simulation	< 1.0%	< 0.5%	< 0.5%	negl.
radiative corrections	< 0.5%	< 0.5%	0.1%	0.1%
fragmentation model dep.	< 1.0%	negl.	negl. ?	negl.
background subtraction	negl.	negl.	negl.	negl.
luminosity monitoring	≤ 5.0%	< 1.0%	2.0%	0.2%
total	5.0%	1%	2.0%	0.3%
machine energy $\Delta E_{CM}/E_{CM}$	$5\ 10^{-4}$	$\leq 5\ 10^{-4}$	$2\ 10^{-5}$	$\leq 2\ 10^{-5}$

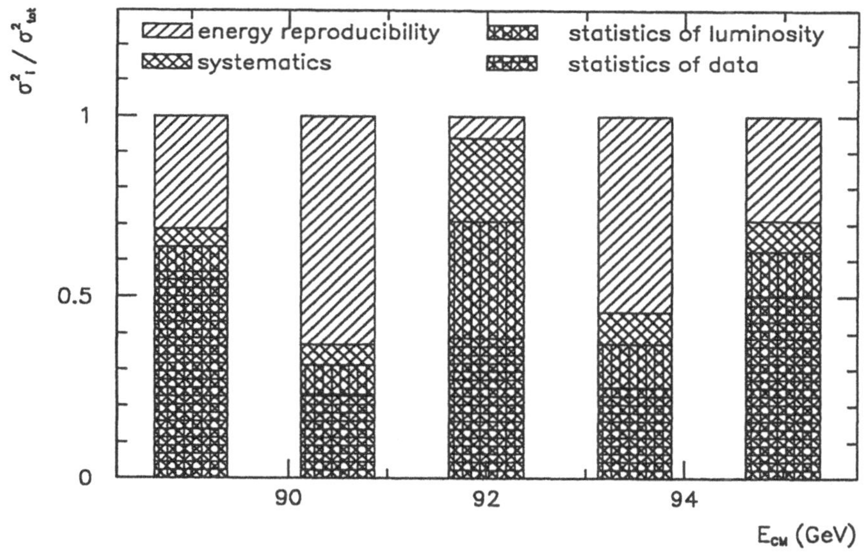

Figure 2:   Relative contributions to the point to point error for an integrated luminosity of 200 nb⁻¹ per point (m_z = 92 GeV). The diagram shows the individual errors σ_i^2 arising from the statistics of the data, the statistics of the luminosity monitoring (based on an effective Bhabha cross-section of 36 nb), the sum of all systematic errors (1%) and the energy reproducibility normalised to the total error σ_{tot}^2 at each point.

5. The trade-off between luminosity and accuracy

The interest in a scan of the Z^0 line shape can be divided into 3 categories

i)   Find the Z^0 peak as quickly as possible to reach high signal rates. Many proposals are circulating which promise an accuracy of $\Delta m_Z = 200 - 500\,MeV$ for an integrated luminosity of O(10) nb^{-1}. One of these proposals might have become reality at the time you read this paper and we will not detail on this further.

ii)   Get a precision measurement on m_Z and Γ_Z and the total hadronic cross-section. The anticipated performance of LEP in '89 foresees a peak lumininosity of $2\ 10^{30}$ cm^{-2} s^{-1} with an efficiency factor of 0.5 over a 20 hours run. These figures would allow to reach an O(50) MeV accuracy on both m_Z and Γ_Z in the near future and we will devote most of the discussion to this scenario.

iii)   Get sensitive to non-photonic corrections and virtual effects of new physics. The boundary between scenario ii) and iii) is fluent. Virtual effects of new physics may already show up in ii). In any case they will have to be distangled from the influence of a massive t quark. We may however hope that by the time we start the ultimate precision scan the t quark mass is known to a much better precision than we will ever get it from the line shape measurement, which could increase the sensitivity of this scan.

To treat the dependence of line shape parameters on the luminosity we have to make sure that we place the scan points in the region of highest sensitivity. In principle this is an optimisation problem with many parameters:

- the total luminosity: L_{tot}
- the number of scan points: N_{scan}
- N_{scan} energies: E_i
- $N_{scan} - 1$ partial luminosities: L_i

To get a stable procedure we have reduced the parameter set by requiring that the scan points are equally spaced and that their partial luminosity is proportional to the cross-section raised to some power α. We will justify this reduction of the parameter set at the end of this section. The remaining parameters are then

- the total luminosity: L_{tot}
- the number of scan points: N_{scan}
- the spacing of scan points: δE
- the offset of points w.r.t. $\sqrt{s} = m_Z$: E_{off}
- the luminosity weighting factor: α

The offset E_{off} is defined such that points are taken at

$$\sqrt{s} = E_i = E_0 + i\,\delta E \quad , i = 1, N_{scan} \quad \text{with } E_0 = m_Z - E_{off} - ([N_{scan}/2])\,\delta E .$$

This offset can only be specified with a certain preknowledge on m_Z, for which we assume 200 MeV for scenario ii) and 50 MeV for scenario iii) respectively. The procedure gives worst case results for a variation of E_{off} within the prespecified accuracy on m_Z. The partial luminosities are chosen such that

$$L_i = \frac{L_{tot}}{\sum_j \sigma(E_j)^\alpha} \sigma(E_i)^\alpha \ .$$

For a given set of scan parameters the covariance matrix of the data can be calculated with elements

$$V_{ii} = (\Delta\sigma_i^{p\varphi})^2 + (\Delta\sigma_i^N)^2 \qquad V_{ij} = \Delta\sigma_i^N \Delta\sigma_j^N \qquad \text{for } i \neq j \ .$$

$\Delta\sigma_i^{p\varphi}$ stands for the Gaussian sum of all point to point errors including the statistical error of the data, the statistical error of the luminosity (based on an effective Bhabha cross-section of 36 nb^{-1}) and the systematic errors in Table 2. The impact of the energy reproducibility on the cross-section has been taken into account as $\Delta\sigma^E = \partial\sigma/\partial E \,\Delta E$. $\Delta\sigma_N$ represents the overall normalisation error. We then choose a line shape parametrisation and calculate the covariance matrix $\tilde{V}$ of its parameters. In the consideration below we treat as free parameters an overall normalisation constant A_N, m_z and Γ_z.

We first investigate the luminosity dependence for the initial systematic errors in Table 2. Figure 3 a) shows the luminosity dependence of the error on m_z (open symbols) and on Γ_z (solid symbols) for two sets of strategies: one optimises Δm_z (dashed lines) the other $\Delta\Gamma_z$ (solid lines). Figure 3 a) clearly demonstrates that optimising Δm_z is on the expense of $\Delta\Gamma_z$ and vice versa. In optimising m_z we have to keep in mind the overall scale uncertainty of the LEP energy. There is no point in accepting the deterioration of $\Delta\Gamma_z$ for a strategy that pushes Δm_z far below this line. On the other hand optimising $\Delta\Gamma_z$ requires at least 1 pb^{-1} to reach $\Delta m_z = 50$ MeV. The value of Δm_z depends quite strongly on the energy reproducibility. Figure 3 b) compares the luminosity dependence of Δm_z and $\Delta\Gamma_z$ for an energy reproducibility corresponding to $\Delta E_{CM}/E_{CM} = 5 \ 10^{-4}$ and a perfect reproducibility. It has been derived for strategies optimising $\Delta\Gamma_z$. Keeping in mind that an energy reproducibility of $5 \ 10^{-4}$ is probably pessimistic, the strategy optimising $\Delta\Gamma_z$ is close to the one which one would like to adopt. Independent of the energy reproducibility it becomes clear that an increase of accuracy becomes very expensive in terms of luminosity beyond an integrated luminosity of 1 pb^{-1}.

As an outlook Figure 4 finally gives the precision on m_z and Γ_z for the assymptotic resolutions in Table 2. Here the accuracy levels off only beyond a luminosity of 4 pb^{-1}.

Choosing the energies and partial luminosities for the scan we don't have to become slaves of the optimisation procedure but ask how much freedom one can gain in the scan parameters on the expense of a small fraction ε on the optimum accuracy. It can be shown that the dependence on the number of scan points is negligible and therefore the choice of N_{scan} represents a compromise between experimental simplicity favoring a low number and the sensitivity to unexpected phenomena favoring a high number of scan points. Figure 5 visualises the range of nearly optimal energy settings and partial luminosities for a 5 point scan with a total luminosity of 1 pb^{-1}. Shown are contours of equal accuracy on $\Delta\Gamma_z$ in the $E_{eff} - \delta E$, $\alpha - \delta E$ and the $\alpha - E_{eff}$ plane. The star symbol represents the optimum scanning strategy, the first contour refers to parameter settings leading to $\Delta\Gamma_z = \Delta\Gamma_z^{opt} + 10 MeV$ – i.e. they surround the preferred region of scanning parameters – and the second contour represents scanning strategies which are considerably off the optimum with $\Delta\Gamma_z = \Delta\Gamma_z^{opt} + 35 MeV$. From Figure 5 we conclude that there is a lot of freedom in the choice of scanning parameters. It is therefore not necessary to go beyond our simple parametrisation of scanning strategies in terms of N_{scan}, δE, E_{eff} and α. To summarise, a good scan should cover an energy range of at least 6 GeV, place points fairly symmetric around the peak and give about equal luminosity to all points.

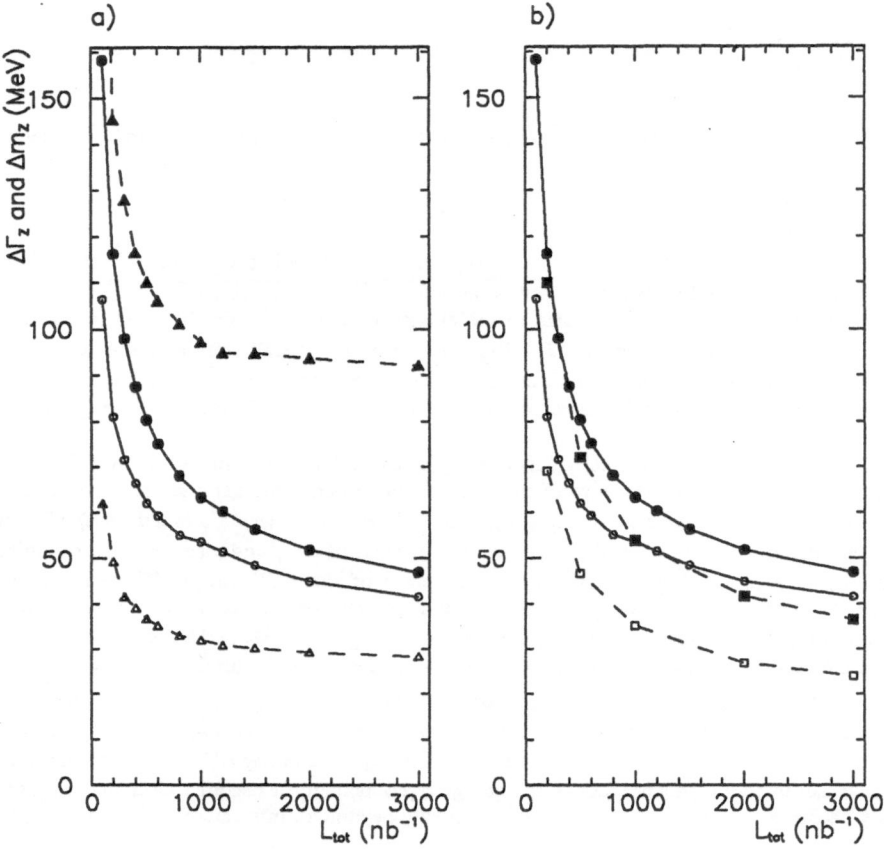

Figure 3:   Luminosity versus accuracy in m_Z and Γ_Z for the near future resolutions in Table 2. Solid symbols denote $\Delta\Gamma_Z$ open symbols Δm_Z. a) Comparison of strategies optimising Δm_Z (dashed lines) and strategies optimising $\Delta\Gamma_Z$ (solid lines). b) Influence of the energy reproducibility on the optimum accuracy. Dashed lines represent a perfect energy reproducibility, solid lines refer to $\Delta E_{CM}/E_{CM} = 5\ 10^{-4}$.

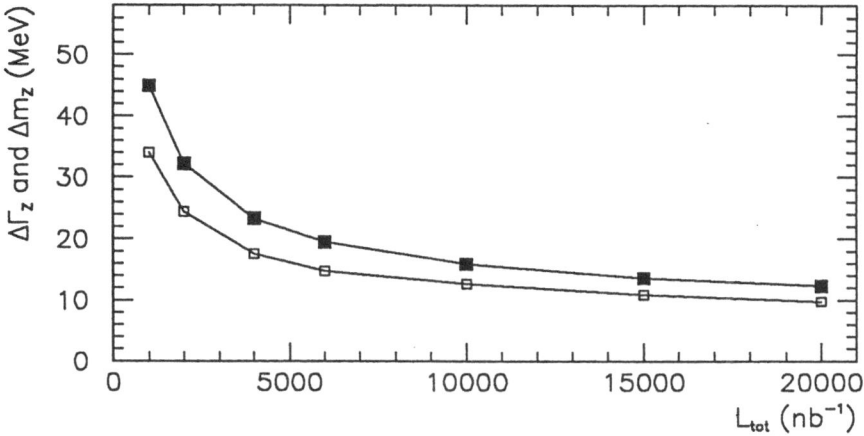

Figure 4:   Luminosity versus accuracy in m_z and Γ_z for the ultimate resolutions in Table 2. Solid symbols denote $\Delta\Gamma_z$, open symbols Δm_z.

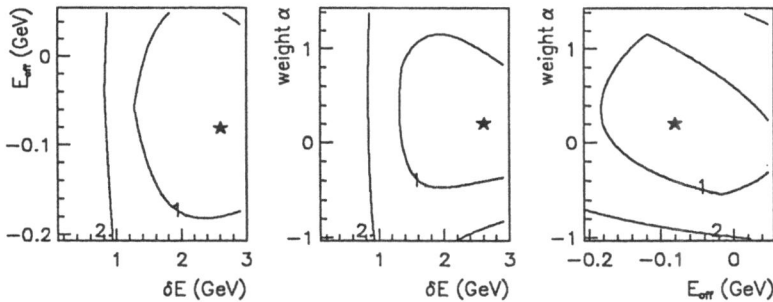

Figure 5:   Range of scan parameters which result in an accuracy on Γ_z which is close to the optimised scan (indicated by the star). The lines represent contours of equal accuracy on Γ_z. The first contour surrounds the preferred region of $\Delta\Gamma_z \leq \Delta\Gamma_z^{opt} + 10MeV$, the second represents $\Delta\Gamma_z \leq \Delta\Gamma_z^{opt} + 35MeV$.

6. Prospects for the near future and beyond

To give an impression of the power of the Z^0 line shape measurement we have created two model data sets with the two estimates of systematic errors given in Table 2. The statistical errors correspond to an integrated luminosity of 1 pb^{-1} for the initial and 4 pb^{-1} for the ultimate systematic errors respectively. The hadronic cross-section has been calculated with the line shape program ZHADRO which is based on the full one loop calculation of [7] including $Z\gamma$ mixing and weak vertex corrections. Initial state radiation is taken into account as described in [8]. As input parameters we have chosen

$$m_Z = 92 \ GeV \quad m_t = 150 \ GeV \quad m_H = 100 \ GeV \quad \alpha_s = .12 \ .$$

With the input parameters above one can calculate $\sin^2\theta_w = .2189$ and $\Gamma_Z = 2.580 \ GeV$.

As mentioned in the introduction the line shape measurement serves as a test of the SM. In a first step we will therefore parametrise the data in a model independent way. The simplest choice is a Breit Wigner with an s-dependent width $\Gamma_Z(s) = \frac{s}{m_Z^2}\Gamma_Z$

$$\sigma = \sigma_Z^{pole} \frac{s\Gamma_Z^2}{(s - m_Z^2)^2 + \frac{s^2}{m_Z^2}\Gamma_Z^2} \tag{1}$$

where the normalisation constant σ_Z^{pole} represents the cross-section and Γ_Z the width at the pole at $s = m_Z^2$. To account for photonic corrections this expression has to be convoluted with the initial state radiation spectrum before fitting it to the data

$$\sigma_{obs}(s) = \int_{s_0}^{s} f(s,s') \, \sigma(s') \, ds' \ . \tag{2}$$

For the radiator f(s,s') in (2) we adopt the calculation of [8] which is based on a second order treatment with exponentiation of soft photons. For the model data sets we get from a fit

$$\sigma_Z^{pole} = 40.2 \pm 2.2nb \quad m_Z = 92.01 \pm .03 \pm .05 \ GeV \quad \Gamma_Z = 2.60 \pm .06 \pm .005 \ GeV \tag{3}$$

at $\chi^2 = 0.9$ for the 1 pb^{-1} sample and

$$\sigma_Z^{pole} = 40.4 \pm 0.9nb \quad m_Z = 92.00 \pm .01 \ GeV \quad \Gamma_Z = 2.59 \pm .03 \ GeV \tag{4}$$

at $\chi^2 = 2.1$ for the 4 pb^{-1} sample. The second error in (3) takes into account the error in the overall energy scale. If the χ^2 for our real data assumes a similar value this would indicate that the QED photonic corrections are understood.

The accurate value on m_Z will give rise to many papers which will combine m_Z with the expected high precision measurement of $m_Z - m_W$ at the hadron colliders and $\sin^2\theta_w$ from neutrino scattering to extract new limits on extended gauge and Higgs structures. In this survey we restrict ourselves to a discussion of results directly derived from the line shape measurement. As the global correlation coefficient of m_Z is small, we can perform an analysis of models by looking at contours in the σ_Z^{pole}-Γ_Z plane. These are displayed in Figure 6 for both data sets. The symbols show the prediction for specific models. The bars through these symbols show the range of these predictions when the top mass is varied within its present uncertainty of $50 - 200$ GeV. The uncertainty in the Higgs mass is well accounted for by the size of the symbols. The solid circle represents the expectation for the minimal Standard Model with 3 generations which fits well within the $1\,\sigma$ contour for both fits. As the solid circle has been calculated with the input parameters to the model data sets, one can conclude that the fit which is only an approximation to the Standard Model has not produced relevant parameter shifts. The open circle shows the expectation for an additional 4th generation neutrino. From the first precision scan it can be excluded nearly at the $3\,\sigma$ level, the ultimate scan would exclude it with almost certainty. The effect of an additional heavy quark b' with a mass $m_{b'} = 40\,GeV$ is indicated by the open box symbol. The first precision scan can exclude such an object at the $2\,\sigma$ level. By the time of the ultimate scan the study of event topologies will have established many limits on massive particles. The remeasurement of the line shape however represents an independent confirmation of the limits given. The ρ parameter at present is constrained by many experiments, a global fit to existing data [9] gives $\rho = 0.998 \pm .0086$. In Figure 6 the triangular symbols represent the expectation for a ρ parameter deviating from 1 at tree level by $\pm 2\%$. Whereas the first precision scan will only be sensitive to such a deviation on the 1σ level, the ultimate scan could compete as a single measurement with todays global analysis of data. The solid star in Figure 6 illustrates the maximum effect that can be expected from an additional Z' of the most general E_6 origin. We have adopted the calculation of [10] where the mixing angle of the Z' with the conventional Z^0 is constrained to lie in the experimentally allowed domain of [9]. The appearance of such an additional Z' is quite unique, as it reduces the observed Z^0 total width below the minimal value expected from the SM. The relative reduction of the leptonic width is even stronger which results in a negative shift of σ_Z^{pole}. From Figure 6 it becomes clear that the measurement of the Z^0 line shape is a valuable tool to constrain E_6 models.

The model independent analysis will allow to restrict the range of models considerably. In a second step one would choose a specific model to test nonphotonic radiative corrections. If nature would reproduce something similar to our model data set we would like to test the data against the full calculation of electroweak corrections within the framework of the Minimal Standard Model with 3 generations. In this case Γ_Z and the overall normalisation are no longer free parameters but can be calculated provided the couplings and all gauge boson, scalar and fermion masses are known. Taking into account α and G_μ from low energy measurements and extracting the contribution of light quarks to Δr via dispersion relations from the data $e^+e^- \rightarrow$ hadrons, the only unknowns in the Minimal Standard Model are m_Z, m_t and m_H. As the fit is not sensitive to m_H its value has been fixed to $m_H = 100$ GeV. A fit based on calculations of [7] gives

$$m_Z = 92.01 \pm .03 \pm .05\,GeV \quad m_t = 116 \pm 135\,GeV \quad at \ \chi^2 = 0.9 \tag{5}$$

for the first precision scan and

$$m_Z = 92.00 \pm .01\,GeV \quad m_t = 107 \pm 82\,GeV \quad at \ \chi^2 = 2.0 \tag{6}$$

for the ultimate scan. The value of the mass is identical to the value obtained with the model independent fit. In both cases the limit on m_t is not very interesting as compared to existing ones [9]. As there is also no significant change in χ^2, when going from the model independent fit to the SM constrained fit, we conclude that we are not sensitive to deviations of the line shape from a Breit Wigner with an s-dependent width caused by non-photonic corrections.

284

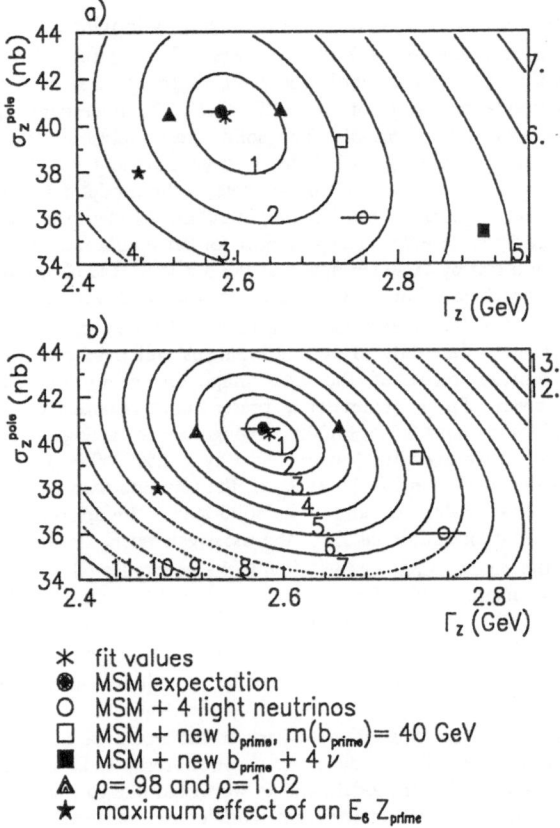

*   fit values
⊛   MSM expectation
○   MSM + 4 light neutrinos
□   MSM + new b_{prime}, $m(b_{prime})$= 40 GeV
■   MSM + new b_{prime} + 4 ν
△ ρ=.98 and ρ=1.02
★   maximum effect of an E_6 Z_{prime}

Figure 6:   Confidence contours for the model independent fit in the Γ_z-σ_z^{pole} plane a) for an integrated luminosity of 1 pb^{-1} and near future systematics and b) for 4 pb^{-1} and assymptotic systematic errors (c.f. Table 2). The numbers refer to n-σ contours. The bars indicate the possible variation of parameters if m_t is varied in the range 50 GeV $\leq m_t \leq$ 200 GeV.

7. Conclusions

The measurement of the Z^0 line shape with hadronic events will be an early, experimentally easy and very sensitive probe of the SM. The expected background is negligible, the main sources of systematic point to point errors will be the luminosity monitoring and the detector modelling under varying machine background conditions. In the early periods of running an integrated luminosity of 1 pb $^{-1}$ represents a reasonable compromise between the gain in accuracy and the cost of running time. At this integrated luminosity scanning strategies are of minor importance, a scan should however cover an interval of 6 GeV to obtain a good accuracy on Γ_z. With 1 pb^{-1} one can reach $\Delta m_z = 50-60$ MeV, limited by the knowledge of the beam energy, and $\Delta\Gamma_z = 60$ MeV, limited to about equal parts by statistics and by initial systematics. The accuray above will allow to exclude an additional neutrino close to the 3 σ confidence level. Many phenomena can broaden the SM width and therefore any positive indication of an additional neutrino needs cross checks. However we also might observe a Z^0 width which is significantly smaller than the SM expectation and therefore have a glance at new physics that lies beyond the energy range of LEP/SLC in the early phase of running.

Finally I would like to thank the organisers of this workshop who succeeded in creating a stimulating atmosphere for the exchange of knowledge and ideas.

References

1.   e.g. G. Altarelli in Physics at LEP, CERN 86−02, Vol. 1 (1986),p. 3.
2.   Plenary Meeting of the LEP Physics Workshop at CERN, February 20−21 1989, transparencies J. Ludwig.
3.   S. Jadach, R.G. Stuart, B.F. Ward and Z. Was, KORALZ Version 3.4, Long write-up (1989), unpublished.
4.   J.E. Campagne and R. Zitoun, LPNHEP 88.08 (1988).
5.   G.A. Bonvicini and L. Trentadue (1989), unpublished.
6.   G. Burgers; Yellow Report on Polarization at LEP, CERN 88−06,Vol. 1, p. 121 (1988).
7.   M. Böhm, H. Spiesberger and W. Hollik; Fort. der Phys. 34, 687 (1986).
        W. Hollik; DESY 88−188 (1988).
        G. Burgers and W. Hollik; Yellow Report on Polarization at LEP, CERN 88−06,Vol. 1 (1988),p. 136.
8.   F.A. Berends, W.L. Van Neerven, G.J.H. Burgers; Nucl. Phys. B297, 429 (1988).
9.   U. Amaldi et al.,Phys. Rev., D36, 1385 (1987).
10.   F.M. Renard and C. Verzegnassi, CERN−TH 5372/89 (1989).

Radiative Corrections for e⁺e⁻ Collisions   Editor: J.H. Kühn
© Springer-Verlag Berlin, Heidelberg 1989

The First Three Months

R. Settles

Max–Planck–Institut für Physik und Astrophysik
Föhringer Ring 6, D–8000 München 40

Abstract

A scenario is described for simple but fundamental tests of the standard model using the data taken with ALEPH during the first three months of LEP operation. Examples chosen for the scenario are the determination of the number of neutrino types N_ν and the strong coupling constant α_s from the measurements of the hadronic and the muonic production cross sections on the Z° peak. If things go well with the experimental start–up, then we should know N_ν to $\pm$ .3 and α_s to $\pm$ .08 (from this experiment only) by the end of 1989. Data from the different channels will give important cross checks on the internal consistency of the standard model.

1. Introduction

The title of this talk reminds one of "The First Three Minutes", the famous book [1] by Steven Weinberg which gives a popular interpretation of the big bang theory. However, we wish here not to describe what happened during the first three *months* after the big bang, but try to guess what will happen during the first three months after the start–up of LEP. Since this author is a member of the ALEPH collaboration [2], this detector will be used in the speculations. The goal will be to perform basic tests on the standard model as soon as possible. The examples given here are the determination of the member of neutrino types N_ν and the strong coupling constant α_s from measurement the Z° line shape as seen in the hadronic $e^+e^- \rightarrow Z^\circ \rightarrow q\bar{q}$ and muonic $e^+e^- \rightarrow Z^\circ \rightarrow \mu^+\mu^-$ channels. N_ν and α_s measured in several ways with the first data will test the internal consistency of the standard model. Cross–section measurements are *counting* experiments which are used here since they do not need full understanding of detector idiosyncrasies.

There is of course a connection between the big bang and the start–up of LEP, because the former [3] sets one of the limits on N_ν: arguments based on cosmology and the abundance of helium in the universe say that there should be $< \sim 4$ neutrino types. Particle physics experiments [3] set $N_\nu < \sim 6$ ($p\bar{p}$ colliders) and $N_\nu < \sim 7$ (e^+e^- colliders). As shown below, with a bit of luck in commissioning LEP and ALEPH, we should know N_ν to $\delta N_\nu \simeq \pm 0.3$ by the end of this year, assuming the standard model is correct.

In the following are described relevant standard–model predictions in section 2, the start–up of LEP operation and of ALEPH in section 3, the measurements of N_ν and α_s in section 4, and the hoped–for measuring accuracy in section 5.

2. The Standard Model

The standard model predictions for $e^+e^- \to f\bar{f}(\gamma)$ are now very familiar, but a few formulae relevant to the present scenario will be given here. On the Z° peak, the Breit–Wigner form for the cross–section in the Born approximation is

$$\sigma_f(s = M_Z^2) = \frac{12\pi}{M_Z^2} \frac{\Gamma_e \Gamma_f}{\Gamma_{tot}^2}, \tag{1}$$

where Γ_f is the rate of decay of the Z° into fermion + antifermion. This $\sigma_f(M_Z^2)$ is reduced by about 30% due to initial–state radiation. Theoretical uncertainty in the absolute value of the cross sections is now less than 1% after full electroweak corrections, as this workshop has impressively demonstrated, so that experiments have solid numbers with which to compare.

The Born peak cross section is good enough to understand the determination of standard model parameters. We can consider measuring four basic quantities: σ_{had}, σ_μ, $R' = \sigma_{had}/\sigma_\mu$ and Γ_{tot}, whereby $\sigma_{had} = \sum_{quarks} \sigma_q$. In terms of decay rates,

$$\sigma_{had} = \frac{12\pi}{M_Z^2} \frac{\Gamma_e \Gamma_{had}^\circ (1 + \frac{\alpha_s}{\pi} + \cdots)}{\left[3\Gamma_\mu + N_\nu \Gamma_\nu + \Gamma_{had}^\circ (1 + \frac{\alpha_s}{\pi} + \cdots)\right]^2}$$

$$\sigma_\mu = \frac{12\pi}{M_Z^2} \frac{\Gamma_e \Gamma_\mu}{\left[3\Gamma_\mu + N_\nu \Gamma_\nu + \Gamma_{had}^\circ (1 + \frac{\alpha_s}{\pi} + \cdots)\right]^2} \tag{2}$$

$$R' = \frac{\sigma_{had}}{\sigma_\mu} = \frac{\Gamma_{had}^\circ (1 + \frac{\alpha_s}{\pi} + \cdots)}{\Gamma_\mu}$$

$$\Gamma_{tot} = 3\Gamma_\mu + N_\nu \Gamma_\nu + \Gamma_{had}^\circ (1 + \frac{\alpha_s}{\pi} + \cdots)$$

For completeness we write down that

$$\Gamma_f = \frac{1}{48} \alpha M_Z N_C (v_f^2 + a_f^2), \tag{3}$$

where

$$v_f = \frac{2I_{3f} - 4Q \sin^2 \Theta_W}{\sin \Theta_W \cos \Theta_W}$$

$$a_f = \frac{2I_{3f}}{\sin \Theta_W \cos \Theta_W}$$

are the vector and axialvector coupling constants in the standard model, and $N_C = 1$ for leptons and $= 3$ for quarks. Using $\sin^2 \Theta_W = .23$, then $\Gamma_\nu = 170$ MeV, $\Gamma_e = 86$ MeV, $\Gamma_{u,c}^\circ = 293$ MeV, and $\Gamma_{d,s,b}^\circ = 377$ MeV.

Looking at (2) again we see that universality has been used and that M_Z, N_ν and α_s are the parameters left. To simplify the discussion here, we assume that SLC will have measured M_Z by the time LEP starts. This leaves $N\nu$ and α_s to be determined from the four measured quantities in (2).

Before going into the measuring procedure, we discuss the LEP machine schedule in order to estimate how much data the first three months may yield.

3. The Commissioning of LEP and of ALEPH

The original commissioning schedule drawn up at the beginning of 1989 foresaw injection into the LEP ring on the 200th Bastille Day followed by 26 "super days" (i.e., days during which everything goes as planned without need of trouble–shooting and repair) of tuning at 20 GeV, the injection energy from the SPS. This period included a pilot run at 20 GeV beam energy to test the detectors. Then energy–ramping tests would start until beams at 45 GeV were achieved, and the first real data–taking run would start mid–October, after a total of 47 "super days" = 85 real days (i.e., safety factor = 1.8 for trouble–shooting).

The advent of Carlo Rubbia as CERN DG was accompanied by a reshuffling of the schedule and resulted in the scrapping of the 20 GeV pilot run and some beam tuning time spent at 20 GeV, so that the first pilot run will be a scan of the Z° peak after only 20 "super days". For this scan, however, the background conditions and luminosity will not be optimal. The real physics run is still foreseen for mid–October, but we now have a chance of getting a few Z° events by mid–August. The expected luminosities are about $\sim 6 \cdot 10^{29}$ $cm^{-2}s^{-1}$ for a few days for the pilot run and $\sim 2 \cdot 10^{30}$ $cm^{-2}s^{-1}$ for a few weeks for the physics run. The integrated luminosities could then be .05 pb^{-1} after the pilot run and 1 pb^{-1} after the physics run.

As for the status of ALEPH, at the time of this writing (May '89), all subdetectors except the outer muon chambers were in place and 90% of the cabling–up was finished. The detector was in beam position and connected to the LEP beampipe. The magnetic field had been on in this situation with subdetectors running. The trigger was being installed. A big task will be to get the readout system to working reliably by August '89. The reconstruction program was essentially ready but still in need of debugging.

4. The Determination of N_ν and α_s

To understand the dependence of N_ν and α_s on the four quantities of (2) measured on the peak, one calculates

$$
\begin{aligned}
\frac{\delta \sigma_{had}}{\sigma_{had}} &= -2 \; \frac{\Gamma_\nu}{\Gamma_{tot}} \; \delta N_\nu(\sigma_{had}) = -\frac{1}{\pi} \; \frac{\Gamma_{had} - \Gamma_L}{\Gamma_{tot}(1 + \frac{\alpha_s}{\pi} + \cdots)} \; \delta \alpha_s(\sigma_{had}) \\[4pt]
\frac{\delta \sigma_\mu}{\sigma_\mu} &= -2 \; \frac{\Gamma_\nu}{\Gamma_{tot}} \; \delta N_\nu(\sigma_\mu) = -\frac{2}{\pi} \; \frac{\Gamma_{had}}{\Gamma_{tot}(1 + \frac{\alpha_s}{\pi} + \cdots)} \; \delta \alpha_s(\sigma_\mu) \\[4pt]
\frac{\delta R'}{R'} &= \quad 0.0 \quad\; \delta N_\nu(R') = \frac{1}{\pi} \; \frac{1}{1 + \frac{\alpha_s}{\pi} + \cdots} \; \delta \alpha_s(R') \\[4pt]
\frac{\delta \Gamma_{tot}}{\Gamma_{tot}} &= \quad \frac{\Gamma_\nu}{\Gamma_{tot}} \quad \delta N_\nu(\Gamma_{tot}) = \frac{1}{\pi} \; \frac{\Gamma_{had}}{\Gamma_{tot}(1 + \frac{\alpha_s}{\pi} + \cdots)} \; \delta \alpha_s(\Gamma_{tot})
\end{aligned}
\tag{4}
$$

where $\Gamma_L \equiv 3\Gamma_e + N_\nu \Gamma_\nu$ and $\Gamma_{had} \equiv \Gamma_{had}^\circ(1 + \frac{\alpha_s}{\pi} + \cdots)$. The symbol $\delta N_\nu(\sigma_{had})$ denotes the variation in N_ν in the expression for σ_{had}, etc. Putting in numbers, $\Gamma_{had} \simeq .7\Gamma_{tot}$, $\Gamma_L \simeq .3\Gamma_{tot}$, $\Gamma_\nu \simeq .07\Gamma_{tot}$, one finds

$$\frac{\delta\sigma_{had}}{\sigma_{had}} \simeq -.13 \ \delta N_\nu(\sigma_{had}) \simeq -.12 \ \delta\alpha_s(\sigma_{had})$$

$$\frac{\delta\sigma_\mu}{\sigma_\mu} \simeq -.13 \ \delta N_\nu(\sigma_\mu) \simeq -.42 \ \delta\alpha_s(\sigma_\mu)$$

$$\frac{\delta R'}{R'} \simeq 0.0 \ \delta N_\nu(R') \simeq +.31 \ \delta\alpha_s(R')$$

$$\frac{\delta\Gamma_{tot}}{\Gamma_{tot}} \simeq +.07 \ \delta N_\nu(\Gamma_{tot}) \simeq +.21 \ \delta\alpha_s(\Gamma_{tot})$$

(5)

So one sees for example that one extra neutrino generation, $\delta N_\nu = 1$, causes a 13% decrease in σ_{had}, so that one has to measure $\frac{\delta\sigma_{had}}{\sigma_{had}}$ to $\sim 4\%$ in order to know δN_ν to $\sim .3$. Taking into account that $\sigma_{had} \simeq 30$ nb and $\sigma_\mu \simeq 1.5$ nb on the Z° peak, these numbers show that it is best to measure N_ν using σ_{had} because of larger statistics and because the N_ν determination is less sensitive to α_s. It is better to measure α_s using R', since, although α_s is more sensitive to σ_μ, its determination using R' is independent of N_ν. The nice thing (4) shows is that the determination of N_ν and α_s using σ_μ and Γ_{tot} will give sensitive checks on the internal consistency of the standard model.

To demonstrate what goes into a cross section measurement, we take σ_{had} as an example:

$$\sigma_{had} = \frac{1}{L\epsilon_{tr}\epsilon_{had}} \ (n_{obs} - \sum_i n_{bcki})$$

(6)

where

n_{obs}	= number of observed events in the sample
L	= integrated luminosity
ϵ_{tr}	= trigger efficiency for hadronic events
ϵ_{had}	= selection efficiency for hadronic events
n_{bcki}	= number of background events from source i to be subtracted from the sample
	= $L\sigma_{bcki}\epsilon_{bcki}$
σ_{bcki}	= cross section for background source i
ϵ_{bcki}	= efficiency by which background source i gets into the observed sample

Using error propagation

$$\left(\frac{\delta\sigma_{had}}{\sigma_{had}}\right)^2 = \left(\frac{\delta n_{obs}}{n_{sig}}\right)^2 + \left(\frac{n_{obs}}{n_{sig}}\right)^2\left(\frac{\delta L}{L}\right)^2 + \left(\frac{\delta\epsilon_{tr}}{\epsilon_{tr}}\right)^2 + \left(\frac{\delta\epsilon_{had}}{\epsilon_{had}}\right)^2$$
$$+ \sum_i\left\{\left(\frac{n_{bcki}}{n_{sig}}\right)^2\left[\left(\frac{\delta\sigma_{bcki}}{\sigma_{bcki}}\right)^2 + \left(\frac{\delta\epsilon_{bcki}}{\epsilon_{bcki}}\right)^2\right]\right\}$$

(7)

where

$n_{sig} = n_{obs} - \sum_i n_{bcki}$ is the number of signal events remaining in the sample after background subtraction.

So in addition to the statistical error, we have to have good estimates for the errors in luminosity, trigger efficiency, selection efficiency, and background subtraction efficiency.

These last three errors above can be lumped together to one "systematic error" for the purpose of the calculations to be done here. Detailed simulations using the ALEPH detector [4] show that these errors added quadratically give a systematic error of 0.7% in the case of hadrons. Another complication comes from the fact that the first data will be gathered at several energies across the Z^0 resonance, and not only at the peak as assumed to simplify the discussion here. The changing conditions of the machine, detector, and background from one energy to the next give additional contributions to the systematic error when measuring the line shape. Here we shall assume an overall systematic error of 1% in the measurement of the hadronic or muonic cross section and allow the luminosity error to vary. The accuracy with which one knows his luminosity will play a central role in getting physics out of the first data.

5. The Measuring Accuracy with the First Data

Figure 1 shows the result of using (4) and (6) to calculate the accuracy with which N_ν and α_s can be determined as a function of luminosity error for different integrated luminosities. For the case of R', the luminosity error drops out, so the systematic error is varied.

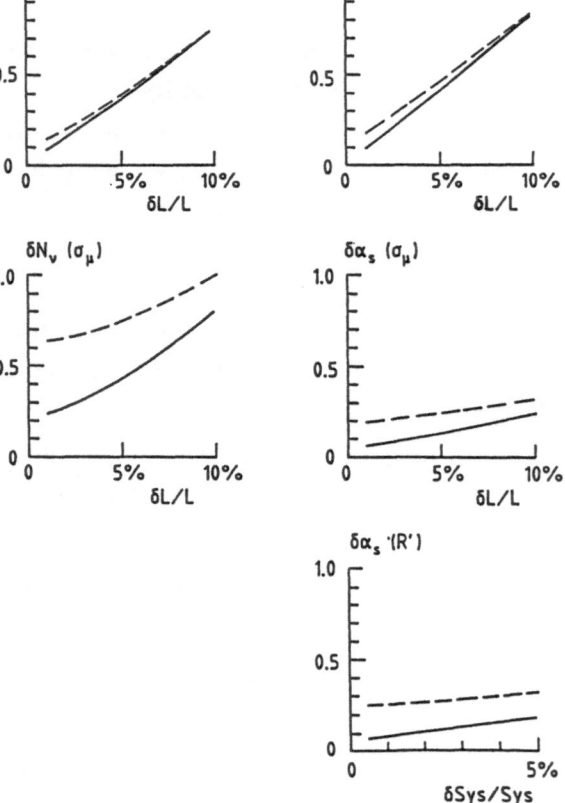

Fig.1. Errors in determining N_ν and α_s on the peak for .1 pb^{-1} (dashed line) and .1 pb^{-1} (solid line) of data.

Table 1 lists the results of the calculations for two scenarios for the first three months: if things go well and we have an equivalent of 1 pb^{-1} of data at the peak, or if things go not–so–well and we have .1 pb^{-1} of data at the peak. These numbers assume a luminosity error of 4% and a systematic error of 1% in measuring each of the cross sections.

	$\delta N_\nu(x)$		$\delta\alpha_s(x)$	
L	.1 pb^{-1}	1 pb^{-1}	.1 pb^{-1}	1 pb^{-1}
$x = \sigma_{had}$	.35	.31	.37	.33
$x = \sigma_\mu$	.70	.37	.23	.11
$x = R'$	∞	∞	.27	.10

Table 1: Uncertainties in N_ν and α_s assuming .1 and 1 pb^{-1} integrated luminosity, luminosity error of 4%, and systematic error of 1% for each cross section measurements.

We have not gone into the measurement of Γ_{tot}. Here the additional errors coming from the changing conditions of machine and detector during the scan must be included. Studies of this measurement are underway, but suffice it to say that the results are similar to those using σ_μ in Table 1.

In summary, the bottom line is that if thing go well (1 pb^{-1}), we should know from ALEPH alone N_ν to $\lesssim$ .3 and α_s to $\lesssim$ .08 (after combining measurements in Table 1) by the end of 1989.

REFERENCES

[1] S. Weinberg, <u>The First Three Minutes</u>, Basis Books, Inc., Publishers, New York (1977)

[2] ALEPH Collaboration: Annecy, Barcelona, Bari, Beijing, CERN, Clermont-Ferrand, Copenhagen, Demokritos, Ecole Polytechnique, Edinburgh, Firenze, Florida, Frascati, Glasgow, Heidelberg, Imperial College, Innsbruck, Lancaster, Mainz, Marseille, MPI–München, Orsay, Pisa, RAL, Royal Holloway College, Saclay, Sheffield, Siegen, Trieste and Wisconsin.

[3] Particle Data Group, *Phys. Lett.* **B 204** (1988) 146–147

[4] L. Bauerdick, M. Bardadin–Otwinowska, C. Benchouk, P. Cattaneo, G. Cowan, A. Ealet, C. Geweniger, M. Kasemann, H. Kühn, E. Lange, M. Pepe, G. Rudolph, M. Schmelling and R. Settles, Physics Workshop on QCD, ALEPH Plenary Meeting at Barcelona, May 1988.

Radiative Corrections for e⁺e⁻ Collisions   Editor: J.H. Kühn
© Springer-Verlag Berlin, Heidelberg 1989

Results on R in e^+e^- Annihilation

Wim de Boer *

Max–Planck–Institut für Physik und Astrophysik,

Werner–Heisenberg–Institut für Physik, D 8000 Munich 40

Abstract

We discuss the determination of the Standard Model parameters α_s, M_Z, and $\sin^2\theta_W$ from the total hadronic cross section in e^+e^- annihilation. At the highest TRISTAN energies, the tail of the Z^0 resonance is increasing R already by 50%, which allows a direct measurement of M_Z. We find: $M_Z =$ $89.0(88.5, 88.1) \pm 1.0$ GeV for a top mass of 60 (120,180) GeV and fixed $\sin^2\theta_W$ $=0.23$. For the strong coupling constant we find: $\alpha_s(34^2\,GeV^2) = 0.143 \pm$ 0.015, if $O(\alpha_s{}^3)$ QCD corrections are taken into account. We study the effect of changing the renormalization scale of α_s and find that the third order QCD corrections, which are larger than the second order ones for the usual scale $Q =$ $\sqrt{s}$, are smaller than the second order contributions at other scales. A Monte Carlo study shows that with 3 energy scan points around the Z^0 resonance peak and a total of about 100 events, one can determine the Z^0 mass and width with an error of about 250 MeV, if one fits the absolute cross section instead of the shape of the resonance.

1. Introduction

Some time ago the CELLO Collaboration has determined α_s [1] from an analysis of all data available from the PEP and PETRA experiments on the total hadronic cross section in e^+e^- annihilation using a complete error correlation matrix. The importance of such a determination of α_s stems from the fact that it is independent of fragmentation models and that the results can be calculated reliably in QCD. These α_s determinations have been updated last year[2]-[6] with new experimental data and new theoretical progress, namely a third order calculation of the QCD corrections by S.G. Gorishny et al.[7]. The third order contribution turns out to be larger than the second order contribution in the commonly used $\overline{MS}$ scheme, if the strong coupling constant is evaluated at the scale $Q = \sqrt{s}$. However, this is not the typical scale for gluon radiation. Therefore we studied the scale dependence of α_s as

---

*Talk at the Ringberg Workshop on Electroweak Radiative Corrections, April 3-7, 1989.
  Mailing address: DESY FH1K, Notkestraße 85, D 2000 Hamburg 52.
  Bitnet address: user F36WDB at node DHHDESY3

determined from R and find that at other scales the third order contributions are smaller than the second order ones.

New experimental data has become available from TRISTAN experiments up to center of mass energies of 60.8 GeV. At these energies the tail of the Z^0 resonance is increasing R already by 50%, thus allowing a direct measurement of the Z^0-mass. Such a mass determination has been published by the AMY Collaboration[8] and presented at various conferences[9,10,11]. The fit to the newest data yields[11]: $M_Z = 88.9 \pm 1.3$ GeV. This result does not agree too well with direct measurements from $p\bar{p}$ experiments, which yield $M_Z = 92.0 \pm 1.8$ GeV[12]. We have repeated the fit to R after applying the radiative corrections to all data in a consistent way.

At present all data has been corrected only for first order QED corrections, i.e. at most one photon radiated in the initial state. If one worries about third order QCD corrections, one should also worry about higher order QED corrections. Recently the higher order QED initial state radiative corrections have been calculated[14], which for the first time allows a comparison between these exact calculations and other approximate methods. The report has been organized as follows: we first summarize the Standard Model (SM) formulae, then discuss the fitting method and results. We finish with a discussion of the higher order radiative corrections, a Monte Carlo study of expected errors on mass and width of data points around the Z^0 resonance, and a summary.

2. Standard Model Formulae

Here we summarize the formulas used in fitting the hadronic cross section. The normalized cross section R is defined as the ratio

$$R \equiv \frac{\sigma[e^+e^- \to \gamma, Z^0 \to hadrons]}{\sigma[e^+e^- \to \gamma \to \mu^+\mu^-]}$$

The $\mu^+\mu^-$ cross section is the lowest order pointlike QED cross section of massless spin $\frac{1}{2}$ particles, and is equal to $4\pi\alpha^2/3s$, where s is the square of the centre of mass energy, while the hadronic cross section has been defined before.

The total hadronic cross section at the Born level is given by the sum of the following contributions:

$$\sigma_{had}^{\gamma} = \frac{4\pi\alpha^2}{3s} r_{QCD} \sum_{q=1}^{5} e_e^2 e_q^2 \tag{1}$$

$$\sigma_{had}^{\gamma Z} = 8\pi\alpha r_{QCD} \frac{K(s - M_Z^2)}{(s - M_Z^2)^2 + s^2\Gamma_{tot}^2/M_Z^2} \sum_{q=1}^{5} e_e e_q v_e v_q \tag{2}$$

$$\sigma_{had}^{Z} = 12\pi r_{QCD} \frac{K^2 s}{(s - M_Z^2)^2 + s^2\Gamma_{tot}^2/M_Z^2} (v_e^2 + a_e^2) \sum_{q=1}^{5} (v_q^2 + a_q^2) \tag{3}$$

The superscripts indicate the contribution from photon exchange, Z^0 exchange and their interference and the sum is taken over five quark flavours, thus assuming the top quark is too heavy; v and a represent the vector and axial vector couplings of the quarks (subscript q) and electrons (subscript e) and Γ_{tot} is the total width of

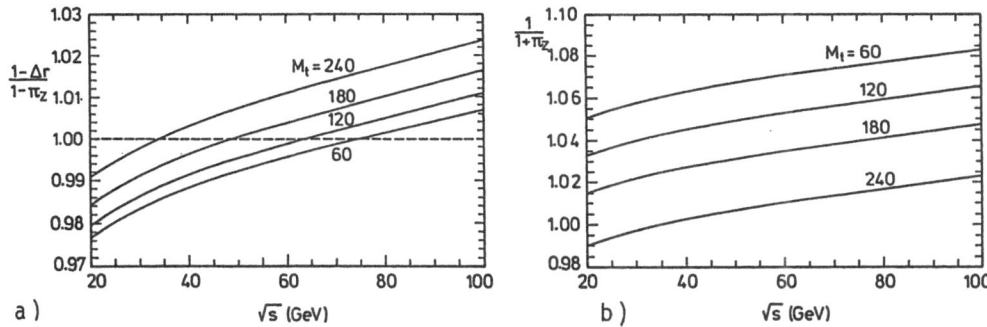

Figure 1: **Electroweak correction factors (κ_1 and κ_2 in text) for the different parametrizations of the Born cross section**

the Z^0. For simplicity we have neglected small mass effects in the formulae above, but they have been taken into account in the analysis, using the formulae in Ref.[1]. The factor r_{QCD} represents the effect from gluon radiation and is given in the $\overline{MS}$ scheme by[7,13]:

$$r_{QCD} = 3\left[1 + \frac{\alpha_s}{\pi} + (1.986 - 0.115n_f)\left(\frac{\alpha_s}{\pi}\right)^2 + (70.985 - 1.2n_f - 0.005n_f^2)\left(\frac{\alpha_s}{\pi}\right)^3\right] \tag{4}$$

The factor 3 on the righthand side accounts for the color of the quarks.

The energy dependence (running) of α_s is given by the 3rd order formula[15]:

$$\alpha_s(s) = \frac{4\pi}{\beta_0 \log(\frac{s}{\Lambda^2})}\left\{1 - \frac{\beta_1}{\beta_0^2}\frac{\log[\log(\frac{s}{\Lambda^2})]}{\log(\frac{s}{\Lambda^2})} + (\frac{\beta_1}{\beta_0^2})^2 \frac{1}{\log^2(\frac{s}{\Lambda^2})}\left[(\log[\log(\frac{s}{\Lambda^2})] - \frac{1}{2})^2 + \frac{\beta_2\beta_0}{\beta_1^2} - \frac{5}{4}\right]\right\} \tag{5}$$

with

$$\begin{aligned} \beta_0 &= 11 - \tfrac{2}{3}n_f \\ \beta_1 &= 2(51 - \tfrac{19}{3}n_f) \\ \beta_2 &= \tfrac{2857}{2} - \tfrac{5033}{18}n_f + \tfrac{325}{54}n_f^2 . \end{aligned}$$

The constant K can be either defined as:

$$K_1 = \frac{\sqrt{2}G_F M_Z^2 \kappa_1}{48\pi} \tag{6}$$

or

$$K_2 = \frac{\alpha\kappa_2}{48\sin^2\theta_W\cos^2\theta_W} \tag{7}$$

Here G_F is the Fermi constant, which is well known from muon decay and $\sin^2\theta_W$ determines the electroweak mixing angle. In the definitions of K we have explicitly included the factor κ which represents the loop corrections to the Z^0 propagator. For example, practically all data from the PEP and PETRA experiments have been corrected with the LUND Monte Carlo program[16], which uses the radiative corrections from Berends et al.[17], thus including the loop corrections for the photon propagator, but not the loop corrections for the Z^0 propagator. In this case the formulae to be fitted to the data should include this κ-factor, which can be written as follows[18,19]:

$$\kappa_1 = \frac{1 - \Delta r}{1 + \Pi_Z(s)} \tag{8}$$

or

$$\kappa_2 = \frac{1}{1 + \Pi_Z(s)} \tag{9}$$

where

$$1 - \Delta r = \frac{\alpha(0)}{\alpha(M_W)} + \delta_r(M_t, M_H) \tag{10}$$

and

$$1 + \Pi_Z(s) = \frac{\alpha(0)}{\alpha(M_Z)} + \delta_\Pi(M_t, M_H, s) \tag{11}$$

Here Δr represents the electroweak corrections to the charged gauge boson exchange in muon decay and $\Pi_Z(s)$ represents the electroweak loop corrections to the neutral gauge boson exchange. One sees that the first term in both cases is given by the running of the QED coupling constant coming from the light fermion loops in the photon propagator (hence the indication of the scale in α). For top quark masses below the gauge boson masses this term is dominant in both expressions. E.g. for a top mass of 70 GeV Δr is about 7 % and 6% is coming from the first term alone. However, for a top mass of 230 GeV the latter term is as large as the first term, but of opposite sign, so the total correction Δr is about zero. $\Pi_Z(s)$ shows a similar behaviour, so that the ratio in κ_1 is much less dependent on the top mass and furthermore close to 1 (see Fig. 1a). This is the advantage of the parametrization with K_1: one can neglect the electroweak corrections to a large extent and the results are insensitive to the unknown top mass. This was the reason why in previous fits to data on R this parametrization has been used, e.g. to determine the strong coupling constant [1]. What was considered further as an advantage compared to the K_2 parametrization was the insensitivity to the Z^0-mass at PETRA energies: the dominant term in both the numerator and denominator in Eq. 3 is proportional to M_Z^6, thus largely canceling the uncertainty in M_Z. However, at TRISTAN-energies one observes the tail of the Z^0-resonance and it becomes possible to make a direct measurement of the Z^0-mass. In this case one obtains much more sensitivity with K_2, since one can measure the pure propagator effect without the compensation from the M_Z^6 factor in the numerator. However, with the K_2 parametrization the electroweak corrections cannot be neglected anymore (κ_2 in this case, see Fig. 1b), since the correction to the total hadronic cross section is of order 3% at 60 GeV, as will be discussed below.

From the definitions of K_1 and K_2 one can deduce the following well known relation between $\sin^2 \theta_W$ and M_Z:

$$\sin^2 \theta_W = \frac{1}{2} \left[1 - \sqrt{1 - \frac{4\pi\alpha}{\sqrt{2} G_F M_Z^2 (1 - \Delta r)}} \right] \tag{12}$$

3. Analysis method

Combining the data from different experiments is always a delicate procedure. It requires that:

- all data are have been corrected to the same level and their errors have a similar meaning;

- correlations between the data points within the same experiment and eventually between different experiments, must be considered.

Correlated errors between measurements can be taken into account by defining the χ^2 via an error correlation matrix[1]:

$$\chi^2 = \Delta^T V^{-1} \Delta \qquad (13)$$

Here Δ is a column vector containing the residuals between the measurements R_i and its estimators R_i^*; V is the NxN error correlation matrix between N measurements. The elements of V can be estimated as follows: Assume the true R values deviate from the measured values by a common normalization factor f, which causes a correlation between the measurements and will make the off-diagonal elements of V nonzero. In this case the best estimator R_i^* from the fit including the correlations will deviate from the best estimator excluding the correlations - called r_i^*- by the same factor f, so $R_i^* = f\, r_i^*$. If the estimator is efficient, r_i^* will just be the averaged R value. The variance of f around 1 is called σ_n^2, where σ_n is the relative normalization error, so $< (f-1)^2 > = \sigma_n^2$. Then

$$
\begin{aligned}
V_{ii} &= \; < (R_i - R_i^*)^2 > \\
&= \; < (R_i - r_i^*)^2 > \; + \; < (r_i^* - f r_i^*)^2 > \\
&= \; \sigma_i^2 + \; < (1-f)^2 > \; (r_i^*)^2 \\
&= \; \sigma_i^2 + \; \sigma_n^2 \, (r_i^*)^2 \\
V_{ij} &= \; < (R_i - R_i^*)(R_j - R_j^*) > \\
&= \; < (R_i - r_i^* + (1-f)r_i^*)(R_j - r_j^* + (1-f)r_j^*) > \\
&= \; < (1-f)^2 > \; r_i^* \, r_j^* \\
&= \; \sigma_n^2 \, r_i^* \, r_j^*
\end{aligned}
$$

All terms containing $< (R_i - r_i^*) >$ have not been written, since they yield 0 and $\sigma_i^2 = < (R_i - r_i^*)^2) >$ contains the uncorrelated part of the error, which is the sum of both the statistical error - and point to point systematic error squared, but excludes the overall normalization error. Note that σ_n is the error on f, so it is the relative normalization error, which has to be multiplied by r_i^*, while σ_i is the absolute error on R_i.

It should be noted that the procedure of taking correlated errors into account via an error correlation matrix has distinct advantages over a likelihood method, in which the correlations are taken into account by fitting a renormalization constant to the data, as is done e.g. in Ref. [2]: in the first case one has only the physical parameters as free parameters, in the latter case one would have to fit 16 additional parameters in our problem. But what is more important, with a correlation matrix one can define a correlation between every pair of experimental points, thus taking correctly the energy dependence of the correlations into account, which exists for some experiments Furthermore one can study the effects of possible correlations between experiments, e.g. correlations from uncertainties in Monte Carlo programs or radiative corrections. This can be done by setting matrixelements connecting different experiments nonzero. It was found that there is practically no change in the fit results if one includes an overall correlation at the percent level[1]. We have not included the uncertainty from higher order QED radiative corrections in the covariance matrix, since we believe that treating it in a probabilistic way is uncorrect. We prefer to quote the variation of the final results for a given assumption on this correction, which will be discussed in the last chapter.

For some experiments the separation into point-to-point and common error was not explicitly given. In these cases it was checked that the numerical values of the fitted parameters were very stable against even large variations of their splittings.

4. Results

The 3 parameters in the total hadronic cross section we would like to determine are: α_s, M_Z, and $\sin^2\theta_W$. At present energies the Z^0 width does not play a role and we fixed its value at 2.5 GeV. We have tried several strategies to determine the other parameters:

- The most trivial way is a three parameter fit assuming no connection between the couplings and the mass.

- Make a two parameter fit of M_Z and α_s assuming relation 12 to hold or taking for $\sin^2\theta_W$ the value obtained in neutrino scattering, also taking into account the dependence on the unkown top mass[22,23].

- Determine α_s by using the G_F parametrization, which makes the results insensitive to M_Z and the unkown top mass due to the compensation in κ_1. One could use for $\sin^2\theta_W$ the world average or fit $\sin^2\theta_W$ as a free parameter. In this case $\sin^2\theta_W$ is mainly determined by the vector coupling of the electron, since the sum over all quark couplings is rather insensitive to $\sin^2\theta_W$ (the different electric quark charges cancel largely in the sum: $\sum(v_q^2 + a_q^2) = 10 - 1.33\sin^2\theta_W$). Since the vector coupling of the electron has been tested in various reactions, we will not put the priority on this and use for $\sin^2\theta_W$ the world average.

We first discuss the fit results from the different strategies and then proceed with a discussion of the higher order radiative corrections.

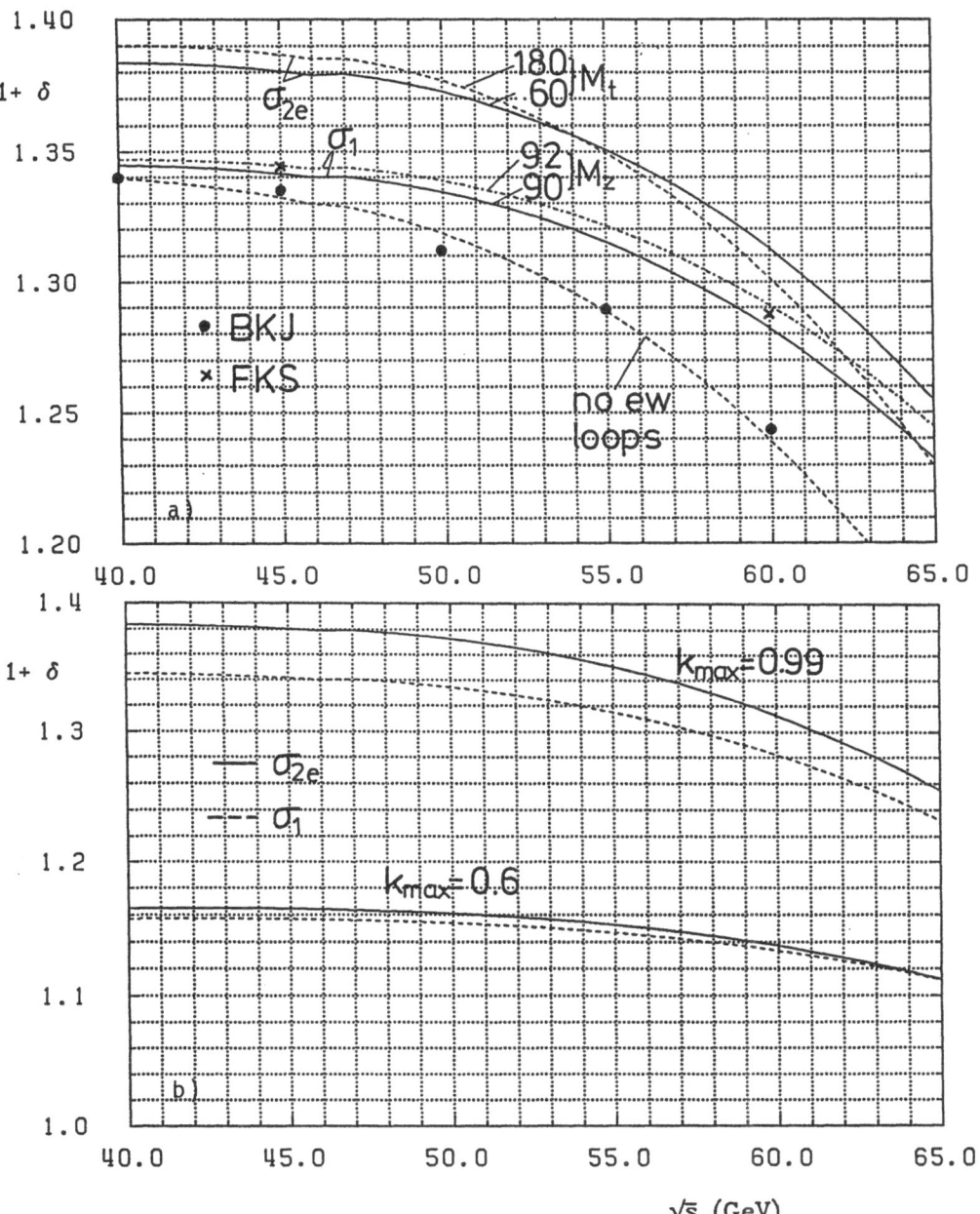

Figure 2: a) Radiative correction factors calculated in first and second order for $k_{max} = 0.99$. The M_Z mass is 90 GeV and the top mass 60 GeV, unless indicated differently. The lowest curve excludes the Z^0 selfenergy and is comparable with the BKJ radiative corrections [17] as implemented in the Lund Monte Carlo (black dots). The first order full electroweak corrections from Fujimoto et al.[21] (FKS) (crosses) are in good agreement with the results from Burgers and Hollik (curves). b) as in a), but now for $k_{max} = 0.6$

$M_t(GeV)$	$\sin^2\theta_W\ fixed$	$\sin^2\theta_W = f(M_Z)$	$\sin^2\theta_W\ free$
60	$M_Z = 89.0 \pm 1.0\ GeV$ $\alpha_s = 0.138 \pm 0.017$ $\sin^2\theta_W = 0.231$	$M_Z = 89.3 \pm 1.3\ GeV$ $\alpha_s = 0.134 \pm 0.018$ $\sin^2\theta_W = 0.250 \pm 0.014$	$M_Z = 89.4 \pm 1.3\ GeV$ $\alpha_s = 0.142 \pm 0.018$ $\sin^2\theta_W = 0.220^{+0.025}_{-0.020}$
120	$M_Z = 88.5 \pm 1.0\ GeV$ $\alpha_s = 0.138 \pm 0.017$ $\sin^2\theta_W = 0.230$	$M_Z = 88.7 \pm 1.3\ GeV$ $\alpha_s = 0.135 \pm 0.018$ $\sin^2\theta_W = 0.248 \pm 0.014$	$M_Z = 88.9 \pm 1.3\ GeV$ $\alpha_s = 0.142 \pm 0.018$ $\sin^2\theta_W = 0.221^{+0.024}_{-0.020}$
180	$M_Z = 88.1 \pm 1.0\ GeV$ $\alpha_s = 0.139 \pm 0.017$ $\sin^2\theta_W = 0.230$	$M_Z = 88.2 \pm 1.2\ GeV$ $\alpha_s = 0.135 \pm 0.018$ $\sin^2\theta_W = 0.246 \pm 0.014$	$M_Z = 88.4 \pm 1.2\ GeV$ $\alpha_s = 0.142 \pm 0.018$ $\sin^2\theta_W = 0.221^{+0.025}_{-0.020}$

Table 1: M_Z and α_s values from various fits.

4.1 Determination from M_Z

TRISTAN data up to centre of mass energies of 60.8 GeV have the potential of a Z^0-mass determination better than the one from the $p\bar{p}$-collider. However, the results are rather sensitive to radiative corrections, which -unfortunately - have not been applied yet in a consistent way. For example, data presented at the Munich Conference up to 57 GeV[10] included data corrected with and without electroweak loops and with different Z^0-masses. These differences are rather important as shown in Fig. 2. The curves, which give the ratio of cross section including radiative corrections and the Born cross section, have been calculated with the programs from Burgers and Hollik[19]. The differences originate from the following physics:

- the lower curve corresponds to initial state and vertex corrections for both γ and Z^0 exchange, but only loop corrections for γ exchange, thus excluding the self energy of the Z^0 propagator. The Z^0-mass was taken to be 90 GeV, the Higgs mass 100 GeV and the top mass 60 GeV. Second order QCD corrections have been included using $\alpha_s = 0.12$). Only first order QED graphs have been taken into account ($O(\alpha^3)$). The maximum photon energy allowed corresponds to $k_{max} = E_\gamma/E_{beam} = 0.99$, except for the b-quark, where k_{max} corresponds to the kinematical limit assuming $m_b = 4.7$ GeV. For the other light quarks u, d, s, and c we assumed masses of 0.04, 0.065, 0.3, and 1.5 GeV, respectively. These quark masses have been used to calculate the vacuum polarization. Since the vertex corrections for the Z^0-propagator are small in the on-shell renormalization scheme of Böhm et al.[20], this curve is very close to the results from the well known Berends, Kleis, Jadach (BKJ) results[17] as implemented in the Lund Monte Carlo. These BKJ results, indicated in the figure as solid dots, ignore the Z^0 self energy too.

- The two middle curves for two Z^0 masses (90 and 92 GeV) include in addition the loop corrections from the Z^0-propagator. The sensitivity comes mainly from the graphs with initial state radiation, which depend on the cross section at the energy after radiation and are therefore sensitive to the shape of the cross section, i.e. to the Z^0 mass. The results agree with the calculations from Fujimoto et al.[21], which have been indicated as crosses in the figure. These crosses were calculated for M_Z=91.9 and m_t=45 GeV and include final state radiation. It should be noted that in these programs final state radiative corrections are large and cannot be neglected, as is the case for the calculations of Hollik et al.[18]. The reason is the different wave function renormalization, which shifts part of the electroweak loop corrections into the vertex corrections; however, the final corrections in both calculations are very close.

- The upper two curves to second order exponentiated cross sections for two different top masses. One observes that the higher order corrections are important, if one integrates over all possible photon momenta. However, experimental cuts require typically a visible energy of $0.4\sqrt{s}$, which corresponds to $k_{max} \approx 0.6$. For such a cut off the difference between first- and second order becomes much smaller as shown in Fig. 2b.

The fit results with both M_Z and α_s as free parameters are presented in Table 1. Here we included the data as presented at the Topical Conference at KEK[24]. These data, which are of a preliminary nature and the results in Table 1 should be considered accordingly, included full electroweak corrections. However, the results are in excellent agreement with older data, as presented at the Munich Conference[10] and after taking the different radiative correction factors into account [25].

The radiative corrections were handled in the following way: the data is corrected only for QED radiative corrections, i.e. we have undone the electroweak corrections to the TRISTAN data (the difference between the lowest and middle curves in Fig. 2a). This has two advantages:

- The TRISTAN data is now corrected to the same level as the PEP and PETRA data, so they can be treated equally in the fits.

- The data corrected in such a way do not depend anymore on the top mass, so we can perform a fit with a function including the top mass dependence (through κ_2 in Eq. 7).

As shown in Fig. 2, the radiative correction factors depend on the Z^0 mass and top mass. Therefore, we have iterated the fit each time by applying the correct correction factor for the best fitted Z^0-mass (and using the top mass given in the table). In addition to the changes in radiative corrections as function of M_Z, we have considered possible changes in the detection efficiency. Using the Lund Monte Carlo, we find that varying the Z^0 mass from 92 to 89 GeV increases the detection efficiency

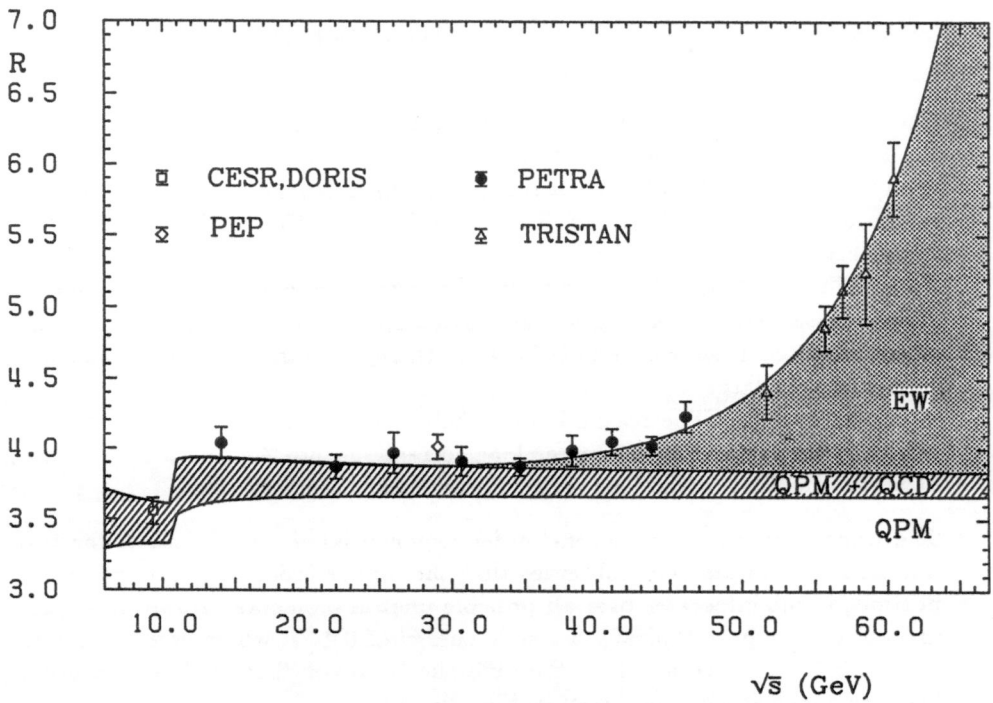

Figure 3: **Experimental data and result of a 3 parameter fit for $M_t = 60$ GeV yielding $M_Z = 89.4 \pm 1.3$, $\alpha_s = 0.142 \pm 0.018$, and $\sin^2 \theta_W = 0.220^{+0.025}_{-0.020}$**

about 0.3 % at $\sqrt{s}$=60 GeV for typical experimental cuts. Without iterations we would find $M_Z \approx 90$ GeV for M_t=60 GeV, which is about 1 GeV higher than the comparable values quoted before[11].

It should be noted that the main sensitivity to M_Z comes from the propagator effect in the TRISTAN energy range, while at PEP/PETRA energies the sensitivity came only through the couplings and the use of relation 12.

The top mass dependence found for M_Z comes from the radiative corrections to the Z^0 exchange present in the κ_2 parametrization. This cannot be avoided by the use of the κ_1 parametrization, since in this case the sensitivity to the Z^0 mass is reduced (the fit would give an error on M_Z of around 3 GeV). Table 1 shows the result from the three parameter fit too. The correlation coefficients between the 3 parameters are (independent of the top mass): $\rho(\sin^2 \theta_W, M_Z) = -0.64$, $\rho(\Lambda^{(5)}_{\overline{MS}}, \sin^2 \theta_W) = -0.42$ and $\rho(\Lambda^{(5)}_{\overline{MS}}, M_Z) = 0.44$. The χ^2/DF is 70/101; this excellent χ^2 comes mainly from the fact that the common normalization errors might have been overestimated. If we calculate the χ^2 only from the diagonal elements of the matrix V^{-1}, thus ignoring the correlations, but including the complete errors, we find χ^2_D/DF is 95/101. The fit result is shown in Fig. 3. For clarity we have averaged the data points within certain energy bins in the following way: we have fitted a constant value to the data points within a certain energy bin using the complete error correlation matrix (we

Data	Energy range	$O(\alpha_s^2)$	$O(\alpha_s^3)$
PEP, PETRA	$14 - 47\ GeV$	$\alpha_s = 0.168 \pm 0.025$	$\alpha_s = 0.151 \pm 0.020$
		$\Lambda_{\overline{MS}}^{(5)} = 590^{+470}_{-340}\ MeV$	$\Lambda_{\overline{MS}}^{(5)} = 320^{+240}_{-180}\ MeV$
PEP, PETRA,	$14 - 57\ GeV$	$\alpha_s = 0.170 \pm 0.025$	$\alpha_s = 0.152 \pm 0.019$
TRISTAN		$\Lambda_{\overline{MS}}^{(5)} = 620^{+460}_{-340}\ MeV$	$\Lambda_{\overline{MS}}^{(5)} = 330^{+230}_{-180}\ MeV$
CESR, DORIS, PEP, PETRA, TRISTAN	$7 - 57\ GeV$	$\alpha_s = 0.158 \pm 0.020$	$\alpha_s = 0.143 \pm 0.015$
		$\Lambda_{\overline{MS}}^{(5)} = 440^{+300}_{-230}\ MeV$	$\Lambda_{\overline{MS}}^{(5)} = 240^{+150}_{-120}\ MeV$

Table 2: α_s and $\Lambda_{\overline{MS}}^{(5)}$ fitted with the G_F parametrization for for $\sin^2 \theta_W = 0.231$ ($M_t = 60$ GeV) and $M_Z = 89.3$ GeV (from Ref. [25]).

have checked that this procedure exactly corresponds to make a weighted average, taking correctly into account independent and correlated errors). So the error bars represent the total errors including the correlation and the data have not been renormalized.

4.2   Determination of α_s

In Table 2 we give the fit results with the G_F parametrization. The values of α_s obtained from the different energy regimes are consistent. Comparing the $O(\alpha_s^2)$ and $O(\alpha_s^3)$ fits in Table 2, one observes a systematic reduction of the third order α_s values by $11 - 12\%$ for all energy regimes in agreement with the estimates given originally in Ref. [7]. This reduction varies between 6.6% and 16.5% in Ref. [2] for different energy regimes, which is strange, although the variations are within errors.

In the definition of α_s we have made the usual choice $Q^2 = \sqrt{s}$, since this is the only large scale in the process. However, one may argue that gluon radiation occurs at a much smaller scale and there is no reason to use such a large scale. A large amount of literature exists with arguments for the choice of scale[26,27]. The main arguments are based on the size of the higher order corrections and/or the sensitivity to the renormalization scheme. Since the third order QCD contributions are large for R, we have studied these contributions as function of scale in contrast to previous results for a few specific scales[28]. This can be done easily as follows. Suppose a variable is given in a certain renormalization scheme and for a given Q^2 scale to be:

$$R = r_1\alpha_s + r_2\alpha_s^2 + r_3\alpha_s^3 + O(\alpha_s^4) \tag{14}$$

If we choose the coupling at a different scale, we get:

$$R' = r_1'\alpha_s' + r_2'\alpha_s'^2 + r_3\alpha_s'^3 + O(\alpha_s'^4) \tag{15}$$

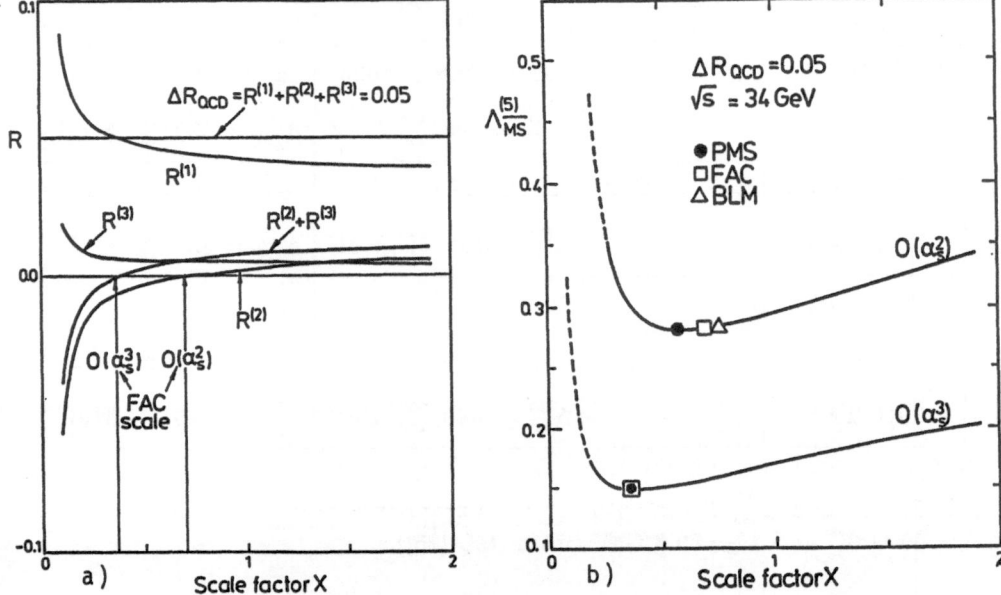

Figure 4: Contributions to R (a) and Λ determinations in second - and third order (b) as a function of the renormalization scale factor x

If one neglects the terms of order $O(\alpha_s^4)$, then

$$R' - R = dR = r_1 d\alpha_s + \alpha_s dr_1 + \alpha_s^2 dr_2 + 2\alpha_s r_2 d\alpha_s + \alpha_s^3 dr_3 = 0 \qquad (16)$$

This can only be zero, if coefficient for each power of α_s equals zero, which yields 3 equations for the 3 unknowns r_1', r_2', r_3'. After calculating $d\alpha_s$ from the renormalization group equation:

$$\mu \frac{\partial \alpha_s(\mu)}{\partial \mu} = -\beta_0 \alpha_s^2(\mu) - \beta_1 \alpha_s^3(\mu) + O(\alpha_s^4), \qquad (17)$$

we find for the coefficients at a different scale $Q' = xQ$:

$$
\begin{aligned}
r_1' &= r_1 \\
r_2' &= r_2 + r_1 \beta_0 ln\ x \\
r_3' &= r_3 + r_1 \beta_1 ln\ x + 2 r_2 \beta_0 ln\ x
\end{aligned}
\qquad (18)
$$

The β-factors are renormalization scheme independent and given by:

$$
\begin{aligned}
\beta_0 &= \frac{1}{6\pi} [33 - 2n_f] \\
\beta_1 &= \frac{1}{12\pi^2} [153 - 19n_f]
\end{aligned}
$$

The various QCD contributions to R as function of the renormalization factor x are shown in Fig. 4a, assuming the total contribution to be constant ($R_{QCD} = 0.05$). One observes that at $x = 1$ the third order contribution is indeed larger than the second order contribution, but at small and large x the absolute value of the second order contribution is larger. However, at all scales the first order contribution is dominant. We have indicated the scales at which the second order or the second plus third order contributions become zero. These are the so-called FAC scales (Fastest Apparent Convergence)[29]. After recalculating the coefficients at a new scale, one can redetermine the corresponding α_s from the measured R-value and recalculate the corresponding Λ value. The result is shown in Fig. 4b. The minimum in this curve is the PMS scale corresponding to the point of minimal sensitivity, a concept introduced by Stevenson[27]. One observes that at all scales the Λ values in third order are roughly a factor two below the Λ values in second order, indicating that for this reaction there is nothing like an optimum scale, where the higher orders are not important.

This clearly indicates that all the heated discussions about choosing a certain scale mean nothing more than betting on the future: you only can say something seriously about higher order contributions by calculating them, not by fiddling with renormalization scales or schemes.

It is interesting to determine the QCD contribution independently from the definition of α_s. If we assume a linear energy dependence within our energy range, we find this contribution to be $f_{QCD}(\sqrt{s} = 34 GeV) = 1.057 \pm 0.008$. An extrapolation of this value to the LEP/SLC energy range yields $f_{QCD}(\sqrt{s} = 91 GeV) = 1.046 \pm 0.005$, which is an experimental number for the infinite series $(1 + \alpha_s/\pi + ...)$.

5. Higher order QED radiative corrections

R is calculated from the number of multihadron events - N_{MH} - and the number of Bhabha events - N_{BB} - in the following way:

$$R = \frac{\sigma_{BB}}{\sigma_{\mu\mu}} \frac{N_{MH}}{N_{BB}} \frac{\epsilon_{BB}(1+\delta)_{BB}(1+\delta_{VP})_{BB}}{\epsilon_{MH}(1+\delta)_{MH}(1+\delta_{VP})_{MH}}$$

Here ϵ is the detection efficiency, δ_{VP} is the vacuum polarization correction, δ is the correction for initial and final state radiation and vertex graphs, and σ_{BB} and $\sigma_{\mu\mu}$ are the Born cross sections for Bhabha scattering and μ-pair production.

We have factorized the effects of loop corrections and other radiative corrections; this yields a higher order contribution $\delta\delta_{VP}$, which should be neglected in first order, but can be large for multihadrons, since typically $\delta = 0.2$ and $\delta_{VP}=0.1$. Such a contribution, representing the radiative corrections to graphs including a fermion pair in the propagator, yields the main difference between the first and second order curves in Fig. 2a. For $k_{max} = 0.6$ δ is small, so one expects the higher order contributions to be smaller too, as proven in Fig. 2b.

In order to estimate more precisely the effect of higher order radiative corrections, one should consider possible changes in ϵ, since the energy loss distribution is changed considerably by the higher order corrections, as shown in Fig. 5.

We have calculated the change in efficiency for a wide range of experimental cuts and center of mass energies using the LUND fragmentation program[16] after implementing the higher order radiative corrections[30]. The result is that indeed

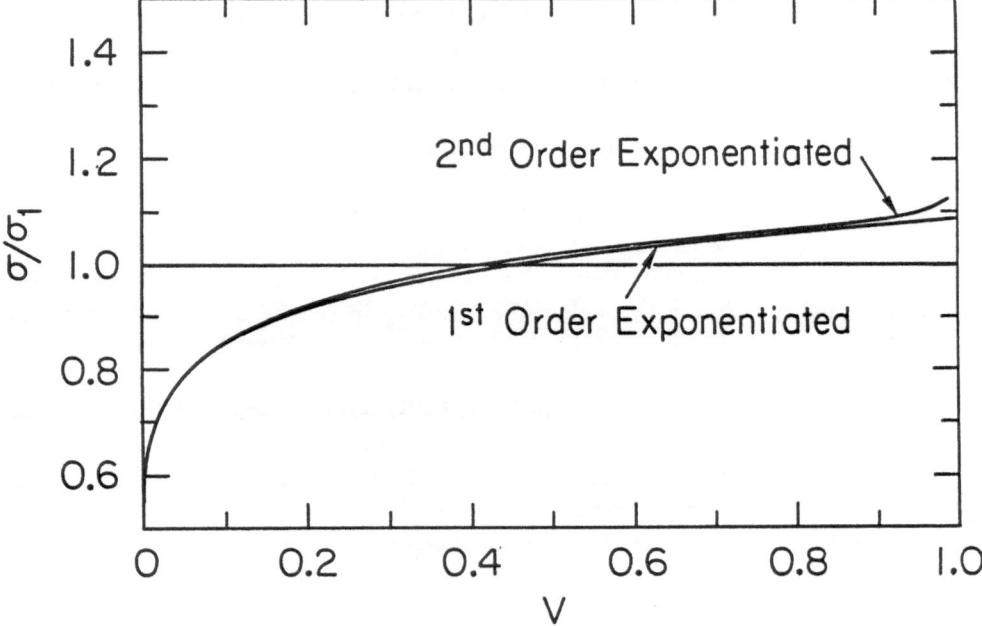

Figure 5: **The first and second order exponentiated cross section normalized to the first order non-exponentiated cross section as function of** $v = 1 - s'/s$**, where s and s' are the invariant masses squared of the initial and final state, respectively. In first order, v is the photon energy normalized to the beam energy. One sees, that the higher order calculations are about 30 % below the first order calculation for small v-values, but 10 % above for large v-values. Exponentiating the first order calculation (horizontal line) gives already results close to the second order calculation; exponentiating the second order calculation hardly changes it**

the product of efficiency and $1 + \delta$ hardly changes by the higher orders, as expected from the small differences in the lower curves of Fig. 2b.

For Bhabha scattering the complete second order calculation has not been done. We have estimated the higher order radiative corrections by applying the exponentiation procedure of Ref[14] to the Bhabha generator[31]; this exponentiation procedure agrees very well with the exact second order calculation in case of μ-pair production[32]. Different experiments usually make quite different cuts on acolinearity and visible energy to select the Bhabha sample. It turns out that the higher order corrections are somewhat sensitive to the various cuts, since one integrates over quite different regions of phase space and different parts are affected differently as shown by the curves in Fig. 5. It was found that the product $\epsilon_{BB}(1 + \delta)_{BB}$ drops between 0 and 1%, if one considers the various cuts. The experiment dependence makes it difficult to correct the R-values. Instead we use these values to make an estimate. If we vary all experimental points between the limits, the refitted value of α_s drops between 0 and 11%. Estimating the error to be half this range or less, we see that the uncertainty from higher order radiative corrections are smaller than the

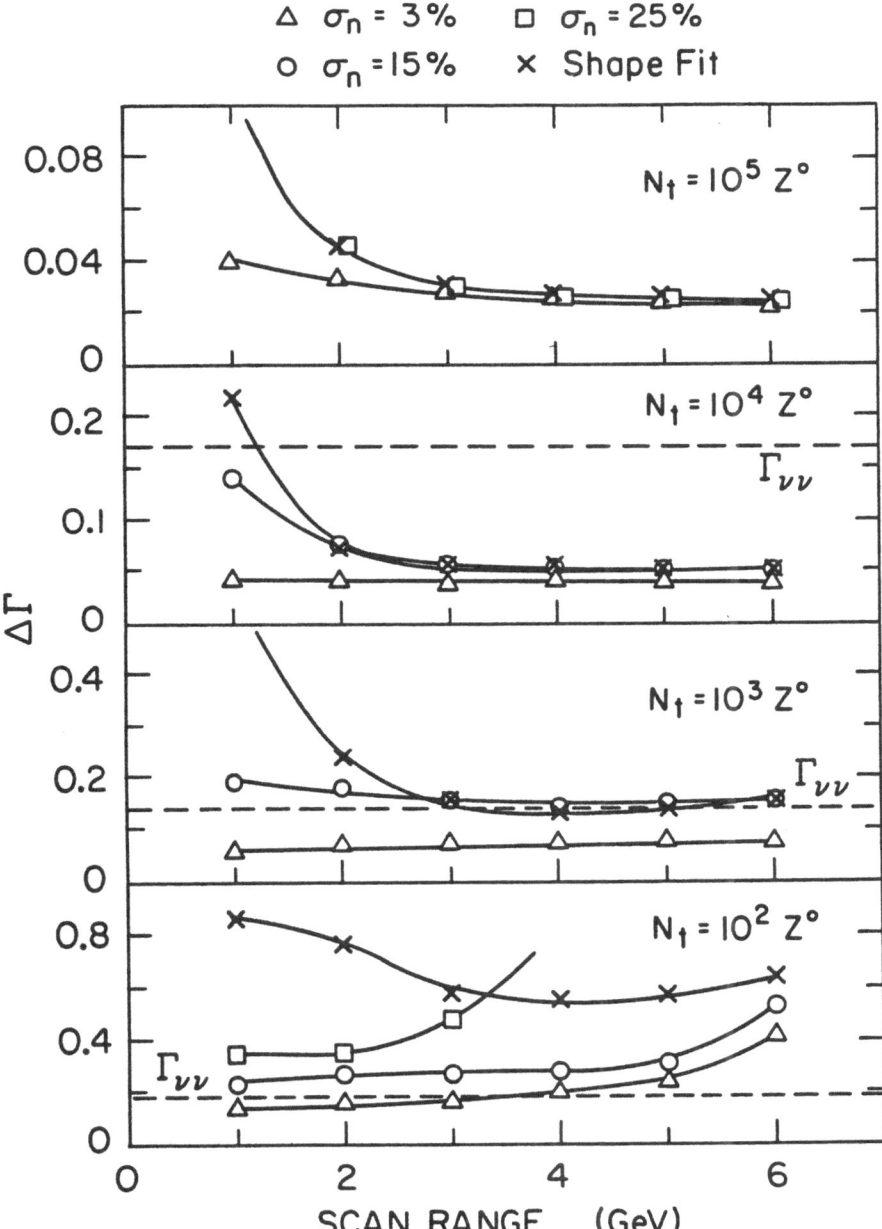

Figure 6: The expected error on the width as function of scanning range assuming 3 scan points and event samples as indicated by N_t (the number of events which would have been obtained if all the luminosity would have been taken at the peak of the resonance). The different assumptions on the normalization errors have been indicated by the different symbols (crosses imply no absolute normalization measured, so only shape fit). From Ref. [32]

quoted experimental error of 11 %.

The effect of higher order corrections on M_Z is appreciably smaller, since the energy dependence is rather smooth, so the shape of R versus energy is not changed significantly.

6. Future M_Z and Γ_Z determinations

Applying the previous analysis method to three hypothetical scan points around the Z^0 resonance would quickly improve the precision of the mass measurement as well as provide a measurement of the width. The expected error on the width as function of scanning range (maximum difference in cm energy between the 3 points) is shown in Fig. 6 for event samples varying between 10^2 and 10^5 Z^0's. The horizontal dashed line indicated the contribution to the width of one new massless neutrino generation. One observes that with 100 Z^0's one can measure the width already with a precision of about 250 MeV, if the absolute normalization error is below 15 %. Similar errors on M_Z can be obtained[32]. For 10^5 Z^0's the precision is limited to about 25 MeV by the systematic poin-to-point uncertainty, which we assumed (optimistically) to be 1% (including the uncertainty from the error on the relative beam energy and reproducibility). Note that if one only fits the shape (crosses in Fig. 6), thus ignoring the knowledge from the absolute normalization (but simultaneously avoiding correlations from the normalization errors), the results are always worse than fits including the normalization. The effect is especially important with low statistics, as expected, since in that case the shape is not well defined, but the width is mainly determined by the events near the peak: the peak cross section is proportional to one over the width squared, so a 10 % error on the peak cross section gives a 5% error on the width (contrary to an eye-fit: the peak cross section determines the width, while events on either side of the peak determine the mass!). For high statistics both fits become equivalent, since then the shape is well determined and the normalization error allows a simultaneous shift of all points without changing the shape. More details can be found in Ref. [32].

7. Summary

We have determined the Standard Model parameters α_s, M_Z, and $\sin^2 \theta_W$ from a 3 parameter fit to the data on R between 7 and 60.8 GeV. The result is:

$$M_Z = 89.4 \pm 1.3 \text{ GeV}$$
$$\alpha_s = 0.142 \pm 0.018$$
$$\sin^2 \theta_W = 0.220^{+0.025}_{-0.020}$$

Here we assumed a top mass of 60 GeV and a Higgs mass of 100 GeV. For a top mass of 120(180) GeV M_Z is lowered to 88.9(88.4) GeV. There is little dependence on the Higgs mass. The value of $\sin^2 \theta_W$ is in good agreement with recent values from deep inelastic neutrino scattering[22]: $\sin^2 \theta_W = 0.231 \pm 0.005$. If we use this value for $\sin^2 \theta_W$, we find:

$$\alpha_s = 0.143 \pm 0.015.$$

Note that the small error comes partly from the fact that we fit over a large energy range, in which case a reduction of all R values by 1% translates into a reduction in α_s of 11% only. Here we have taken the third order QCD corrections into account, which lower the α_s values about 10% (see Table 2) with respect to the second order one. The QCD series $1 + \alpha_s/\pi + ...$ has been determined from a direct fit to the data in a model independent way. Extrapolating to the Z^0 region, we find this factor to be 1.046 ± 0.005.

References

[1] CELLO Coll., H.J. Behrend et al., Phys. Lett. **183B** (1987) 400.

[2] R. Marshall, RAL-Preprint-88-049, submitted to Z. Phys. **C**.

[3] G. d'Agostini, W. de Boer, and G. Grindhammer, Contr. to 24th Int. Conf. on High Energy Physics, Munich (1988)

[4] G. d'Agostini, Proc. Renc. de Moriond (1987), to be published.

[5] W. de Boer, SLAC-Pub 4482, Proc. of the Xth WARSAW Symposium on Elementary Particle Physics, Kazimierz, Ed. Z. Ajduk, Poland, (1987), p. 503

[6] W. de Boer, Proc. of 24th Int. Conf. on High Energy Physics, Munich (1988), Eds. R. Kotthaus and J.H. Kühn, p. 905.

[7] S.G. Gorishny, A.L. Kataev, and S.A. Larin, Hadron Structure '87, Proc., Smolenice, Czechoslovakia, Physics and Applications, Vol. **14** (1988) 180, and preprint JINR, E2-88-254

[8] AMY Coll., T. Mori et al., Phys. Lett. **218B** (1989) 499.

[9] T. Kamae, UT-HE-Preprint-88-05, published in Proc. of 24th Int. Conf. on High Energy Physics, Munich (1988), Eds. R. Kotthaus and J.H. Kühn, p. 156.

[10] T. Tauchi, KEK-Preprint-88-39, to be published in Proc. of Multiparticle Dynamics, Arles, France (1988).

[11] T. Nozaki, presented at Renc. de Moriond, Les Arcs, France (1989).

[12] UA1 Coll., C. Albajar et al., CERN-EP/88-168, November 1988, to be published in Z. Physik **C**,
UA2 Coll., R. Ansari et al., Phys. Lett. **186B** (1987) 440.

[13] M. Dine, J. Sapirstein, Phys. Rev. Lett. **43** (1979) 668
K.G. Chetyrkin et al. , Phys. Lett. **85B** (1979) 277
W. Celmaster, R.J. Gonsalves, Phys. Rev. Lett. **44** (1980) 560

[14] F. A. Berends, G. J. H. Burgers, and W. L. van Neerven, Phys. Lett. **185B** (1987) 395

[15] W.J. Marciano, Phys. Rev. **D29** (1984) 580

[16] T. Sjöstrand, Comp. Phys. Comm. **27** (1982) 243, ibid. **28** (1983) 229
T. Sjöstrand and M. Bengtsson, Comp. Phys. Commun. **43**(1987)367

[17] F.A. Berends, R. Kleiss, S. Jadach, Comp. Phys. Commun. **29** (1983) 185

[18] W. Hollik, DESY 88-188, December 1988, and references there in.

[19] G. Burgers, CERN-TH/5119/88, G. Burgers and W. Hollik, CERN-TH5131/88, both published in the Yellow Book on Polarization at LEP (CERN- 88-06, Vol. 2)

[20] M. Boehm, W. Hollik, H. Spiesberger, Fort. der Physik **34**(1986)687

[21] J. Fujimoto, K. Kato and Y. Shimizu, Prog. Theor. Phys. **79** (1988) 701,
J. Fujimoto, Y. Shimizu, Mod. Phys. Lett. **A3** (1088) 581.

[22] G.L. Fogli and D. Haidt, Z. Physik **C40** (1988) 379.

[23] J. Ellis and G.L. Fogli , Phys. Lett **B213** (1988) 526.

[24] AMY Coll., H. Sagawa et al., Phys. Rev. Lett. **60** (1988) 93, Phys. Lett. **218B** (1989) 499, and G. Kim, Topical Conference, KEK (1989);
TOPAZ Coll., I. Adachi et al., Phys. Rev. Lett. **60** (1988) 97
and S. Suzuki, Topical Conference, KEK (1989);
VENUS Coll., H. Yoshida et al., Phys. Lett. **198B** (1987) 570
and K. Ogawa, Topical Conference, KEK (1989)

[25] G. d'Agostini, W. de Boer, and G. Grindhammer, DESY 89-57, to be published.

[26] D.W. Duke, R.G. Roberts, Phys. Reports **120** (1985) 275

[27] P.M. Stevenson, Phys. Rev. **D23** (1981) 2916

[28] C.J. Maxwell and J.A. Nicholls, Phys. Lett. **B213** (1988) 217
A.P. Contogouris and N. Mebarki, Phys. Rev. **D39** (1989) 1464

[29] G. Grunberg, Phys. Lett. **B95** (1980) 70

[30] The higher order calculations of the Ref.[14] were done for μ-pairs. They have been implemented in the standard μ-pair generator Ref. [17] by J.P. Alexander (private communication) and adapted for quarks by W. de Boer, Mark-II Note 210.

[31] F. A. Berends and R. Kleiss, Nucl. Phys. B228(1983) 537. This Bhabha generator was first exponentiated by M. Levi (private communication) according to a prescription from Tsai, SLAC Pub 3129 (1983). We modified it in order to have the same exponentiation for hadrons and Bhabhas; we know that this exponentiation procedure agrees with the exact second order calculation in case of μ-pair production. In the Bhabha case both the initial state and final state have been exponentiated, as well as their interference following the formulae from M. Greco, Riv. Nuovo Cim. 11(1988) 5:1 and Phys. Lett. 177B(1986) 97. See also Berends and Komen, NP B115(1976) 114.

[32] W. de Boer, SLAC-Pub 4682, to be published in Nucl. Instr. Meth. (1989).

Radiative Corrections for e$^+$e$^-$ Collisions Editor: J.H. Kühn
© Springer-Verlag Berlin, Heidelberg 1989

PRECISION MEASUREMENTS OF QUARK ASYMMETRIES

J. Drees

Fachbereich Physik, University of Wuppertal

Abstract

Recent studies concerning the measurement of quark asymmetries at LEP are presented. It seems possible to achieve accuracies adequate for a precision test of the Standard Model.

1. Introduction

The aim of this contribution is to discuss up to which level of accuracy measurements of quark forward-backward asymmetries can be performed during the first years of LEP operation. If polarized e$^+$, e$^-$ beams are available one can directly determine the couplings of the Z^0 to the final state fermions. This case has been investigated in [1], [2]. Here I would like to concentrate on measurements with unpolarized beams. At the Z peak, in Born approximation, the forward-backward asymmetry is given by

$$A^q_{FB} = 3/4 \ A^e A^q ,$$

and thus highly sensitive to physics phenomena causing a shift of $\sin^2\theta_w$:

$$
\begin{aligned}
\Delta A_{FB} &= -1.9 \ \Delta\sin^2\theta_w \ \text{for} \ \mu,\tau \\
&= -4.4 \ \Delta\sin^2\theta_w \ \text{for} \ u, c \\
&= -5.6 \ \Delta\sin^2\theta_w \ \text{for} \ d, s, b
\end{aligned}
$$

assuming $\sin^2\theta_w = 0.23$.

Let us further note that the "known" electroweak radiative corrections (mainly QED) are small. Consequently uncertainties, originating from the numerical evaluation of radiative corrections are smaller for quark than for muon forward-backward asymmetries. Writing

$$A_{FB} \ (\text{radiatively corrected}) = A_{FB,Born} \ (1 + \delta_{rad})$$

and evaluating δ with the program EXPOSTAR for $M_Z = 92$ GeV, $m_t = 60$ GeV and $M_H = 100$ GeV yields [3]:

$$
\begin{aligned}
\delta_{rad} &= -0.81 \ \text{for} \ \mu, t \\
&= -0.15 \ \text{for} \ u, c \\
&= -0.065 \ \text{for} \ d, s, b .
\end{aligned}
$$

QCD corrections are also small. Writing again:

$$A_{FB} \text{ (1st order QCD corrected)} = A_{FB} (\alpha_s = 0) (1 + \delta_{QCD}),$$

and assuming $\alpha_s(M_Z^2) = 0.12$, one finds for light quarks [3]:

$$\delta_{QCD} = -0.038 \quad \text{no cuts applied to data}$$
$$-0.011 \quad \text{selection of 2-jet events}.$$

For c and b quarks the correction is even smaller.

Obviously the statistics of $q\bar{q}$ events is very high, e.g. for an integrated luminosity of $L = 10 \text{ pb}^{-1}$ (12 days at $10^{31} \text{ cm}^{-2} \text{ sec}^{-1}$) one will take about 300 000 $q\bar{q}$ events but only 15 000 $\mu^+\mu^-$ events.

The reason why measurements of quark asymmetries are often looked at with some scepticism is that the results depend, at least to some extent, on the use of fragmentation models. Let me recall, however, that large apparative investments have been made to enable detailed investigations of quark fragmentation. High statistics measurements of the hadronic final state will improve our understanding of fragmentation. Therefore estimates of systematic errors of A^q_{FB} based on results obtained by different fragmentation models as existing today may not be too optimistic.

In the following I would first like to discuss the measurement of a quark asymmetry which can be performed in the early days of LEP operation. Chapter 3 contains a summary of studies of A^q_{FB} measurements using methods of flavour tagging.

2. Charge Asymmetry of Hadronic Events A_{ch}

Let us imagine one could identify the number of events with a negative quark travelling in the forward direction N_- and the number of events with a negative quark travelling in the backward direction N_+. Defining

$$A_{ch} = (N_- - N_+)/(N_- + N_+)$$

we find in the Born approximation at $\sqrt{s} = M_Z$

$$A_{ch} = 3/4 \; A^e A^{<q>}$$

with

$$A^{<q>} = 2 \left(3 \, a_d v_d - 2 \, a_u v_u\right) / \left(3 \, (a_d^2 + v_d^2) + 2 \, (a_u^2 + v_u^2)\right)$$

for light u, d, c, s and b quarks. For $\sin^2\theta_w = 0.23$, $A_{ch} = 0.047$ and $\Delta A_{ch} = -2.2 \, \Delta \sin^2\theta_w$. The sensitivity with respect to $\sin^2\theta_w$ is comparable with that of the muon forward-backward asymmetry.

The average jet charge has been measured, for instance, in deep inelastic scattering [4]. The charge asymmetry of hadronic events has been measured at PEP [5] and at PETRA [6] applying two different methods. From this work we know that jet charge measurements prove reliable for tagging the quark charge and that the leading charged hadrons have a high probability of containing the parent quark. It was also shown that jet generators reproduce the data.

The MAC collaboration determined A_{ch} at $\sqrt{s} = 29$ GeV by analyzing the jet charge of one hemisphere relative to the thrust axis

$$q_{jet} = \Sigma \; q_i \eta_i{}^{\gamma}, \;\; \gamma = 0.2$$

where q_i and η_i are the charges and the rapidities of the particles in the hemisphere. From A_{ch} they obtain a value for the product of the axial-vector coupling constant of the electron with the axial-vector coupling constant averaged over all quarks:

$$a_e \, a_{<q>} = -1.36 \pm 0.24 \pm 0.20$$

where the second error is systematic.

The A_{ch} analysis of the JADE collaboration is based on a weight technique [6]. They find

$$a_e \, a_{<q>} = -1.09 \pm 0.18 \pm 0.23.$$

It should be noted that the systematic error is about 20% in both cases.

The question arises whether a measurement of A_{ch} at $\sqrt{s} = M_Z$ can yield useful information on $\sin^2\theta_w$. Following the JADE method we evaluated the z distributions of the three most energetic charged particles in each hemisphere using the Lund JETSET Monte Carlo generator ($z = -q \, p_L/E_b$, q is the particle charge, p_L the momentum along the sphericity axis, E_b the beam energy). The weight function is determined from the differential multiplicity distributions of z_1, z_2, z_3 in the forward jet and of z_1, z_2, z_3 in the backward jet for events where the negative quark travelled in the forward direction respectively in the backward direction. Let us call these distributions $f_-((z_1,z_2,z_3)_f \, (z_1,z_2,z_3)_b)$ and $f_+((z_1,z_2, z_3)_f(z_1,z_2,z_3)_b)$. Figure 1 shows the projection of f_- onto the z_1, z_2 and z_3 axes for $\bar{u}, d, s, \bar{c}$ and b quarks and for all quarks. Negative particles have $z_i > 1$. The cuts applied are sphericity less than 0.1 and summed charged momentum larger 0.1 E_b per jet. 82% of all $q\bar{q}$ events survive these cuts. The difference between f_- for negative and positive particles is most pronounced for $\bar{u}$ quarks but there is also a clear indication of a negative charge excess

314

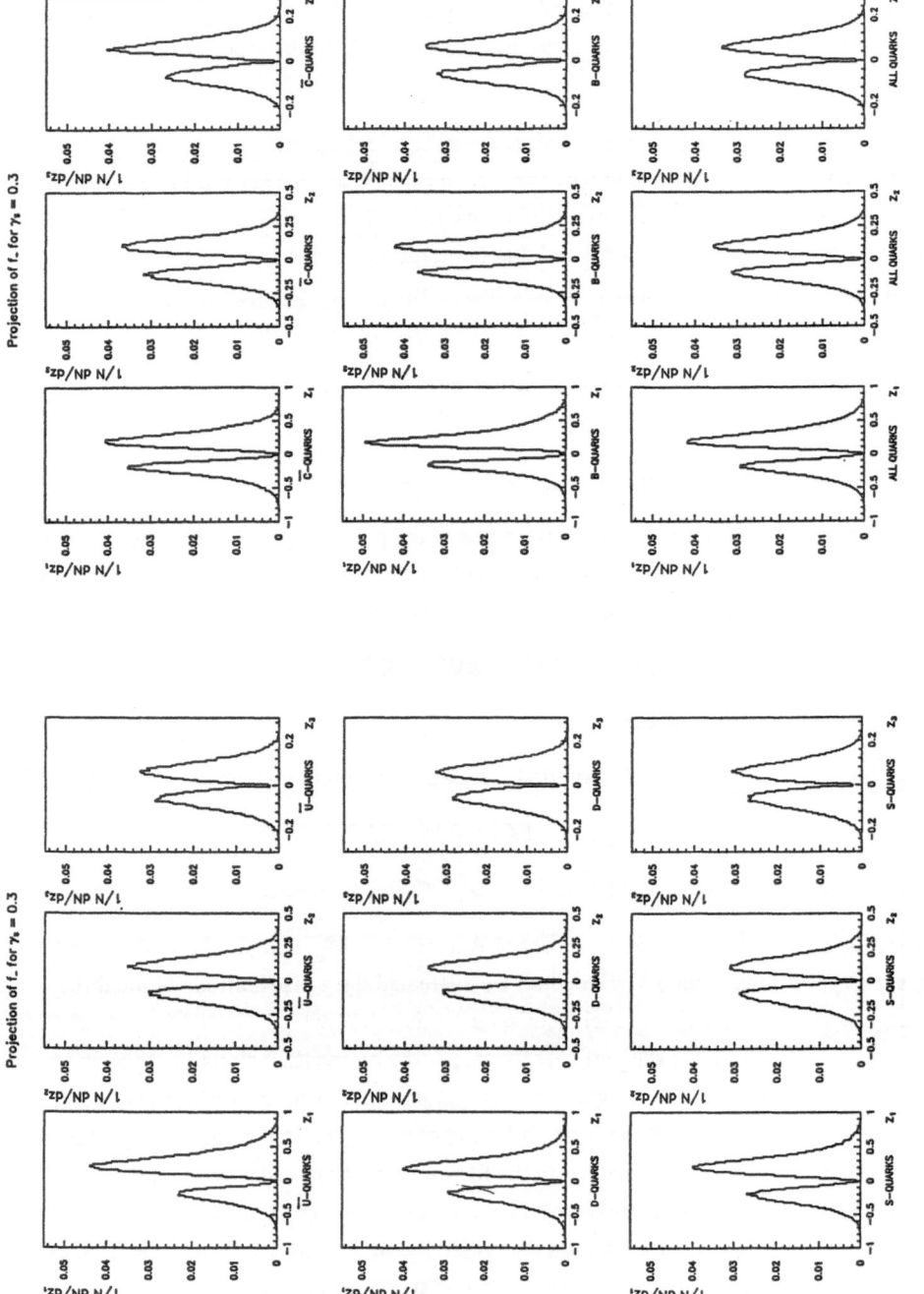

Fig. 1 Projection of f_- generated by JETSET 6.3 onto the z_1, z_2, z_3 axis, $|z_1|>|z_2|>|z_3|$

for all other negative quarks.

With the weight function

$$W(\overline{z}^i) = f_-(\overline{z}^i) / (f_+(\overline{z}^i) + f_-(\overline{z}^i))$$

one can evaluate the mean weight $\overline{W}$ of the data. Then

$$N_-/N = (\overline{W} - \overline{w}_+) / (\overline{w}_- - \overline{w}_+)$$

where $\overline{w}_-$ and $\overline{w}_+$ are the weighted means of f_- and f_+, and $N = N_- + N_+$. Since N_- and N_+ are highly correlated, the statistical error of A_{ch} is given by

$$\Delta A_{ch} = \pm ((1 - A_{ch}^2 + 4\,X/N)\,/N)^{1/2}$$

with $X = (N_-(\overline{w_-^2} - \overline{w_-}^2) + N_+(\overline{w_+^2} - \overline{w_+}^2))/(\overline{w}_- - \overline{w}_+)^2$.

From the Monte Carlo we find $4\,X/N = 3.4$ and $\Delta A_{ch} = \pm 2.1/\sqrt{N}$ at $\sqrt{s} = M_Z$ compared to $\Delta A_{ch} = \pm 1.6/\sqrt{N}$ quoted by JADE at $\sqrt{s} = 35\ \text{GeV}$.

The experimental value of A_{ch} will depend on the fragmentation model used to compute the distributions $f_\pm$. In the JADE analysis the parameter found to have the strongest influence is γ_s, the ratio of s quarks relative to u and d quarks produced from the vacuum during the fragmentation process. To study the influence at the Z peak the distributions $f_\pm$ have been evaluated with $\gamma_s^f = 0.3$ and used to determine A_{ch} for events generated with different values of γ_s. The resulting values of A_{ch} are shown in figure 2.

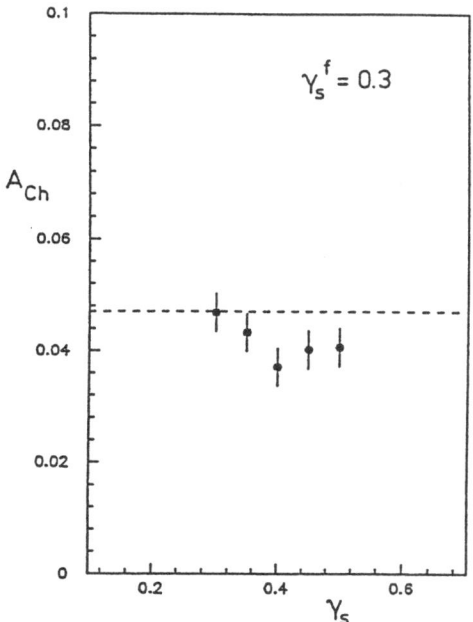

Fig. 2 A_{ch} for event samples generated with JETSET 6.3 for various γ_s, $\sin^2\theta_w = 0.229$

From figure 2 as well as from further investigations using events generated with the parton shower program HERWIG 3.0, one can conclude that the systematic error due to the use of fragmentation models is about 20% or $\Delta A_{ch} = \pm 0.01$.

Systematic errors originating from apparative effects, for instance, from the difference of the acceptance for positive or negative particles are small, $\Delta A_{ch} = \pm 0.001$.

The QED correction is $\delta_{rad} = -0.013$, its error is negligible. Also the QCD correction is small and the error is negligible.

For an integrated luminosity L of $10\ pb^{-1}$ the total error of A_{ch} is dominated already by systematics, the expected accuracy of $\sin^2\theta_w$ is ± 0.005. For the same luminosity the same accuracy of $\sin^2\theta_w$ can be obtained from a measurement of the muon forward-backward asymmetry, the total error of $\Delta A_{FB}^\mu = \pm 0.01$ still being dominated by statistics. For smaller values of L the measurement of A_{ch} should give a more accurate value of $\sin^2\theta_w$ than the measurement of A_{FB}^μ.

3. Forward-Backward Asymmetries of Identified Quarks

Higher precision in $\sin^2\theta_w$ can be reached by measuring forward-backward asymmetries employing flavour tagging methods. The experimental tools are provided by the LEP detectors:

- Full reconstruction of charged particles and electromagnetic showers in the barrel and end-cap regions,

- Charged lepton identification using muon chambers and electromagnetic calorimeters,

- Pion, kaon and proton identification using ring-imaging Cherenkov counters and dE/dx techniques,

- Secondary vertex analysis with micro-vertex detectors.

Examples of the apparative power are given in figures 3 and 4. Figure 3 shows the probability to identify a pion, kaon or proton produced in a $q\bar{q}$ event with the DELPHI barrel RICH as a function of the hadron momentum [8]. The efficiency for track reconstruction is included. Even at momenta of about 25 GeV/c the probability to identify a charged hadron inside a jet is high. Figure 4 illustrates the ability to distinguish $c\bar{c}$ or $b\bar{b}$ events from light $q\bar{q}$ events with the DELPHI microstrip silicon vertex detector [9]. Plotted are the distributions of the sums of the three largest impact parameters for $s\bar{s}$, $c\bar{c}$ and $b\bar{b}$ events for particles with p > 2 GeV/c. A simple cut at 110 μm removes 95% of the light quarks, about 60% c quarks,

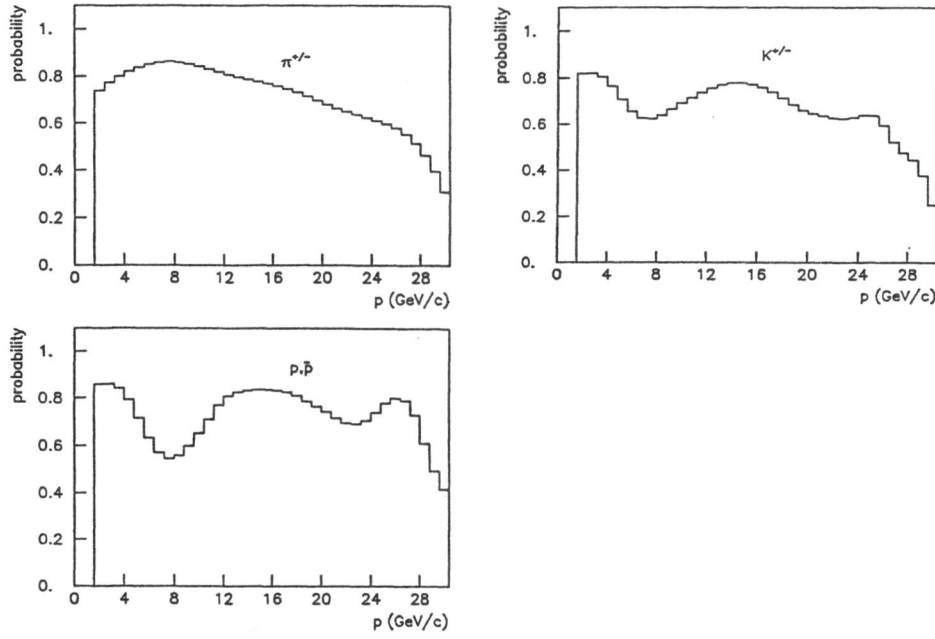

Fig. 3 Probability to identify charged hadrons inside jets with a ring imaging Cherenkov counter

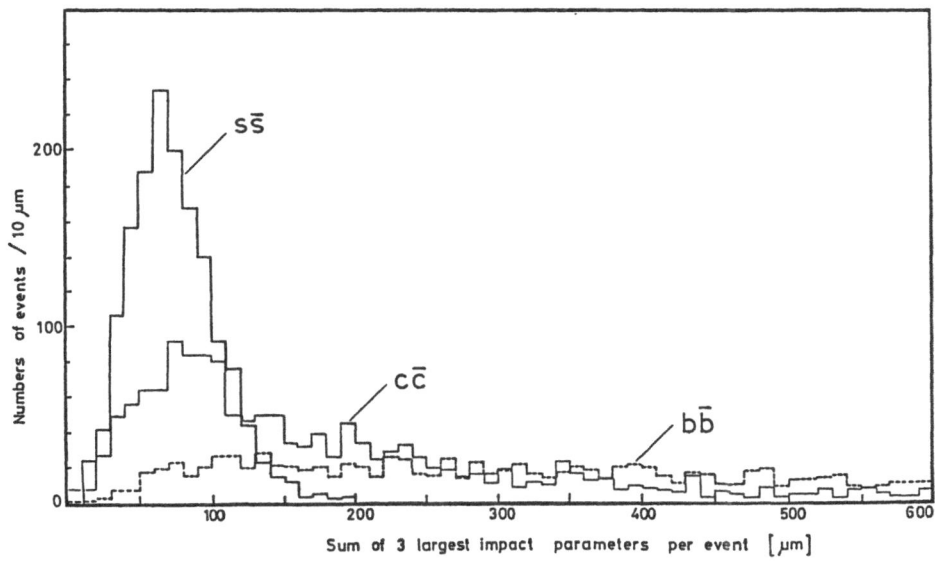

Fig. 4 Heavy quark tagging with a microstrip silicon vertex detector, the rφ resolution is 5 μm

but only 25% b quarks.

With such instruments jet fragmentation will be measured with very high statistics. All inclusive differential hadron distributions for $\pi^{\pm}, K^{\pm}, p, \bar{p}, \Lambda, K_s^o, K^{*o}, D^{*+}$, etc., all two particle correlation functions, e.g. for $p\bar{p}$, will be accurately known. This will allow an update of fragmentation models, for instance, of the baryon sector of parton shower models. One should also note that for flavour tagging only high momentum particles matter, it will not be necessary to understand the soft fragments in detail.

The procedures proposed to identify b, c, s and u quarks can be found in [2]. Let me summarize the status of the DELPHI Monte Carlo studies. In all cases only events with thrust values larger 0.75 are investigated.

i) b-tagging

Using the full detector information including the micro-vertex detector ($45^o < \theta < 135^o$), a rather clean sample of $b\bar{b}$ events can be obtained from a multidimensional analysis [9]. The purity of the $b\bar{b}$ sample will be 90% with a tagging efficiency of nearly 90%.

For a measurement of A_{FB}^b, however, the b charge must be identified. The best results obtained so far are from a multivariate discriminant analysis with a discriminant function

$$F = 1 * PTLE - 0.330 * PTSUM + 0.242 * PKLP$$

where PTLE is the maximum lepton transverse momentum with respect to the thrust axis, PTSUM is the sum of all transverse momenta, and PKLP is the geometrical mean $\sqrt{p_k^{\pm} p_e^{\pm}}$ of the momenta of a same sign kaon-lepton pair inside one jet. Including detector simulation, one achieves a purity of the b sample of 89% with 11% efficiency. 5% background originates from $\bar{b}$ jets. At this stage neutral B mixing is not yet included.

As a consequence of $B^o\bar{B^o}$ mixing, the measured asymmetry is reduced compared to the Standard Model value $A_{FB}^b{}_{,SM}$:

$$A_{FB}^b{}_{,meas} = (1 - 2\chi) A_{FB}^b{}_{,SM}$$

with

$$\chi = R_d^o \chi_d + R_s^o \chi_s,$$

Here χ_d and χ_s are the mixing parameters,

$$\chi_d = \Gamma(B_d^o \to \overline{B_d^o}) / \left(\Gamma(B_d^o \to B_d^o) + \Gamma(B_d^o \to \overline{B_d^o})\right)$$

with an equivalent definition of χ_s. The quantities R are the ratios $b \rightarrow B_d^0, \overline{B_d^0}$ respectively $b \rightarrow B_s^0, \overline{B_s^0}$ to all final states produced in the fragmentation of b quarks. From a measurement of the ARGUS collaboration we know $\chi_d = 0.17$ [10]. Assuming maximal $B_s^0, \overline{B_s^0}$ mixing $\chi_s = 0.5$. The effective mixing parameter χ has been evaluated with the JETSET Monte Carlo. Including effects of the experimental tagging procedure $2\chi = 0.28$.

The increase of the statistical error of A_{FB}^{b},$_{meas}$ due to mixing is studied in [11] for the case that b tagging is performed with high p_T charged leptons. If χ is determined from the measured ratio of like sign to all dilepton events

$$r = (N_{++} + N_{--}) / (N_{++} + N_{--} + N_{+-} + N_{-+}) = 2\chi(1-\chi)$$

the error is given by

$$\Delta A_{FB,SM}^{b} = \pm \sqrt{\frac{1-A_{FB,meas}^2}{1-2r} \frac{1}{N_b} + A_{FB,SM}^2 \frac{r(1-r)}{(1-2r)^2} \frac{1}{N_{ll}}}$$

$$\cong \pm \sqrt{12+1} \cdot 10^{-3} \text{ for } L = 200 \text{ pb}^{-1} .$$

The error originating from the statistics of the observed number N_{ll} of dilepton events is small compared to the error due to the statistics of the number N_b of tagged b events. In the following table only the first term of the above equation is taken into account in order to calculate the statistical error.

ii) c-tagging

The tagging of c quarks can be performed in two steps. The first consists of an analysis of opposite charged kaon-lepton pairs with cuts on the geometrical mean of the momenta and on the angle between the kaon and lepton momenta. An efficiency of 2.1% can be achieved.

In the second step fast D^{*+} mesons are analyzed via their decays into $D^0 \pi^+$. The D^0 is reconstructed from the invariant mass of the $K^- \pi^+$, $K^- \pi^+ \pi^0$ and $K^- \pi^+ \pi^- \pi^+$ final states. If the RICH information is used in the veto identification mode, i.e. for the invariant mass calculation only those particle combinations are rejected which are definitely identified as being wrong, an additional efficiency of 1.8% with a purity better than 70% can be reached.

iii) s-tagging

Again a two-step procedure is proposed. First, events with a high momentum charged kaon in one hemisphere as well as an additional high momentum kaon in the opposite hemisphere are selected. The second kaon must be either neutral or of opposite charge. Further cuts include a veto on a high momentum lepton and a veto on kaon pairs in one hemisphere. Figure 5 shows the background versus the efficiency achieved [12].

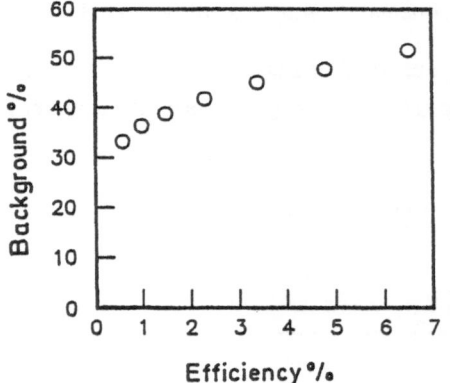

The second step consists of a $K^{0*} \rightarrow K^+\pi^-$ analysis using veto identification. An efficiency of 2.4% with a purity of about 55% is obtained.

Fig. 5   Background of s-tagging using fast kaons versus efficiency

iv) u-tagging

The method is based on the assumption that high momentum protons originate mainly from u and d quarks [13]. After selecting events containing a proton or an antiproton with momentum larger 10 GeV/c in one hemisphere a cut on the average jet charge of the opposite hemisphere reduces $d\bar{d}$ events. The evaluated tagging efficiency and purity depend at present on the fragmentation model used. A much better understanding of baryon fragmentation is needed before reliable results can be obtained.

Table 1 summarizes the results. The values quoted for the b quark asymmetry include effects of neutral B mixing.

Table 1

Efficiencies and backgrounds of b, c, s and u quark tagging, predicted value of $A_{FB,meas}$ for $\sin^2\theta_w = 0.23$ and statistical error, total error of $A_{FB,SM}$ including the systematic error, accuracy of $\sin^2\theta_w$. In the Born approximation $A_{FB} = 0.080$ for u, c and $A_{FB} = 0.112$ for d, s, b quarks. An integrated luminosity of 200 pb^{-1} is assumed.

	b	c	s	u
Efficiency %	11	4	6	1.4
Background %	11	34	45	30 ?
$A_{FB, meas}$	0.065	0.073	0.072	0.070
Statistical error	0.0026	0.005	0.004	0.008
Total error of $A_{FB,SM}$	0.0055	0.007	0.007	0.014
$\Delta\sin^2\theta_w$	0.0010	0.0015	0.0012	0.003

The precision in $\sin^2\theta_w$ expected from the measurement of b, c and s quark asymmetries is rather high. In comparison, a measurement of the muon forward-backward asymmetry will result in an accuracy of $\sin^2\theta_w$ of about 0.0017 for 200 pb^{-1}.

4. Conclusions

Quark forward-backward asymmetries are affected by different types of systematic errors than lepton asymmetries, for instance, uncertainties due to QED corrections are much smaller. They therefore provide additional information for a precision test of the Standard Model.

A measurement of the charge asymmetry of hadronic events can yield an early result with an accuracy of $\sin^2\theta_w$ of 0.005 for an integrated luminosity of a few pb^{-1}.

More precise results will be obtained from measurements of b, c and s quark forward-backward asymmetries using tagging methods. The additional error of the b quark asymmetry due to neutral B mixing remains small. If the Standard Model is correct, a combination of the results of all quark asymmetry measurements is expected to give a precision in $\sin^2\theta_w$ of 0.0007.

References

1. D. Treille: CERN 88-06, 265 (1988)
2. J. Drees et al.: CERN 88-06, 317 (1988)
3. F.M. Renard et al.: CERN 88-06, 197 (1988)
4. J.P. Albanese et al.: Phys. Lett. 144 B, 302 (1984)
5. W.W. Ash et al.: Phys. Rev. Lett. 58, 1080 (1987)
6. T. Greenshaw et al.: Z. Phys. C - Particles and Fields 42, 1 (1989)
7. U. Krüner-Marquis et al.: DELPHI Report in preparation
8. DELPHI - Report 84-60, CERN/LEPC 84-16 (1984)
9. M. Burns et al.: CERN-EP/88-82 (1988)
10. H. Albrecht et al.: Phys. Lett 192 B, 245 (1987)
11. P.J. Dornan: CERN 88-06, 344 (1988)
12. S. Überschär: Report WU B 89-4 (1989)
13. P. Mättig: Int. Journal of Mod. Phys. A 3, 1 (1988)

Radiative Corrections for e⁺e⁻ Collisions   Editor: J.H. Kühn
© Springer-Verlag Berlin, Heidelberg 1989

Physics with the Pretzel LEP

D. Treille

CERN, Geneva, Switzerland

A new option was recently proposed for LEP [1]: multibunch operation on the Z, providing a tenfold increase in luminosity. This is known as the pretzel*⁾ scheme [2]. Physics-wise, no thorough study — such as was done for LEP 200 and for polarization on the Z — has yet been carried out. Therefore, these preliminary and personal remarks should be considered with reservation, since they rely more on guesses than on detailed studies.

With the pretzel scheme (see Section 1), if it can be successfully implemented, one could imagine an exposure on the Z leading to ~ 10^8 such particles. Clearly, many experimental and organizational problems would have to be solved in order to cope with such a flux and with the total amount of information. They are briefly discussed in Section 2. The output of physics from 10^8 Z is then explored along three directions:

i)   the improvement of accurate measurements on or near the Z (Section 3);
ii) the detection of rare modes of the Z, with branching ratios $\leq 10^{-5}$ (Section 4);
iii) the physics of beauty that one can hope to perform (Section 5) on the Z resonance.

One could think of other potentially interesting fields as well (e.g. tau physics).

An important remark should now be made (it will be substantiated later): high luminosity on the Z, however interesting it may be, is physics-wise not an alternative to an increase of LEP energy (in particular, to the W-pair domain) or to the exploitation of longitudinal polarization on the Z: these two options contribute specific irreplaceable information.

1.   The Pretzel Scheme

The well-known formula for the luminosity,

$$L = \frac{k_b \, I_b \, (E_0/mc^2)\bar{\xi}}{2er_e\beta_y^*} \, ,$$

exhibits the main parameters governing its value. With

  $k_b \equiv$ number of bunches = 4,
  $I_b \equiv$ current per bunch = 0.75 mA,
  $\beta_y^* \equiv$ beta function at IP = 7 cm,
  $\bar{\xi} \equiv$ beam–beam limit factor = 0.04,

we get the canonical value of L = 1.7×10^{31} cm⁻²s⁻¹ at the Z mass.

To go beyond this value, we can think of increasing the bunch current (by a factor of 2?) and decreasing β_y^* (by a factor of 1.5?), and we can hope that a larger beam–beam limit (a factor of 1.5?) can be tolerated. However, any such attempt has its unknown features.

---

*⁾ This is the English name of a biscuit called Pretze by the Bavarians (and Brezel by the Germans) and that can be thus described: 'ein Backwerk, etwa in Form einer Acht' (i.e. a biscuit, resembling somewhat the shape of an eight).

A radical approach is to increase the number of bunches, whilst avoiding unwanted bunch–bunch collisions. The best idea would be to have a second ring on top of the first, but occupancy of that location is already foreseen! An alternative method is the pretzel scheme: its principle is to give opposite wavy patterns to the e^- and e^+ closed orbits (Fig. 1), so that the bunches avoid each other. These patterns cannot be tolerated in the RF section because synchro-betatron resonances would be excited; this imposes an upper limit on the number of bunches, namely:

$$n_{max} = \frac{\text{circumference}}{2 \times (\text{distance IP to end RF})} \simeq 54 .$$

Furthermore the number of bunches should divide the RF harmonic number of the machine, so that proper acceleration is ensured. And obviously one must have encounters at even intersection points. If $k_b \neq p \times 4$ there are no collisions in the odd pits. This gives the following possibilities:

$$k_b = 2, 4, ..., 36, 40 .$$

For several reasons, the choice would be to separate the beams in the horizontal plane, in particular because there the aperture is larger.

The residual beam–beam tune shift resulting from the close encounter of opposite bunches has been studied [2], using separators (electrostatic or RF magnetic) in place of the last half-cell of RF cavities. With $\ell = 8$ m and $E_x = 0.8$ MV/m, this tune shift would still be acceptable for a 36-bunch solution. The resulting luminosity would be of the order of $L = 1.4 \times 10^{32}$ cm^{-2}s^{-1}.

However, it is necessary to have sufficient RF power, and such a scheme can only be possible if some of the copper cavities are already replaced by superconducting (SC) ones (also, room is needed for the separators); in 1991, 32 SC cavities could be installed.

With an RF power of $P_{beam} \approx 10$ MW, the scheme could be used up to $\sqrt{s} = 100$ GeV; however, it would help less to increase the luminosity at LEP 200, and the problem of obtaining 500 pb^{-1} (the 'Aachen quantum') above the W-pair threshold is still severe.

It is premature to make any prediction about polarization: however, the choice of a horizontal separation does not *a priori* compromise the chances of maintaining the polarization (if there is any) in a pretzel scheme.

2. Experimental Problems

LEP experiments have been prepared having in mind an interval of 23 μs between crossings and for a peak luminosity of 1.4×10^{31} cm^{-2}s^{-1}. The 36-bunch solution would bring this interval to 2.5 μs and increase the luminosity by an order of magnitude: provided the vacuum can be kept at around the same value, the background would probably scale with the number of bunches. Finally, 10^8 Z represents at least an order of magnitude increase in the volume of data foreseen. Let us note first that the Z rate is approximately four per second: therefore physics-wise, all or nearly of the $\sim 360,000$ crossings per second are empty, and the two-photon interaction does not change this picture.

The argument that the existence of long-drift devices (e.g. TPC, RICH) would preclude the use of such a bunch spacing is not a valid one. What has to be done is to change the gating mode from a systematic opening at each crossing to an externally triggered mode: the elements of such a fast trigger have to be provided, but whatever its rapidity, a few centimetres of information will be lost on the edges of the drift volume. It is difficult to see what could prevent us from having such an efficient trigger at a rate of 1 kHz or less.

Another effect of long-drift devices would be to integrate the background of several crossings. For the nominal LEP conditions, with a vacuum chamber of radius r = 8 cm, the background conditions [3] are not severe (after an initial learning phase, probably); the detectors should stand two orders of magnitude more. However, if at the same time we try to decrease the radius of the chamber, things rapidly get worse. In fact, we can act only in the vertical plane. From the point of view of background, a chamber profile as shown in Fig. 2 [4] seems to represent the ultimate tolerable limit for the nominal luminosity. Therefore, it is hardly possible to run the pretzel and at the same time use such a chamber. My guess is that instead of struggling to approach the interaction point below r = 5-6 cm in order to improve the extrapolation accuracy, we should rather act in such a way as to decrease the effect of Coulomb scattering by using thinner microvertex detectors, possibly installed in a pre-vacuum region. We will come back to this discussion in the section devoted to b physics.

More problematic is the fact that the various detectors, because they were prepared for a large interbunch spacing, have, for some aspects of their electronics and readout, been 'taking their time'. None of them would find it a problem to accept eight bunches. Beyond that, non-negligible modifications would have to be performed in some cases. This would take time, effort, and money, but such outlay would anyway remain at the level of a few per cent of the total project investment.

The problem of acquiring about four Z per second (assuming that we know how to avoid triggering on the background) is not a major one. The idea of processing 100 million Z appears to be challenging. As we will see, quite rare and unusual events are being searched for, and therefore we have to avoid the systematic use of fast but biasing filters for speeding up the production.

The problems of mass storage, fast link, and manpower are probably more severe than the question of CPU availability. However, relying on the rapid evolution of techniques, we can envisage that these problems will be manageable in the near future. Furthermore, it is quite likely that a collider such as the LHC will, several years after the pretzel is introduced, require even more computing power, and therefore we should be prepared for this eventuality.

3. Accurate Measurements

The interest of accurate measurements on the Z, to test in depth the Standard Model and possibly reveal new physics, has been studied with great care in the past.

The obvious tendency here is for the statistical errors — which in many cases were still dominating — to fade away. The measurement is thus more exposed to experimental systematic errors and theoretical uncertainties.

Another aspect is that, with abundant statistics, we can also act on the systematic errors in order to decrease them: cuts can be made tighter so as to avoid doubtful kinematic regions, acceptances can be measured by comparing redundant procedures, and our knowledge improves with the statistics, etc.

The $\mu\mu$ (or $\ell\ell$) charge asymmetry (Table 1) will then be dominated by the error due to the uncertainty in the position of the measurement relative to the Z pole: here I have assumed that it is to ± 20 MeV, a value which already implies energy calibration with a resonant depolarization method. The value quoted in Table 1 for the uncertainty in the radiative correction ($\sim$ 5-6% of the correction) can probably be reached, but first there will have to be a lot of progress with respect to the present state of the art, which, for unessential reasons, shows much dispersion of the results between the various programs. The experimental error (built up from the difference between the acceptances for + and − integrated over the detector) is an example of a quantity that can benefit from an increase in the statistics, since it can be measured from the data.

For the τ-polarization measurement, the statistical error, which still dominates at the level of a few 10^6 Z (the signal is only 6‰ of the Z rate if the $\tau \to \pi\nu$ mode is used), disappears, and we are left with the dominating systematic errors: for instance, in the case of $\tau \to \pi\nu$, the confusion with $\tau \to \varrho\nu$. Our improved knowledge of the detectors, the increased statistics that allow tighter cuts to be made, the likely exploitation of $\tau \to \ell\nu\nu$ modes, and the possible combined exploitation of the $\tau \to \pi\nu$ and $\tau \to \varrho\nu$ modes, could finally lead to numbers below those quoted in Table 2.

It is certainly in the exploitation of hadronic modes that the improvement is mostly felt. Indeed, owing to the severity of cuts at the tagging level, the statistics were always considered to be the limiting factor. Moreover, information obtained on the Z peak is essential to our having confidence in the 'stability' of programs describing fragmentation — in particular for quarks such as the s, which were 'minority carriers' at PEP and PETRA, and in deep-inelastic scattering. It is likely that this information, and therefore the quality of tagging, will gradually improve in the LEP era. Expected numbers [5] are given in Table 3. Included in the systematic errors are: QED, with Δm_Z uncertainty; QCD corrections, which seem to be well under control; and the dominating instrumental and tagging errors. We should note that such measurements are not the privilege of detectors with powerful hadron identification. For instance, the promising c-tagging procedure [6] of the third column requires only the good measurement of a 2–3 GeV pion, and its ultimate systematic error, which is still uncertain, will depend on the quality of such a measurement and on the reliability of background subtraction (Fig. 3).

Table 3 shows a set of impressive numbers to be obtained from quark tagging. However, the amount of work and of further understanding that will be necessary in order to reach this level is impressive too!

Therefore the polarization asymmetry A_{LR} measurement offered by longitudinal polarization keeps its invaluable quality. With a modest 50 pb^{-1}, provided the conditions quoted in Table 4 are fulfilled, an experimental accuracy of $\Delta\sin^2\theta = \pm0.0003$ can be obtained [7]. The systematic errors are of a totally different nature compared to the charge asymmetry A_{ch}, and the amount of data to process is quite limited. The experimental accuracy matches well with the theoretical uncertainty of ±0.0004. The polarization option is more interesting than ever.

Another accurate measurement, where abundant luminosity can make a major impact, is neutrino counting. If we consider the classical radiative method, luminosity allows us to go higher above the Z (say around 100 GeV, since the pretzel is still operative there), select photons of higher energy (> 3 GeV) and angle ($> 45°$), keeping a large signal ($\sim$ 3000 events for 100 pb^{-1}, i.e. two weeks running time). This transverse energy (E_T) cut, combined with the quite low angle tagging foreseen with a second-generation small-angle tagger (SAT) ($< 2°$), almost eliminates the eeγ background; if necessary, this can be subtracted, using a measurement below the Z. If an experimental problem of acceptance to $\gamma > 3$ GeV still exists, then by triggering on photons we can measure the ratio of $\nu\nu\gamma$ to $\mu\mu\gamma$, which in principle eliminates such uncertainty, but at the expense of statistical limitations with $\mu\mu\gamma$. Normalization problems traditionally limited to $\pm2\%$ in the SAT should be improved by the use of forward electromagnetic calorimeters counting about one tenth of the Z (300,000 e$^+$e$^-$ for 100 pb^{-1}) and with hopefully much lower systematics. If a month of pretzel is devoted to ν counting, it is difficult to see why an experimental error of a few (5?) per cent on the ν number should not be obtained. We saw that on the theoretical side [8] the situation, within such an acceptance, is also quite favourable. Since there is a growing tendency, fed by various lines of thinking [9], to propose non-integer ν numbers at LEP, we should thus be able to reply to this challenging statement.

On the other hand, 'fast' measurements such as m_Z and Γ_Z do not improve with statistics, whilst R and $\Gamma_{\ell\ell}$ should get somewhat better in the long term, in particular if the absolute normalization error is decreased.

4. Rare Modes of the Z

Non-standard phenomena should lead to the existence of new decay modes for the Z. Such phenomena should also slightly modify the accurate measurements on the Z (or at higher energy). However, whenever a direct effect has a chance to be observed, we should certainly look for it. Here we enter a domain in which, most of the time, detailed Monte Carlo work is missing, and which would deserve to have a specific Workshop devoted to it.

Intuitive arguments tell us, however, that whenever a channel with a branching ratio of 10^{-5} or smaller is proposed, an increase in luminosity can only increase the chances of revealing this channel in a significant way.

Before considering non-standard phenomena, let us see what luminosity could bring to the standard Higgs search. This is shown in Fig. 4. In the absence of background, the cross-over point between pretzel LEP I and LEP II (the mass at which we get the same rate per unit time) is ~ 60 GeV. The cross-section at LEP 200 is less energy-dependent and a nice overlap should be available. The status of the background for the Higgs discovery at LEP I is not clear to me: compared with early studies, the likely absence of $t\bar{t}$ should make the detection easier. On the other hand, new sources of background [10], for both light and heavy Higgses, should be quantified. We should also remember that at LEP 200 (provided $\sqrt{s}$ = 200 GeV and $\int L\, dt$ = 500 pb^{-1} can indeed be reached), a Higgs with $m_H \approx m_Z$ could in principle be revealed as well, if a good tagging of the $b\bar{b}$ final state is available.

4.1 Compositeness

One non-standard topic is compositeness. It has been studied at Aachen [11], with emphasis on LEP 200 but with frequent references to possible effects at the Z. The usual compositeness (we use the ELP notation of Ref. [12]) leads to contact terms and to excited e. With regard to the contact terms, the Z is clearly not the right place to look for their effect (one cannot get the interference of a real contribution with an imaginary resonance): even off the Z, the scaling laws giving the level of sensitivity to Λ (Table 5) show that 'patience', i.e. L, does not 'pay'. The same is true for e^* single production, or indirect evidence of it through the $\gamma\gamma$ final state: here, increasing $\sqrt{s}$ is far more rewarding.

However, if nearby compositeness (with $\Lambda \sim m_Z$) is a reality — a good example is the strongly coupled Standard Model (SCSM) [13] — new rare decay modes of the Z can occur. A well-known example is $Z \to 3\gamma$, which could have a branching ratio of $\sim 10^{-5} Q^4$, where Q is the average charge of the constituents. It has been shown [14] that above the $ee \to \gamma\gamma(\gamma)$ background [15], a signal corresponding to BR = 1.1×10^{-5} (95% CL) could be revealed with an exposure of 100 pb^{-1} on the Z, owing to quite different signal-to-background behaviour. A tenfold increase in L should bring this to $\sim 2 \times 10^{-6}$, a much more convincing number if the above prediction is to be taken seriously. The SCSM predicts, as well, quite strong enhancements [16] of visible Z decay modes, such as into $H\gamma$ (Figs. 5). The limit of 100 events per year for full acceptance with the pretzel scheme is indicated in the figures. The limit of visibility of $b\bar{b}\gamma$ final states corresponding to $H\gamma$ deserves further study.

4.2 Supersymmetry

Supersymmetry (SUSY) also could give rise to rare and interesting decay modes of the Z. The most promising sectors are those of the Higgs and neutralinos. In minimal (and most of the non-minimal) SUSY versions, quite strong predictions for the Higgs exist [17] (see Fig. 6):

i)   at least one neutral scalar should be lighter than the Z;

ii) depending on the ratio of vacuum expectation values (VEVs), v_2/v_1, either this Higgs is produced as a standard one, or the Z decays into pairs of unidentical neutral Higgs, degenerate in mass for large v_2/v_1: if these are light ($\sim 10\,\text{GeV}$), the effect is abundant (a few per cent of the Z decays). If, however, we want to reach heavy Higgses ($\sim 40\,\text{GeV}$), and furthermore if for visibility reasons one of them, at least, is required to decay into a τ pair, the overall branching ratio drops rapidly below the 10^{-4} region. The signals to be searched for are: four b-jets, two b-jets + two τ, four τ events (e.g. quadrileptons), etc.

About neutralinos, which could also be light objects, general studies of $Z \to \chi\chi'$ decay have been performed [18]. The branching ratio and the nature of the subsequent cascading of $\chi' \to \chi + X$ depend on the main SUSY parameters (μ and M of Fig. 7) and on the nature of the neutralinos: X can be a lepton pair, a light Higgs, or a photon; the resulting events, with two disappearing χ, are spectacular. Figure 7 gives the limit of the (μ and M) area, where the $Z \to \chi\chi'$ branching ratio is above 10^{-5}.

4.3   Flavour-changing neutral currents

The existence of a fourth generation and of further new physics (such as two Higgs doublets) predicts flavour-changing neutral currents (FCNC) leading to unconventional final states [19] (Fig. 8):

$$Z \to t\bar{c}$$

$$\to b'\bar{b}$$

$$\text{etc.}$$

The same can happen for leptonic decays [20]:

$$Z \to e\tau$$

$$\to \mu\tau$$

$$\text{etc.}$$

The predicted branching ratios vary widely but may reach accessible values.

For the leptonic modes the most obvious background is $Z \to \tau\tau$, with one τ decaying into $\ell\nu\nu$ and the lepton carrying most of the momentum. The momentum spectrum is nearly linear in that region, and under the width of the peak at the beam energy (given by the momentum resolution ϵ) the background is roughly ϵ^2. Putting in numbers [$\epsilon \approx 2\%$, $\mathrm{BR}(\tau \to \mu\nu\nu) \approx 20\%$], we find that with large statistics ($\sim 10^8$ Z), branching ratios of a few 10^{-6} should be accessible.

For the hadronic-mode $t\bar{c}$ and a heavy t ($m_t \approx 80\,\text{GeV}$), the signature would be

$$(\bar{c}\; b\; \ell\; E^{\text{miss}})$$

with quite specific features. For $b'\bar{b}$, it would depend on the decay mode of the b': either the classical $b' \to c$ cascade, which could be revealed by an excess of charm ($D^* \to \pi D$), or ($b' \to b$) neutral current leading to $b\gamma$, bg, etc., final states. In any case, much simulation work is needed in order to obtain firm numbers on the visibility of such FCNC decays.

5. The Z as a b$\bar{\text{b}}$ Factory

The Z is a democratic source of all fermion pairs. The physics that could be extracted from a few million τ pairs is, for instance, to be studied: whilst information on the ν_τ mass is probably the domain of dedicated τ factories, we should learn much about, for example, the Lorentz couplings of the τ on the Z.

For b$\bar{\text{b}}$ physics, the prospects are the following. From 100 million Z (visible), about 19 million events of the b$\bar{\text{b}}$ type are available, i.e. 38 million 'B' particles. According to the Lund fragmentation, their distribution by population should be as indicated in Table 6.

One can say that a single exposure (to be compared with the situation at threshold b$\bar{\text{b}}$ factories) provides an abundance and a variety of beauty species. Table 6 and Fig. 9 give the main characteristics of b$\bar{\text{b}}$ events [21] according to the Lund program. Various properties allow such events to be tagged in a very efficient manner (leptons, offsets in a microvertex detector, leading particles, etc.). However, the impressive results of Table 7 [22] should be viewed with caution since we will be looking for rare and peculiar b-decay modes, whilst the tagging results are obtained for typical events. One way out would be to tag on one jet only, leaving the other jet free, and the effect of this procedure is still to be evaluated.

A *sine qua non* condition for performing good b physics is the implementation of a highly performing vertex detector. Table 8 compares the characteristics of the present DELPHI vertex detector with those of an ideal one [21]. One should go from two to three layers, from one- to two-dimensional accurate measurements: the inner radius should be decreased (but remember Section 2), and the lever arm possibly increased. In order to limit the Coulomb scattering, the first two layers should be kept as thin as possible. If this is achieved and if the main vacuum chamber represents a large fraction of the remaining thickness, one should try to push it outward (near the third layer): this means that the first two layers should be in a pre-vacuum region, just separated from the ultravacuum of the machine by a thin window. This is quite challenging technically (and may be impossible), since cooling, as well as shielding against electromagnetic noise, has to be achieved. It is also likely that the inner layer, at least, should be providing space points (as in a pixel device) rather than projective information. All these conditions imply a major technological step forward, compared with present achievements. We could then safely expect a longitudinal resolution of typically 150 μm on the decay vertex of B particles, with minimal confusion and wrong attribution of B tracks to the main vertex (Fig. 10), a source of error in the determination of the charge of the decaying beauty particle.

Beauty physics can be obtained from an exposure on the Z at various levels of integrated luminosity. We can give four examples.

1) *The measurement of the lifetimes of individual beauty states.*
Two ways are possible. One is to use semileptonic decays.

$$B^0 \rightarrow \ell\nu D \ , \tag{1}$$

$$B^\pm \rightarrow \ell\nu D \ , \tag{2}$$

$$B_s \rightarrow \ell\nu D_s \ , \tag{3}$$

and to identify the D states produced. The flight length is obtained from the intersection of the lepton and the D trajectories. The rate per million Z should be 200–300 for processes (1) and (2), and 20–30 for process (3).

Another way is to reconstruct exclusive decays; for instance,

$$B^0 \to \psi K \pi \,,$$

$$B^\pm \to \psi K \,,$$

etc.

The rates depend on the branching ratios [23], which in most cases are poorly known. More information should come from CESR (Cornell) before beauty physics really starts at LEP. Ten million Z should already make it possible to resolve the question of the lifetime measurements. They should also provide substantial spectroscopic information on new B states (baryons etc.).

2) *Rare modes of the B*

These 'precious rarities' [24] should give access to the $V(b \to u)$ Kobayashi–Maskawa (KM) matrix element ($B \to \pi\pi$, $p\bar{p}$, etc.), and can open windows on new physics ($b \to s\gamma$, $B \to K\ell\ell$, $B \to \tau\nu$, etc.).

Figure 11 recalls that whilst such branching ratios are small, or even inaccessible, within the Standard Model, new phenomena such as SUSY could boost them to quite observable values.

A threshold B factory has well-known advantages; the main one is its excellent mass resolution of reconstructed channels (of a few MeV), obtained with the help of beam constraints. At LEP, on the other hand, whilst the mass resolution is somewhat worse (typically $\sigma \approx 40$–50 MeV), the natural boost ensures the visibility of the B decay vertex, which is the only secondary vertex for charmless decay modes. It also guarantees that the decay products of the two b's are in opposite hemispheres, with no possible confusion: for a 'meagre' final state such as $B \to \tau\nu$, this is certainly a major advantage, compared with threshold $B\bar{B}$ physics.

Let us quote an exercise done on the $K^*\gamma$ final state [25]. Figure 12 shows that even before using the microvertex, a 3σ effect can be observed for a branching ratio of less than 10^{-4} and 3×10^6 Z. With high statistics and the full power of the detectors, I think that the level of 10^{-5} can be attained. Here again much work is needed. In particular, the challenging $B \to \tau\nu$ mode should be studied.

3) *The mixing of B_s*

One can admit that the mixing of ordinary B's will be well measured at that time. Inspection of the KM matrix (see Table 9 for notation) shows that the measurement of x_s ($x = \Delta M/\Gamma$), once x_d, ϱ, and λ are known, gives access to the phase of the matrix:

$$\frac{x_s}{x_d} = \frac{1}{\lambda^2(1 + \varrho^2 - 2\varrho\cos\delta)} \,.$$

Experimentally (Fig. 13), we know that x_s is larger than ~ 3. Therefore, for B_s states we can expect quite fast time oscillations.

One can readily assert the resolution that is required from a vertex device in order to give access to a measurement of these oscillations. The period is $T = (2\pi/x)\tau$, and it corresponds to a spatial distance of $(2\pi/x)\gamma c\tau$. If we say, for instance, that we should resolve a third of a period, then the required resolution on the flight path (which in fact means the resolution on the secondary vertex, see Table 8) is

$$\Delta x = \frac{4.4}{x} \text{ (in mm)} .$$

Attempts to see B_s oscillations by fully reconstructing B_s final states are too demanding in statistics. Work is currently being carried out along the lines of semi-exclusive processes such as

$$B_s \rightarrow \ell\nu D_s ,$$

where only the charged tracks of the D_s (containing either zero or two K's) are used. In spite of incomplete reconstruction, the intersection of ℓ and D lines of flight gives a proper measure of the secondary vertex: Fig. 14 [21] shows that according to the estimate made above, x-values as large as ~ 20 are accessible. However, we also need the γ factor of the decaying B_s, and incomplete reconstruction could be a problem. In fact, the peaked fragmentation function of the b-quark (Fig. 15) gives, by itself, a good indication of the B momentum: the situation can be improved by scratching away the low-energy tail of this distribution, using the measurement of particles accompanying the B. The remaining uncertainty on γ has no effect for short times, but precludes exploitation of the distributions for flights longer than ~ 2 mm.

It is felt (see Fig. 16) that an x_s of ~ 10 can be measured with a σ of ± 0.25, using the B_s semileptonic decays for an exposure yielding 30 million $b\bar{b}$ [21]. The corresponding information on the phase of the KM matrix can be read from Fig. 17.

4) CP violation in the B system

This is the most ambitious goal of beauty physics.

CP violation in the mass matrix, to be searched for using dileptons, is totally out of reach within the framework of the Standard Model (10^{-3} expected): conversely, an exposure of 10^8 Z could give access to the observation of CP violation with a non-standard level of a few per cent.

The most promising way is to consider decays to a CP eigenstate that are accessible from two different paths, such as

$$B^0 \rightarrow \psi K_s$$
$$\searrow \bar{B}^0 \nearrow$$

Other final states could be considered as well (ψK^*, $D^{*+}D^-$, ...).

A naïve calculation of the number of $b\bar{b}$ needed to give a n^σ effect gives

$$N_{b\bar{b}} = \frac{n_\sigma^2}{A^2} \frac{1}{2BR} \frac{1}{\epsilon_{tag}} \frac{1}{\epsilon_{det}} \frac{1}{\epsilon_{B^0}}.$$

For $n_\sigma = 3$, $A = 0.2$, $BR = 5 \times 10^{-5}$, $\epsilon_{tag} = 0.2$, $\epsilon_{det} = 0.3$, and $\epsilon_{B^0} = 0.35$, we obtain

$$N_{b\bar{b}} \approx 10^8 \text{ events} .$$

Figures 18 show that with a good vertex detector, the time evolution of the effect can be exploited. Figure 18a represents the most favourable asymmetry for the ψK_s^0 mode between B^0 and $\bar{B}^0$:

$$B^0 \rightarrow \psi K^0_S \propto e^{-\Gamma t}\,(1\,+\,\sin\phi\,\sin x\Gamma t)\,,$$

$$\bar{B}^0 \rightarrow \psi K^0_S \propto e^{-\Gamma t}\,(1\,-\,\sin\phi\,\sin x\Gamma t)\,,$$

for the largest possible value of $\sin\phi \simeq 0.6$. Figure 18b indicates that on the tagging side a cut against long times can limit the unwanted mixing.

A more serious evaluation has been done at Snowmass [26], where all types of $b\bar{b}$ factories have been compared. Table 10 shows interesting features:
- The next to last line demonstrates that even within the Standard Model the uncertainty in the expected asymmetry is quite large. Only the most favourable side of the domain can give rise to observable effects.
- Even then, the task is difficult. On the Z without polarization, the required integrated luminosity is $0.5 \times 10^{40}\,cm^{-2}$; this corresponds to a quantum of 1.7×10^8 Z, which is larger than the one we considered. A substantial amount of polarization [27], providing an automatic tagging of b (versus $\bar{b}$), would greatly help. We have seen that such polarization is not *a priori* incompatible with the pretzel scheme, although we enter a domain of wild speculation.
- Nevertheless, the situation at a Z factory is not very different from that at a dedicated-threshold b factory. Both will have a hard time reaching the sensitivity needed to see CP violation at the standard level. On the other hand, the world can be non-standard: new effects can substantially enhance the CP-violation level. Certainly, 10^8 Z give access to a region whose exploration is essential.

6. CONCLUSIONS
The exploitation of high intensity at LEP I, which increases the speed of Z physics and offers the possibility of an exposure of 10^8 such particles, is certainly a very interesting option and one that is worth studying in depth.
- It is not an alternative to LEP 200, nor to the use of polarization at the Z.
- It can provide very accurate measurements of essential quantities, such as the number of neutrinos.
- It will make it possible to improve substantially the accuracy of the tests of the Standard Model through the well-known methods (e.g. lepton and quark asymmetries, τ polarization). However, compared with a measurement of A_{LR} with polarized beams, the volume of work and the number of problems to be solved are much greater for a result of somewhat lower quality.

High luminosity may reveal rare decay modes of the Z that are of considerable interest, and many of these channels should be studied carefully.

The pretzel LEP is a factory for $f\bar{f}$ states. The prospect of identifying $q\bar{q}$ states is good.

In particular, for $b\bar{b}$, the pretzel represents a very promising approach with a large amount of physics output, part of which would already appear at the level of 10^7 Z.

The success of the most ambitious goal, i.e. CP violation, is not guaranteed, and nature will decide. This statement is true for all the facilities being considered at present. I think that the implementation of the pretzel should not preclude the continuing effort to conceive an 'ideal' threshold $B\bar{B}$ program.

Acknowledgements

A lot of the work described here on $b\bar{b}$ physics is due to P. Roudeau [21], to whom I express my appreciation. I want to thank also M. Bosman, M. Dittmar, P. Mattig and G. Wormser for useful discussions.

References

[1]   C. Rubbia: CERN–EP/88–130 (1988).
[2]   J. Jowett: talk to the LEP Management Board (1988) and More bunches in LEP, preprint CERN–LEP–TH/89–17, Talk given at the Particle Accelerator Conf., Chicago, 1989.
[3]   G. von Holtey: CERN/LEP–B1/88–52 (1988).
[4]   V.H. Ritson and G. von Holtey: CERN/DELPHI 88–70 Gen. 83 (1988).
[5]   J. Drees: these proceedings, and in Polarization at LEP, eds. G. Alexander et al. (CERN 88–06, Geneva, 1988), Vol. 1, p. 317.
[6]   G. Wormser: SLAC PUB–4576 (1988) and refs. therein.
[7]   D. Treille: Workshop on Polarization in LEP, Geneva, 1987, eds. G. Alexander et al. (CERN, Geneva), p. 19.
[8]   L. Trentadue: these proceedings.
[9]   S. Glashow: *Phys. Lett.* **187B** 367 (1987).
      G.F. Giudice et al.: Pisa report IFUP–TH 14/88 (1988).
[10] N. Glover and J. Van der Bij: Rare Z^0 decays, to be published in Proc. LEP Physics Workshop (1989).
[11] D. Treille: ECFA Workshop on LEP 200, Aachen, 1986, eds. A. Böhm and W. Hoogland (CERN 87–08, ECFA 87/108, Geneva, 1987), Vol. II, p. 414.
[12] E.J. Eichten, K.D. Lane and M.E. Peskin, *Phys. Rev. Lett.* **50** 811 (1983).
[13] L.F. Abbott and E. Farhi: *Phys. Lett.* **101B** 69 (1981) and *Nucl. Phys.* **B189** 547 (1981).
[14] D. Bloch: Strasbourg report CRN/HE 86–08 (1986).
[15] F.A. Berends and R. Kleiss: *Nucl. Phys.* **B186** 22 (1981).
[16] D.W. Düsedau and J. Wudka: report MIT–CTP 1389 (1981).
[17] G.F. Giudice: *Phys. Lett.* **208B** 315 (1988).
[18] R. Barbieri: Pisa report IFUP–TH 26/88 (1988) and *Riv. Nuovo Cimento* **11** 1 (1988).
[19] W.S. Hou and R.G. Stuart: Munich report MPI–PAE/PTh 55/88 (1988).
[20] See contributions quoted in [10].
[21] P. Roudeau: Talk given at the Rencontres de Moriond, Les Arcs, 1989.
[22] C. de la Vaissière et al.: CERN/DELPHI 89–5 Phys. 35 (1989).
[23] K. Berkelman: *B meson decays,* Lectures given at the SLAC Summer Institute (1988).
[24] I.I. Bigi: ibid.
[25] F.R. Cavallo: CERN/DELPHI 88–5 Phys. 22 (1988).
[26] Report of the Snowmass B Factory Subgroup, Snowmass, 1988.
[27] W.P. Atwood et al.: SLAC–PUB 4544 (1988).

Table 1
Accuracy on $\sin^2\theta_w$ from $A_{ch}^{\mu\mu}$ (for 1000 pb^{-1})

Error	ΔA (%)	$\Delta\sin^2\theta_w$
Statistical	0.1	
Systematic:		
$\Delta m_Z = \pm 20$ MeV	0.16	
Detection efficiency	0.1	
QED radiative correction	0.12	
Total	0.25	0.0012

Table 2
Accuracy on $\sin^2\theta_w$ from P_τ ($\tau \to \pi\nu$ channel only)
(for 1000 pb^{-1})

Error	ΔP_τ (%)	$\Delta\sin^2\theta_w$
Statistical	0.4	0.0005
Systematic: $\varrho\nu$ channel + radiative corrections + ...	0.65	0.0008
Total	0.8	0.00095

Table 3
Accuracy on $\sin^2\theta_w$ from quark asymmetries A_{ch}^{qq} (100 million Z)

Type of quark	b	c	c (π of D*)	s	u
Efficiency of tagging	0.11	0.04	0.15[a]	0.05	0.02
Stat. error on A_{ch}	0.0007	0.0013	0.0015	0.001	0.0018
Syst. error on A_{ch}	0.003	0.004	> 0.0021	0.004	0.004
$\Delta\sin^2\theta_w$	0.00055	0.0009	> 0.00056	0.0007	0.0009

a) Signal/Background = 1/5

Table 4

Typical conditions for obtaining $\Delta \sin^2\theta = \pm 0.0003$
(experimental error) from A_{LR} measurement.
$L_{i,j}$ are luminosities registered with different spin configurations

---

Integrated luminosity: $\int L\, dt = 50\ pb^{-1}$

Polarization: $P \cong 50\%$

Uncertainty on P: $\Delta P = 1\%$

Rate of Bhabha events: $4 \times$ rate of Z events

Relative error on luminosity: $\Delta(L_i/L_j)_{syst.} = 1.5\%_0$

Table 5

Scaling laws for compositeness tests

Contact terms:
  Magnitude $c \propto s/\alpha\Lambda^2$
  Scale $\Lambda \propto \sqrt[4]{\int L\, dt}\ \sqrt[4]{s}$

Search for e^* from anomalous behaviour of $ee \rightarrow \gamma\gamma$:
  $(\text{Mass } e^*)^4 \propto \alpha\, s^{3/2}\, \sqrt{\int L\, dt}$ a)

---

a) The usual parameter λ in the Lagrangian is taken to
   be equal to 1.

Table 6
Some characteristics of $Z \rightarrow b\bar{b}$

Cross-section: $\sigma_{b\bar{b}} = 6.5$ nb.

Percentage, relative to the hadronic Z modes:
$\sigma_{b\bar{b}}/\sigma_{had} = 0.22$.

Percentage, relative to the visible Z modes:
$\sigma_{b\bar{b}}/\sigma_{vis} = 0.19$.

Population of various species, from 100 million Z:

$15.5 \times 10^6\, B^0$	$15.5 \times 10^6\, B^+$
$4.5 \times 10^6\, B^0_s$	$1.7 \times 10^6\, '\Lambda'_b$
	$0.35 \times 10^6\, '\Xi'_b$

Mean number of charged particles per B: ~ 5.

Mean number of charged particles at the primary vertex: ~ 10.

Mean flight path of B: ~ 2.2 mm.

Table 7
The percentage of well-classified events is $80.3 \pm 0.9\%$.
Example of a classification of 2000 $Z \rightarrow q\bar{q}$ events between four classes:
L(q = u, d, s); C (q = c); B(q = b); T(q = t). Events are classified
according to a 'class likelihood' derived from 15 variables, seven of
which use the measurement of impact parameters given
in the barrel region by the microvertex detector.

Classified	Purity	Class generated			
		L	C	B	T
L	90.8%	1012	87	14	2
C	47.0%	173	187	38	0
B	87.4%	9	41	348	0
T	66.3%	14	4	12	59
Loss		16.2%	41.4%	15.5%	3.3%

Table 9
The KM matrix and the various sources of information on its elements, from B physics

$$\begin{pmatrix} V_{ud} & V_{us} & V_{ub} \\ V_{cd} & V_{cs} & V_{cb} \\ V_{td} & V_{ts} & V_{tb} \end{pmatrix} \equiv \begin{pmatrix} 1-\dfrac{\lambda^2}{2} & \lambda & A\lambda^3\varrho e^{i\delta} \\[2mm] -\lambda & 1-\dfrac{\lambda^2}{2} & A\lambda^2 \\[2mm] A\lambda^3(1-\varrho e^{-i\delta}) & -A\lambda^2 & 1 \end{pmatrix}$$

V_{ub} comes from B → no charm

V_{cb} comes from τ_B, B_{SL}

V_{td} comes from $B^0_d\bar{B}^0_d$ oscillations

V_{ts} comes from $B^0_s-\bar{B}^0_s$ oscillations

Table 8
Vertex detector: present and future

Microvertex 1 (present one of DELPHI)

R_1 (cm)	9		
R_2 (cm)	11		
σ (μm) (1 coordinate)	5		
Si thickness (μm)	400		
Al thickness of vacuum chamber (mm)	1		
θ_{min} (°)	42		

p (GeV/c)	3	4.7	14
Transverse accuracy (μm)			
without microvertex	136	90	41
with microvertex	54	41	27
σ along flight path (μm) (secondary vertex)	250		

Microvertex 2 (ideal)

R_1 (cm)	5		
R_2 (cm)	8		
R_3 (cm)	11		
σ (μm) (2 coordinates)	5		
Si thickness (μm)	200		
Be thickness of vacuum chamber (mm)	1		
θ_{min} (°)	20		

p (GeV/c)	3	4.7	14
Transverse accuracy (μm)	27	19	11.6
σ along flight path (μm) (secondary vertex)	150		

Table 10
Comparison of B-factory techniques

Factor	Case				
	Asymmetric Υ(4S)	Symmetric Υ(4S) +	$\sqrt{s}$ = 16 GeV	Z P = 0	Z P = 0.9 (P = 0.45)
$b\bar{b}$ cross-section (nb)	1.2	0.3	0.11	6.3	6.3
Fraction of B^0	0.43	0.34	0.35	0.35	0.35
ψK_S reconstruction efficiency	0.61	0.61	0.61	0.46	0.46
Tagging efficiency (and method)	0.48 (ℓ, K)	0.48 (ℓ, K)	0.30 (ℓ, D)	0.18 (ℓ, D)	0.61 (A$_{FB}$)
Wrong tag fraction	0.08	0.08	0.08	0.08	0.125 (0.27)
Asymmetry dilution	0.71	0.63	0.52	0.52	0.71
$\int$ L dt needed for 3σ effect (10^{40} cm^{-2})[*]	0.3–12	2.2–78	14–490	0.5–19	0.1–3.6 (0.3–9.6)
Relative $\int$ L dt needed	1.0	6.4	40	1.5	0.3 (0.8)

[*] The peak luminosity needed in units of 10^{33} cm^{-2} s^{-1} for 10^7 seconds of fully efficient running at peak luminosity.

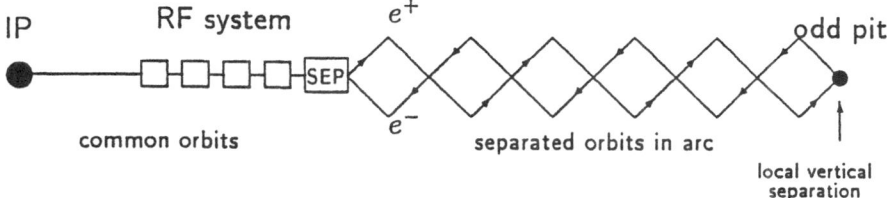

Fig. 1 Basic idea for a pretzel scheme (showing one octant) (from [2])

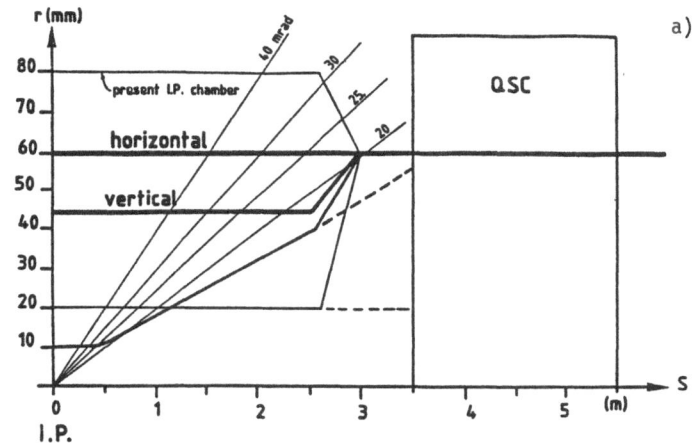

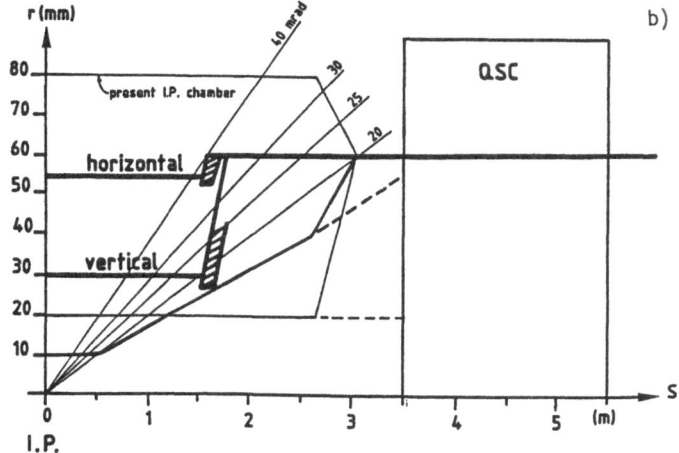

Fig. 2 Minimum radius foreseen for the vacuum chamber at LEP (from [4]): a) no mask, b) with masks.

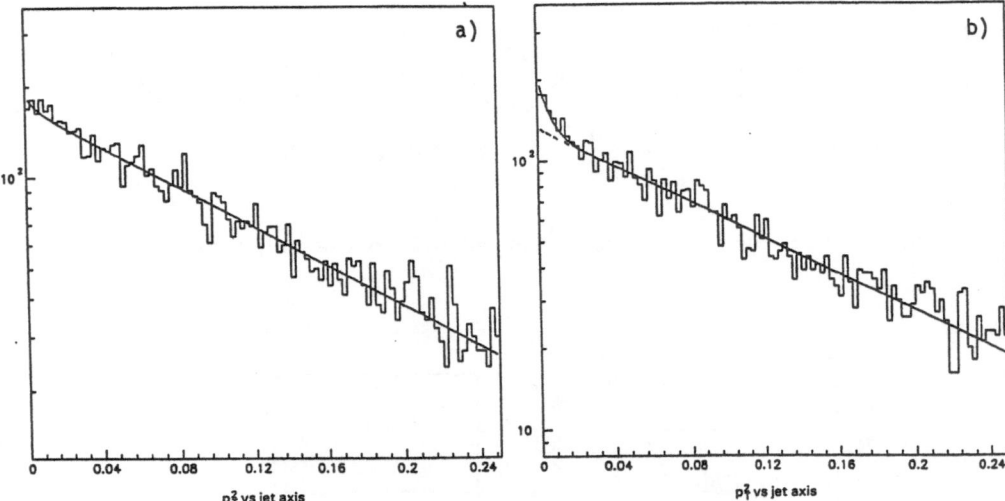

Fig. 3 Spectrum of the p_T^2 of single π relative to the jet axis: a) non-charm, b) all events (from [5] and [6]).

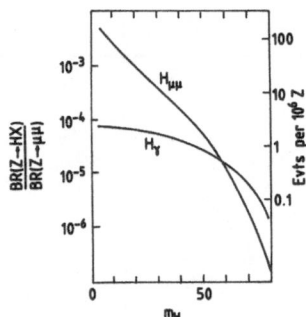

Fig. 4 Standard Higgs cross-section (from [10])

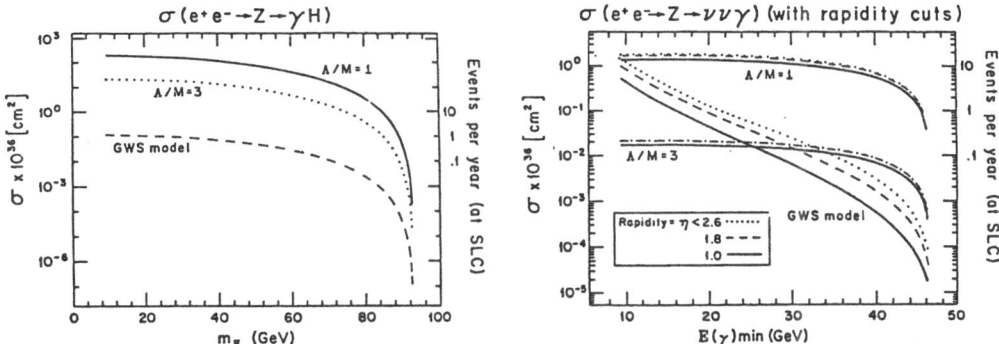

Fig. 5 Enhancement of cross-sections in the SCSM (from [16]); 1 event per year at the SLC (L = 10^{30} assumed) means 100 events per year at the pretzel LEP.

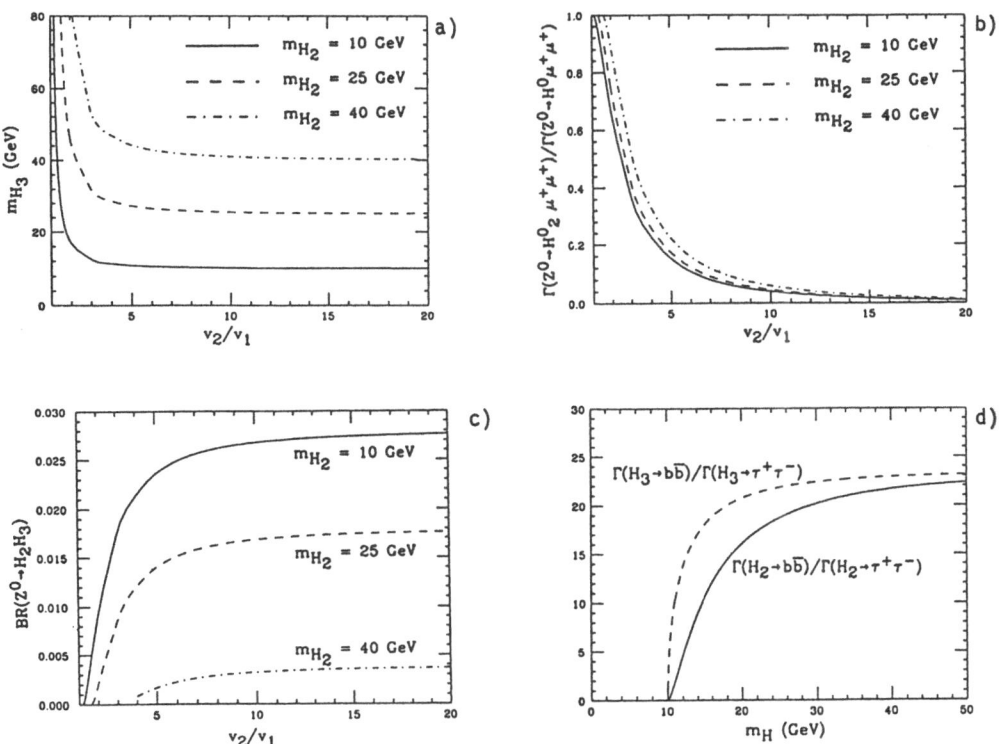

Fig. 6 The neutral Higgs system in minimal SUSY (from [17]): a) m_{H_3} versus v_2/v_1; b) BR of the $Z \to H_2\mu\mu$ versus v_2/v_1; c) BR of the $Z \to H_2H_3$; d) decays of H_2H_3.

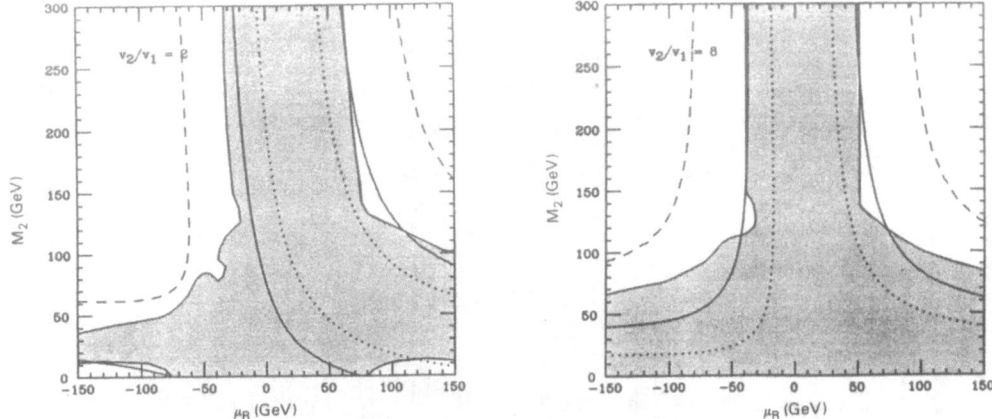

Fig. 7 Contours of neutralino production with a BR > 10^{-5}; M_2 and μ_R arer close to the usual M,μ parameters of SUSY (from [18]).

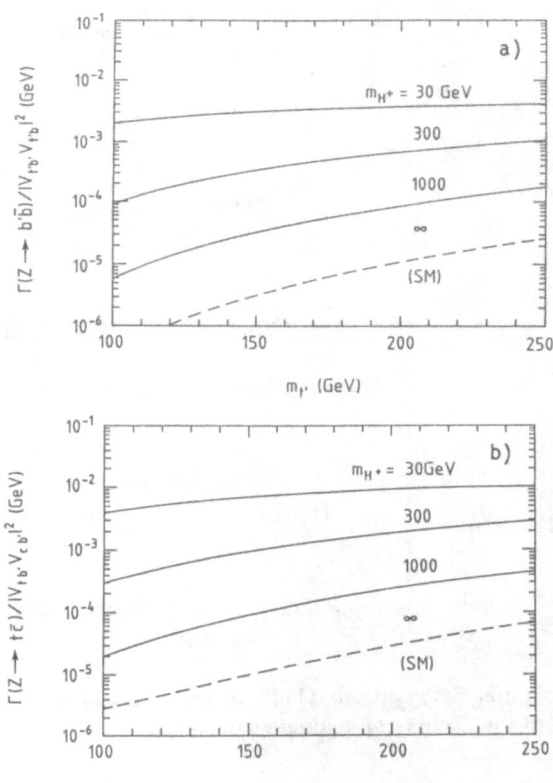

Fig. 8 FCNC cross-sections for the extended Higgs model (from [19])

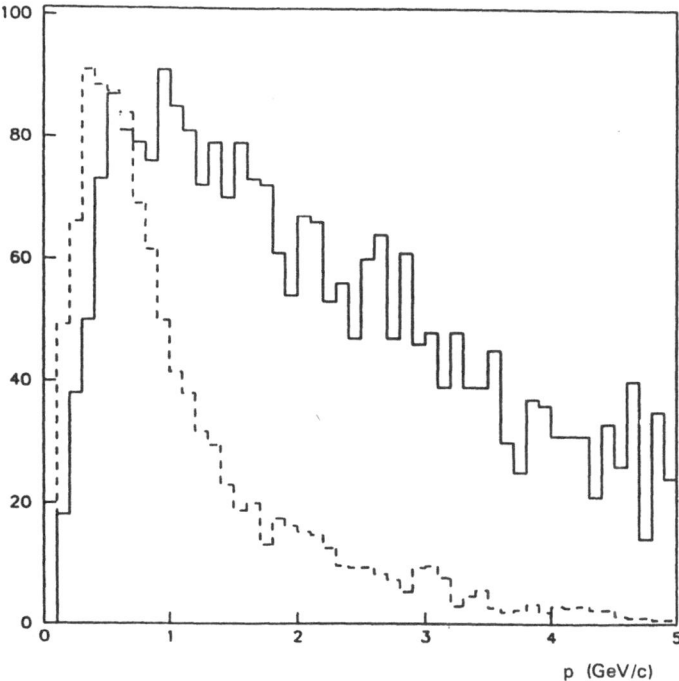

Fig. 9 Momentum of tracks from b̄b events at the Z: —— from B, B̄; --- at the main vertex (from [21]).

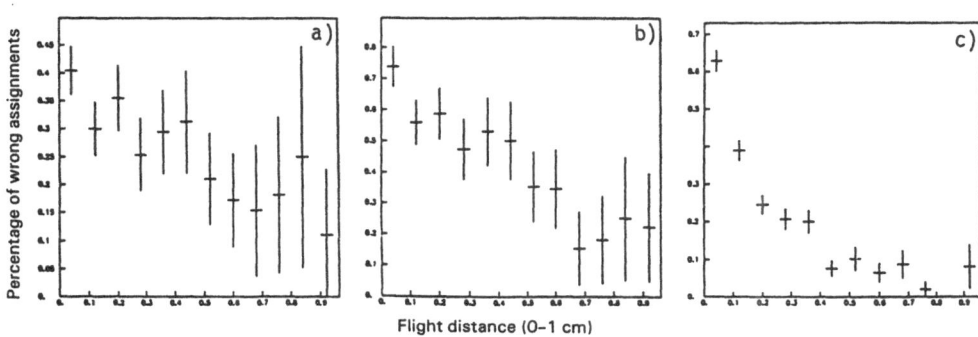

Fig. 10 Probability of wrong assignment of B tracks to the main vertex as a function of flight distance: a) microvertex 1 of DELPHI; b) microvertex 2 with one coordinate; c) ideal microvertex 2 (from [21]).

344

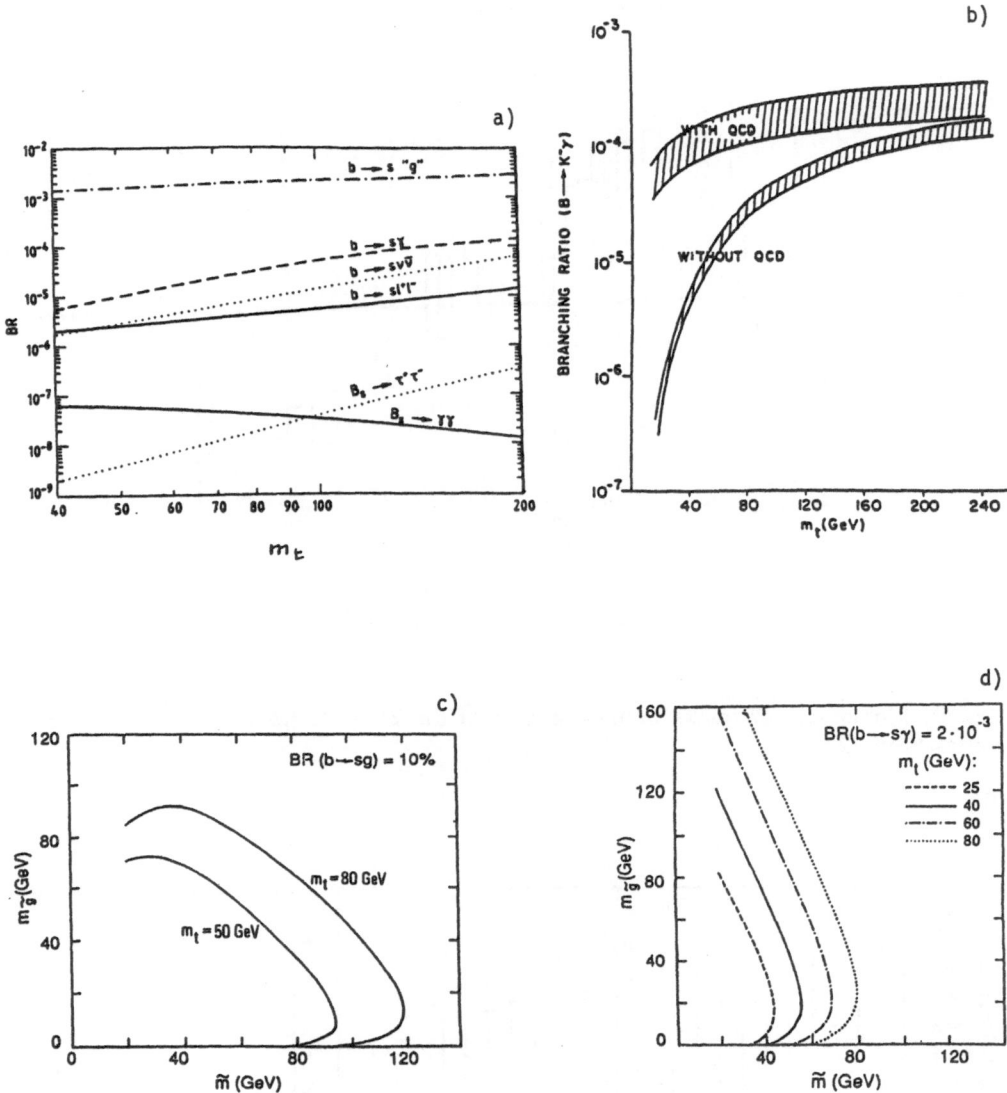

Fig. 11 SM predictions for rare B decays (a, b) and possible deviations due to SUSY (c, d)

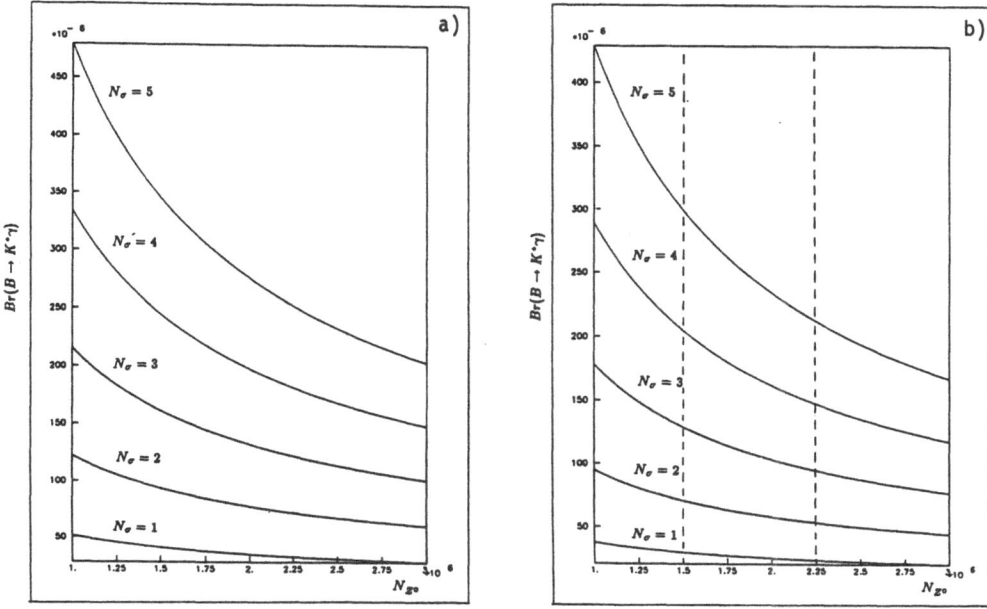

Fig. 12 Number of standard deviations for observing B($\to$ K$^*\gamma$) versus the number of Z analysed (from [25]): a) all events, b) full tag of b$\bar{\text{b}}$ (100% efficient).

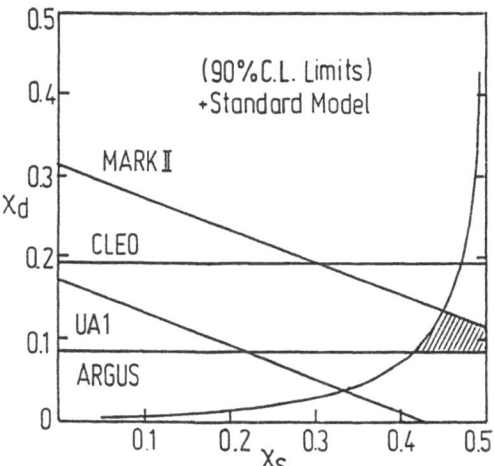

Fig. 13 Limits on χ_s versus χ_d; $\chi = r/(1 + r)$, where $r = x^2/(2 + x^2)$.

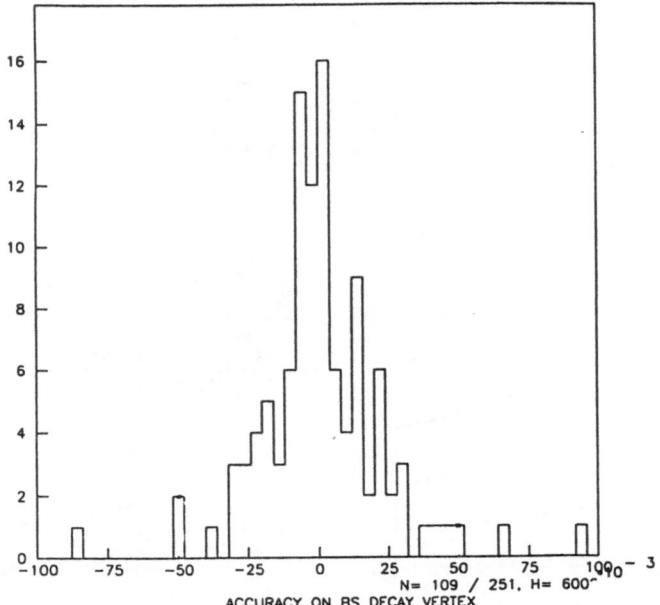

Fig. 14 Longitudinal accuracy on the B_s decay vertex (see text), from ref. [21] (10 units = 100 μm)

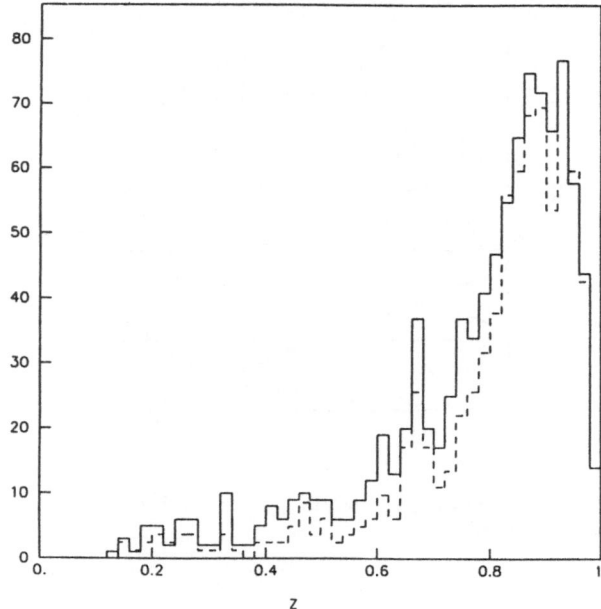

Fig. 15 The Z distribution in B production: —— all B; --- cos jj < −0.95 (from [21]).

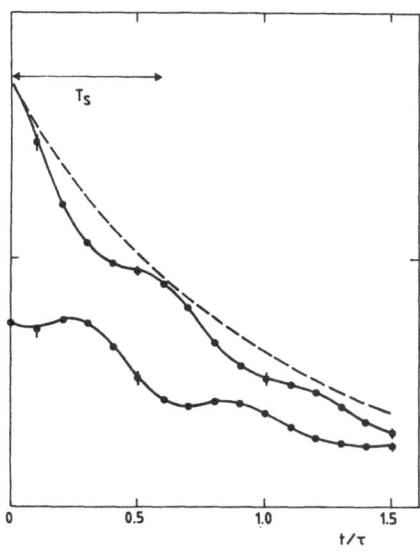

Fig. 16 The expected oscillation pattern for B_s (3×10^7 $b\bar{b}$ events) from [21] (top curve: $B_{tagged}\bar{B} + \bar{B}_t B$, bottom curve: $B_t B + \bar{B}_t\bar{B}$).

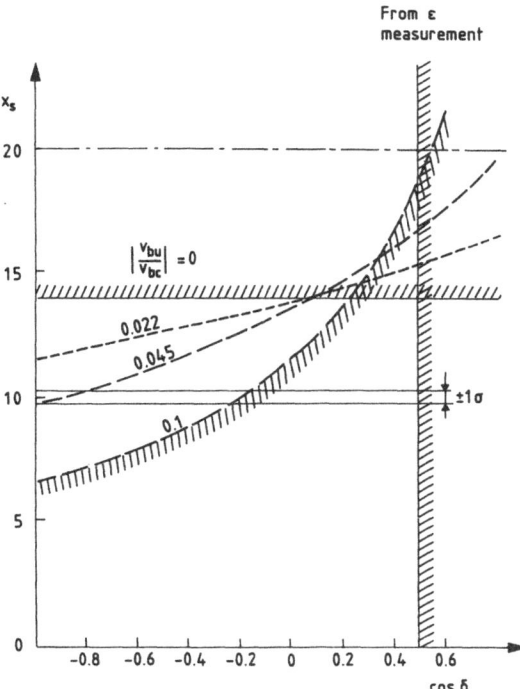

Fig. 17 Information on cos δ from various measurements, including $x_s = 10$ (from [21]).

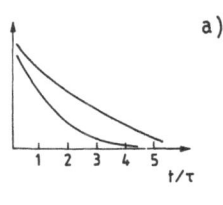

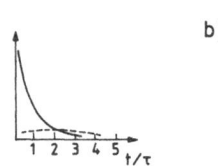

Fig. 18 a) B_d (top) and $\bar{B}_d$ (bottom) decay into ψK_S^0 (sin δ = 0.6); b) $B^0 \rightarrow$ ($\ell^- X$) (full line) and B^0 ($\ell^+ X$) (dotted line).

List of Participants

F. Berends, University of Leiden (Netherlands)
W. Blum, MPI Munich (FRG)
W. De Boer, MPI Munich (FRG)
A. Buras, MPI Munich/Technical University Munich (FRG)
M. Böhm, University Würzburg (FRG)
H. Burkhardt, CERN, Geneva (Switzerland)
G. Cowan, MPI Munich (FRG)
A. Denner, MPI Munich (FRG)
N. Dombey, University of Sussex, Brighton (United Kingdom)
J. Drees, University Wuppertal (FRG)
F. Dydak, CERN, Geneva (Switzerland)
R. Guth, MPI Munich (FRG)
W. Hollik, University Hamburg (FRG)
S. Jadach, University Cracow (Poland)
F. Jegerlehner, SIN, Zurich (Switzerland)
M. Jezabek, University Cracow (Poland)
R. Kleiss, CERN, Geneva (Schweiz)
B. Kniehl, MPI Munich (FRG)
H. Kroha, MPI Munich (FRG)
J. Kühn, MPI Munich (FRG)
W. van Neerven, University of Leiden (Netherlands)
G. Passarino, University of Torino (Italy)
E. Papageorgiou, MPI Munich (FRG)
P. Rankin, University of Colorado, Boulder (USA)
T. Riemann, Academy of Sciences DDR, Zeuthen (GDF)
R. Rückl, MPI Munich/University Munich (FRG)
C. Salazar, MPI Munich (FRG)
D. Schaile, CERN, Geneva (Switzerland)
R. Settles, MPI Munich (FRG)
A. Sirlin, University of New York(USA)
U. Stiegler, MPI Munich (FRG)
D. Treille, CERN, Geneva (Switzerland)
L. Trentadue, University of Parma (Italy)
C. Verzegnassi, CERN, Geneva (Switzerland)
Z. Was, MPI Munich (FRG)